Plant Disease
AN ADVANCED TREATISE

VOLUME III
How Plants Suffer from Disease

Advisory Board

Plant Disease

AN ADVANCED TREATISE

VOLUME III

How Plants Suffer from Disease

Edited by

JAMES G. HORSFALL
The Connecticut Agricultural
Experiment Station
New Haven, Connecticut

ELLIS B. COWLING
Department of Plant Pathology
and School of Forest Resources
North Carolina State University
Raleigh, North Carolina

ACADEMIC PRESS   New York   San Francisco   London   1978

A Subsidiary of Harcourt Brace Jovanovich, Publishers

ACADEMIC PRESS, INC.
111 Fifth Avenue, New York, New York 10003

United Kingdom Edition published by
ACADEMIC PRESS, INC. (LONDON) LTD.
24/28 Oval Road, London NW1 7DX

Library of Congress Cataloging in Publication Data
Main entry under title:

Plant disease.

    Includes bibliographies and index.
    CONTENTS: v. 1.   How disease is managed.
                        v. 3.   How plants suffer from disease.
1.   Plant diseases.   I.   Horsfall, James Gordon, Date
II.   Cowling, Ellis Brevier, Date
SB731.P64   632   76–42973
ISBN 0–12–356403–4   (v. 3)

To those who hand on the torch,
to those who transmit excellence,
to the teachers of Plant Pathology,
To you we dedicate this volume.

Contents

CHAPTER 1 PROLOGUE: HOW PLANTS SUFFER FROM DISEASE
ELLIS B. COWLING AND JAMES G. HORSFALL

CHAPTER 2 HOW HEALTHY PLANTS GROW
T. T. KOZLOWSKI

CHAPTER 3 THE DYNAMIC NATURE OF DISEASE
DURWARD F. BATEMAN

CHAPTER 4   THE CAPTURE AND USE OF ENERGY BY DISEASED PLANTS

TSUNE KOSUGE

CHAPTER 5   DYSFUNCTION IN THE FLOW OF FOOD

MARTIN H. ZIMMERMANN AND JANET McDONOUGH

CHAPTER 6   DYSFUNCTION OF THE WATER SYSTEM

P. W. TALBOYS

CHAPTER 7   DISTURBED MINERAL NUTRITION
DON M. HUBER

CHAPTER 8   ALTERATION OF GROWTH BY DISEASE
JOHN P. HELGESON

CHAPTER 9   CROWN GALL—A UNIQUE DISEASE
DONALD J. MERLO

CHAPTER 10   PLANT TERATOMAS—WHO'S IN CONTROL OF THEM?
CHARLES L. WILSON

CHAPTER 15 DISEASE ALTERATIONS IN PERMEABILITY AND MEMBRANES

HARRY WHEELER

CHAPTER 16 CHANGES IN INTERMEDIARY METABOLISM CAUSED BY DISEASE

JOSEPH KUĆ

CHAPTER 17 TRANSCRIPTION AND TRANSLATION IN DISEASED PLANTS

D. J. SAMBORSKI, R. ROHRINGER, AND W. K. KIM

CHAPTER 18 SENESCENCE AND PLANT DISEASE
G. L. FARKAS

CHAPTER 19   RELATION BETWEEN BIOLOGICAL RHYTHMS AND DISEASE
T. W. TIBBITTS

List of Contributors

Numbers in parentheses indicate the pages on which the authors' contributions begin.

DURWARD F. BATEMAN (53), Department of Plant Pathology, Cornell University, Ithaca, New York 14850

G. D. BOWEN (231), CSIRO, Division of Soils, Glen Osmond, South Australia 5064, Australia

ELLIS B. COWLING (1, 299), Department of Plant Pathology and School of Forest Resources, North Carolina State University, Raleigh, North Carolina 27650

G. L. FARKAS (391), Institute of Plant Physiology, Biological Research Center, Hungarian Academy of Sciences, Szeged, Hungary

JOHN P. HELGESON (183), USDA, SEA, FR, NCR Plant Disease Resistance Research Unit, Department of Plant Pathology, University of Wisconsin–Madison, Madison, Wisconsin 53706

JAMES G. HORSFALL (1), The Connecticut Agricultural Experiment Station, New Haven, Connecticut 06504

DON M. HUBER (163), Department of Botany and Plant Pathology, Purdue University, West Lafayette, Indiana 47907

W. K. KIM (375), Agriculture Canada, Research Station, Winnipeg, Manitoba, R3T 2M9, Canada

TSUNE KOSUGE (85), Department of Plant Pathology, University of California, Davis, California 95616

T. T. KOZLOWSKI (19), The Biotron, University of Wisconsin–Madison, Madison, Wisconsin 53706

JOSEPH KUĆ (349), Department of Plant Pathology, University of Kentucky, Lexington, Kentucky 40506

JANET McDONOUGH (117), Harvard Forest, Harvard University, Petersham, Massachusetts 01366

D. E. MATHRE (257), Department of Plant Pathology, Montana State University, Bozeman, Montana 59717

DONALD J. MERLO (201), Departments of Microbiology and Immunology, University of Washington, Seattle, Washington 98195

M. S. MOUNT (279), Department of Plant Pathology, University of Massachusetts, Amherst, Massachusetts 01002

R. ROHRINGER (375), Agriculture Canada, Research Station, Winnipeg, Manitoba, R3T 2M9, Canada

D. J. SAMBORSKI (375), Agriculture Canada, Research Station, Winnipeg, Manitoba, R3T 2M9, Canada

P. W. TALBOYS (141), East Malling Research Station, East Malling, Maidstone, Kent ME19 6BJ, England

T. W. TIBBITTS (413), Department of Horticulture, University of Wisconsin–Madison, Madison, Wisconsin 53706

HARRY WHEELER (327), Department of Plant Pathology, University of Kentucky, Lexington, Kentucky 40506

CHARLES L. WILSON (215), USDA Agricultural Research Service Nursery Crops Research Laboratory, Delaware, Ohio; and Ohio Agricultural Research and Development Center, Wooster, Ohio

MARTIN H. ZIMMERMANN (117), Harvard Forest, Harvard University, Petersham, Massachusetts 01366

Preface

Death rides on every passing breeze,
He lurks in every flower.
Each season has its own disease,
Its peril every hour.

Reginald Heber

In plants, too, death rides on every passing breeze. Plants often suffer, and sometimes die, from their afflictions. For this reason, we have devoted the third volume of this treatise to the dysfunctions that occur in plants when they are ill. Illness implies suffering. Thus, we chose deliberately to include the word *suffer* in the title of this volume—"How Plants Suffer from Disease." Suffer is a verb. It implies action. Plants that are sick are functioning poorly.

Plants comprise an extraordinary set of systems and subsystems, each tied to the others in a complex web. Each functional system, and its interconnections with the others, is a potential target for disruption. Pathogens of many types interfere with the systems that constitute photosynthesis, respiration, transpiration, food transport and storage, mineral nutrition, regulation of growth, symbiosis, reproduction, mechanical support, permeability, transcription, translation, intermediary metabolism, and biological rhythms.

If any of these processes is impaired, the plant suffers a loss of efficiency. If the process is stopped, the plant dies.

This volume, as well as the others in the treatise, is designed for advanced thinkers in the whole field of plant protection, whatever their specialties may be—from entomology through plant pathology to virology, from graduate student to Nobel Laureates.

A major objective of plant pathology is to learn how the vital processes of plants are disrupted and to design methods to prevent and circumvent the damage. This volume is meant to go beyond a comfortable exposition of where we have been. It is intended to tell us where we are going, to point out regions of our science where the prospecting for new knowledge will be most productive, and to emphasize how our profession can grow in usefulness to the society that pays the bills for the groceries.

To increase the success of our prospecting, we have asked the 22 authors of this volume to write provocatively and imaginatively, to make us think, to stretch our minds. Oliver Wendell Holmes put it more beautifully than we can when he said, "Man's mind, stretched to embrace a new idea, never returns to its original dimension." We hope you find it so with this volume.

James G. Horsfall
Ellis B. Cowling

Contents of Other Volumes

Tentative Contents of Other Volumes

Chapter 1

Prologue: How Plants Suffer from Disease

ELLIS B. COWLING AND JAMES G. HORSFALL

I. INTRODUCTION

It is our pleasant duty in this chapter to welcome you to the third of five volumes in An Advanced Treatise on Plant Disease. With this volume we shift from concern about disease in populations of plants to disease in individual plants. This will require a major change in our perspective and patterns of thinking even though the ultimate goal of our analysis remains the same—to understand disease so well that we can manage it efficiently.

In the first two volumes we needed binoculars to broaden our view. Now we need our spectacles for close work so we can see the individual plants and peer into the world of organs, tissues, cells, and metabolites.

We seek now to understand the dynamics of dysfunction due to disease. Thus we turn from concern about the spread of disease in fields and forests to development of disease within individual plants and organs; from concern about the productivity of fields and the population dynamics of pathogens to concern about physiological functions and metabolic alterations; from breeding strategies to mechanisms of tolerance and resistance; from synoptic weather patterns to the microclimate within the boundary layer of a crop; from management with chemicals to mechanisms of parasitism and pathogenesis.

We have taken various precautions to avoid making this shift into a dichotomy. We carefully selected authors who can make their subject

1

relevant to a wide cross section of plant pathologists. Terminology has been selected to be as broadly understandable as possible. Most important to all, perhaps, significant linkages among all five volumes have been built into the general organization of the treatise. By these efforts we hope to encourage greater communication both among all the various specialists within plant pathology and also between plant pathologists and all the other scientists it takes to feed, clothe, and house a hungry and shivering world.

II. THE PROGRESS OF DISEASE IN INDIVIDUAL PLANTS

Disease develops in individual plants by a series of sequential steps beginning with the arrival of inoculum at the plant surface and ending with the terminal stages of pathogenesis. The innumerable stages in between are unique to each particular pathogen, host, and variation of the environment.

To cause disease a pathogen must first penetrate its host. How does it get in? Is ingress accomplished actively, with its own mechanical or chemical power, or passively, by depending on a vector or entering through wounds?

Recognition and compatibility phenomena are the next stages of pathogenesis. They are among the most poorly understood aspects of disease. They are critical because resistance and susceptibility, virulence and avirulence usually are determined at this early stage. A complex series of interacting biochemical events is initiated. The results of this interaction determine if the pathogen and host are compatible or incompatible. If compatible, pathogenesis can continue. If not, the host goes on about its normal functions without further disturbance by the pathogen.

In the third stage of pathogenesis, invasion begins in earnest. Biotic pathogens spread from cell to cell within the plant tissue, living at the expense of the host. To obtain energy and substance for their own nourishment, many pathogens secrete enzymes and other metabolites that break down the cell walls, cell membranes, and other complex organelles and constituents of the host cytoplasm. Some living pathogens secrete toxins which disturb the normal metabolic machinery of host cells. Others produce growth regulatory substances which alter the normal patterns of cellular development and differentiation. Viruses take over the synthetic machinery of host cells and divert it to synthesis of viral nucleic acid and proteins instead of host nucleic acids and proteins. Air pollutants act as metabolic poisons and disrupt the normal processes of synthesis and catabolism.

In all stages of pathogenesis the insults of the pathogen force the host to alter its vital functions. These alterations are manifest in all the affected cells and tissues, the membranes and organelles, the metabolic pathways, and the growth regulating processes of the host.

These alterations themselves constitute invisible symptoms of disease. They also give rise to the characteristic visible symptoms by which we diagnose disease. The invisible symptoms of disease are legion. Respiration almost always increases during the early stages of pathogenesis and decreases in the terminal phases. The expanding lesions become metabolic sinks into which flow nutrients and other metabolites of all sorts. Mechanisms of dynamic defense are elicited. Hypersensitive reactions are initiated. Phytoalexins are produced. Lysosomes release their packets of enzymes and other metabolites. Protective glycosides are cleaved to release their antimicrobial aglycones. Reaction zones, and barrier zones of abnormal tissues are established in efforts to contain the expanding lesions.

The defensive responses of the host vary qualitatively and quantitatively with the particular organ and tissue that is affected as well as with the particular pathogen and environmental conditions that prevail during each stage of pathogenesis. If feeder roots are the site of pathogenesis, uptake of water and soil nutrients as well as secretion of root exudates will be impaired. If structural roots, stems, branches, or petioles are affected, transpiration and/or transport of photosynthetic products will be affected depending on whether the lesion is in the xylem or phloem. If meristems are attacked, growth will be affected—increased or decreased according to the amount of nutrients available and the nature, supply, and ratios among the various growth-regulating substances produced by the host and the pathogen. If flower parts or pollen are affected, reproduction will be impaired. If foliage and green shoots are attacked, photosynthesis will be affected. If pathogens attack storage tissues, mobilization of reserve foods will be disrupted.

During all of these alterations, the host carries on as best it can. The plant continues its efforts to cope with the constantly changing stresses imposed by the environment as well as the additional stresses imposed by the pathogen. Hot days follow cold days. Bright days follow cloudy days. Dry days follow wet days, or dry days continue during protracted drought. Nutrients may be plentiful at times, but they are seldom constant in supply. Competition from weeds and other plants in the field becomes intense. The pathogen becomes aggressive. Other pathogens join in. Insects graze on the canopy. Soil animals chew on the roots. Frost may nip the early buds. Episodes of air pollution come and go. The farmer supplies the wrong spray or adds too little or too much of some essential nutrient. Conditions never stay the same. The struggle between the host and the pathogen(s) continues day after day, week after week, and year after year in perennial plants.

Plants that can cope well with all these variations in environmental stresses prove themselves to be hardy. Those that can exclude virulent pathogens prove themselves to be resistant. Those that produce a good crop despite the insults of pathogens show themselves to be tolerant. Plants that cannot cope, grow poorly, yield poorly, or perhaps are killed by a lethal combination of environmental and pathogenic stresses.

III. A THEORETICAL ANALYSIS OF SUFFERING IN PLANTS

Ernst Gäumann (1946) was a pioneer in establishing a theoretical base for plant pathology. That is why we dedicated Volume II to him. He was a generalist in an age of specialists. Following in Gäumann's footsteps, Horsfall and Dimond (1959) organized certain chapters in Volume I in the original treatise around the physiological functions of plants that are impaired by pathogens—tissue is disintegrated; growth is affected; reproduction is affected; the host is starved; water is deficient; and respiratory patterns are altered.

George McNew (1960) was another early theoretician in plant pathology. He wrote on "The Nature, Origin, and Evolution of Parasitism" in Volume II of the original treatise. In that chapter he recognized that interference with six vital processes provides the basis for a rational analysis of pathogenesis: (1) destruction of food reserves; (2) prevention of seedling metabolism; (3) interference with procurement; (4) interference with upward transport; (5) destruction of food manufacture; and (6) diversion of foodstuffs to abnormal use.

McNew concluded his conception with these simple but elegant ideas:

> If any one of these processes is brought to a stop, the plant succumbs but this is the exception rather than the rule. Ordinarily, the function is injured but not destroyed so only the efficiency of the plant is affected. A major objective of plant pathology is to learn how these processes are injured and to design methods of preventing or circumventing that injury. The science, therefore, has as its applied objective the conservation of foodstuffs and the promotion of efficiency in their synthesis and utilization by the plant.
>
> If the pathogens that disrupt each of these six processes are grouped together irrespective of their taxonomic derivation, a surprising degree of order is obtained. . . . Here, then, is a system for organizing the types of pathogenesis. . . . By so organizing the knowledge on plant diseases into an integrated system depending on function, certain natural laws of pathogenesis can be evolved. . . .
>
> The basic tenets of this system are that: disease is an abnormal physio-

logical process in plants; the physiological processes are constantly in a state of flux; there is an orderly progression of parasitism depending upon which function of the host is under attack; and the system is natural enough to permit development of some general laws and principles for understanding disease and its control.

What a brilliant conception! It is inherently logical. The approach is useful as well as fascinating. But McNew's brilliant ideas have not captured the imagination of plant pathologists generally. Why is this so? We suggest two main reasons. The first has to do with the way plant pathologists are clustered. The second has to do with the way we study pathogenesis. As we shall see below, the two reasons are connected.

A. The Prevalent Patterns of Clustering Inhibit the Generalist

Like "birds of a feather," plant pathologists cluster together in groups according to their own specialty. As we discussed more fully in Chapter 2, Vol. I, common interests and shared experience often have led to mutual reinforcement of ideas and support in the labor of research, but it has also led to some narrowness of view, internal cliques, and even institutional biases.

If you ask a plant pathologist where his interests lie he generally responds as a specialist and not as a generalist: "I work on diseases of vegetable crops"; or "I am a forest pathologist"; or "I am a nematologist"; or "I work on air pollution damage to plants"; or "I work on phytoalexins." As Stakman (1969) points out, excessive concern with disciplinary identity has led to attitudes of exclusiveness and internal self-sufficiency within disciplinary "guilds." In this same sense, concern with particular specialties has tended to inhibit thinking about the whole matrix that constitutes McNew's "Theory of Vital Functions."

For example, histological methods have long been a favorite method of plant pathologists. Thus, for many individual diseases we have a remarkably good understanding of the anatomy and morphology of disease. This base of descriptive data is very valuable in understanding the pathways of penetration and invasion of host tissue. These are very important aspects of pathogenesis. But pathological histology is a very indirect method if our objective is to gain an analytical understanding of the dynamics of dysfunction in growing plants.

Ten Houten (1959) says that plant physiology is the science most closely related to plant pathology. But it is rare that plant pathologists have worked and published together with plant physiologists. Some institutions have created departments of plant pathology and physiology. However, even then joint publications were few and far between. We know of no institutions that have combined departments of biochemistry and plant pathology, although there are a few plant pathologists who work in departments of biochemistry and vice versa.

Among those of us who declare an interest in the dynamics of plant disease, the impact of molecular biology is impressive. If you ask members of this group where their interests lie the prevalent answers are four in number and all are predominately molecular: the pathogen-oriented section of the group will mention "cell wall degrading enzymes" and "host-specific toxins." The members of the host-oriented section of the group will mention "mechanisms of disease resistance" and "phytoalexins." Plant pathologists who study disease as abnormal physiology appear to have specialized in the molecular biology of pathogenesis without first learning about the physiology of whole plants and organs. It appears that many of us took the path of least resistance by electing to study some single isolated component of disease, usually at the cellular or molecular level, rather than attempt to cope with the integrated systems of the host or its vital organs as a whole.

Considering these preferences, it is curious that our profession generally uses the term "physiological plant pathology" when it means "biochemical plant pathology." We have both a major journal and a fine new text by this name (Heitefuss and Williams, 1976). Of course, we can note some exceptions such as *The Biochemistry and Physiology of Infectious Plant Disease* by Goodman *et al.* (1967) and *Biochemical Aspects of Plant–Parasite Relations* by Friend and Threlfall (1976).

B. The Way We Analyze Pathogenesis Inhibits the Generalist

Two strongholds of physiology are visible in the great rush to biochemical plant pathology. These are wilting and abnormal growth. In the first case, the typical approach has been indirect and, in the second, a more adequate conceptual basis for experimental research is needed. These and three other aspects of the way we analyze pathogenesis illustrate the need for more adequate conceptual theories of pathology.

1. The Physiology of Wilting

Wilt is a common symptom of disease in plants. Many different types of pathogens affect the water system—feeder-root and xylem-invading fungi, bacteria, nematodes, and parasitic seed plants. Wilt results whenever the supply of water does not keep up with the demand. This deficiency can develop in at least two different ways: by increasing the rate of loss of water from foliar organs, or by decreasing the supply of water to foliar organs, for example, by inhibiting the uptake of water by roots or by blocking the xylem. These two possibilities can be readily distinguished by direct measurement of changes in the water potential of various organs. Such direct measurements hold the key to understanding dysfunction in transpiration. However, direct measurements of water potential are rare indeed in the literature of plant pathology.

The idea of water potential has provided physiologists with an elegant conceptual basis for experimental analysis of dysfunction in the water system. Two simple methods—the pressure bomb method of Sholander and the older but reliable method of thermocouple psychrometry—permit easy, rapid, and direct analysis of the water status of plants. These two methods are described in detail by Barrs (1968), Brown and van Haveren (1972), and Slovik (1974).

Despite the abundance of diseases that affect transpiration and the availability of reliable methods for measurement of water potential, only a few plant pathologists [for example, Cook and Papendick (1972) and Duniway (1976)] have used direct methods for experimental analysis of mechanisms of wilting. Most of us are accustomed to the methods of pathological histology, and are convinced that we should be able to "see the plugs." Thus, we continue to use indirect methods instead of getting to the heart of the matter by measuring the dynamics of water stress directly (see also Chapters 2, 5, and 6, this volume).

2. The Physiology of Abnormal Growth

Abnormal growth, another common dysfunction due to disease, is induced by a wide array of pathogens including many gall-forming insects, growth-altering fungi such as *Plasmodiophora brassicae* and *Gibberella fujikuroi,* the crown gall and hairy root bacteria (*Agrobacterium tumefaciens* and *A. rhizogenies*), rootknot nematodes such as *Meloidogyne incognita,* the leaf curling viruses, and parasitic seed plants such as the mistletoes.

Plant pathologists are proud of our Japanese colleague Kurosawa (1926), whose curiosity about growth regulation was stimulated by plants showing the "foolish rice disease." He showed for the first time that culture filtrates from *Gibberella fujikuroi* could induce the same growth-promoting effects as the fungus itself. His curiosity led to the discovery and characterization of an entirely new class of powerful growth-regulating compounds. Soon plant physiologists and biochemists such as Yabuta and Sumiki (1938) took up the torch and showed that gibberellins are produced in healthy plants as well as sick ones and that those compounds play an essential role in the regulation of growth and differentiation in many species of healthy plants. This case illustrates the influence of plant pathology on other sciences. It also emphasizes that healthy plants are a special case of the diseased. This latter point provides encouragement for increased cooperation and collaboration between plant pathologists and plant physiologists.

Progress by plant physiologists in understanding the role of various growth regulatory substances in normal plant growth and differentiation inspired plant pathologists to seek understanding of abnormal growth in diseased plants by applying similar methods. Coleoptile tests, callus growth assays, and specific chemical procedures were used to show that many pathogens produce growth-regulating compounds including auxins, cytokinins, gibberellins and ethylene (Sequiera, 1963). These bioassay procedures are elaborate and tedious. Thus, few plant physiologists, let alone plant pathologists, could easily tool up to assay for more than one of these types of compounds at a time.

After years of descriptive analytical research with single compounds, plant physiologists finally concluded it was not the concentration of single compounds that controls growth in plants but rather the dynamic, changing ratios of two, three, or sometimes all four types of compounds. The same "principle of ratios" apparently applies in diseased plants as well as in healthy ones. What a blow! Suddenly, the cost of measuring the amount and activity of plant growth substances doubled or tripled. As a result, research on these aspects of plant pathogenesis will be an expensive and risky business for some time to come. In this case hindsight has shown how badly we needed a more adequate conceptual basis for experimental analysis of abnormal growth (Sequiera, 1973). (See Chapter 8, this volume.)

3. The Descriptive Analysis of Model Systems

Another striking trend in the analysis of disease in individual plants is the consistency with which plant pathologists have selected certain particular sets of experimental materials and methods of approach. It appears that many of us have had a preferred host and pathogen which we liked to call a "model system" on the assumption that it would turn out to be "representative of other major diseases." The "model" typically involved a matrix of four experimental organisms—a virulent and an avirulent genotype of the same pathogen and a resistant and a susceptible variety of the same host. The ideal case was one in which a single gene was believed to control resistance to the pathogen in the host and another single gene was believed to control virulence in the pathogen. Inoculations were made with all four possible combinations among the resistant and susceptible varieties and the virulent and avirulent pathogens. Then, periodic measurements were made of some particular metabolite—an enzyme, a toxin, a phytoalexin, a nucleic acid, a phenol, etc. The four curves resulting from these measurements were then plotted and compared, with the hope of discovering "the key" (note singular) to pathogenesis or "the ultimate cause of disease" in the model system.

This approach has given rise to a large amount of descriptive data which has proved to be of limited general value for two reasons: the assumption of a single controlling factor, and the lack of an adequate conceptual basis for integrating the data from many diverse model systems. Even with some of our most frequently studied pathogens the extent of our ignorance remains profound. In 1970, Bateman wrote of *Rhizoctonia solani*: "In general there appears to be a great void in our knowledge of the physiological characteristics of Rhizoctonia-infected plants." As Bateman also suggests in Chapter 3, this volume, a more substantial investment of time and energy in analysis of a few carefully selected diseases will yield greater benefits of understanding plant disease than continuing study of many different diseases.

4. The Chemical Basis of Resistance

Studies of the chemical basis for disease resistance provide another striking example of the importance of an adequate conceptual basis for experimental analysis of disease. In 1933, Link and Walker published their now classic paper on the resistance of red onions to smudge caused by *Colletotrichum circinans*. They showed that the dead outer scales of red onions contain protocatechuic acid and that this substance, which does not accumulate in yellow onions or in the inner scales of red onions, inhibits germination of *Colletotrichum* spores and thus accounts for the resistance of mature red onions to this disease. This is an elegant piece of work! As Daly (1976) asserted only recently, "it remains as one, if not the only, clear case where the chemical basis for resistance is known."

This achievement stimulated a great flurry of efforts during the next three decades to identify in other plants preexisting chemical factors which would explain resistance in a similarly definitive manner. Unfortunately, almost all these efforts have not been successful (Schönbeck and Schlösser, 1976). Compounds formed prior to infection appear to have importance in a relatively few diseases of which the heart rots of living trees are one important class (Scheffer and Cowling, 1966).

Some years after Walker and Link's classic work, Müller and Börger (1940) enunciated their phytoalexin theory of resistance. This theory stimulated much research by Cruikshank (1963), Kuć (1976), and many others. It holds that specific compounds formed in response to infection account for the resistance of most plants to disease. The present status of this theory will be discussed in depth in Volume V of this treatise.

5. *Confusion Over the Difference between Parasitism and Pathogenism*

As Bateman shows more thoroughly in Chapter 3, in this volume, many plant pathologists have emphasized the nutritional aspects of parasitism and pathogenism. Thus, many of us have tried to explain pathogenesis using a conceptual basis for experimental analysis that is applicable only to hungry parasites rather than irritating pathogens.

To clarify the often subtle distinctions between these two aspects of disease, Horsfall and Dimond (1960) offered the so-called "mother-in-law" analogy. Admittedly with some trepidation we offer this analogy once again. Although the phrasing has been changed the message of this analogy remains the same:

> Living in a commensal relationship with ones' mother-in-law is an uncertain business. If she is a delightful lady who pays the cost of her room and food at the table and never complains about the children, you have symbiosis. If she depends on you for her income and joins with the family for her meals but still never complains, you have parasitism, but not pathogenesis. If she is a constant source of irritation to the family, you have pathogenesis whether she pays her own way or not—if she pays her own way, she is only a pathogen, if she does not pay her own way, you have both parasitism and pathogenesis.

IV. WHERE ARE THE POTENTIALS FOR PROGRESS IN PLANT PATHOLOGY TODAY?

A few general inferences can be drawn from the above analysis. The first and most obvious is that rational analysis of parasitism and pathogenesis requires a more adequate conceptual basis for experimentation than has been used in the past. The second is that direct methods for the measurement of changes in physiological functions are more likely to facilitate understanding than indirect methods. These are the keystones to successful research to determine how plants suffer from disease and the mechanisms of offense by the pathogens and defense by the host. This is the major point of this chapter and also of Chapter 3, this volume.

We see great merit in McNew's "Theory of Vital Functions" and have organized the various chapters of this volume to emphasize the present status of our knowledge and promising methods for the analysis of dysfunction in these vital processes. We have enlarged the list of functions proposed by McNew but the major tenets remain the same:

> Disease is an abnormal physiological process in plants. . . . A major objective of plant pathology is to learn how these processes are injured and to design methods of preventing or circumventing that injury . . . the system

is natural enough to permit the development of some general laws and principles for understanding disease and its control.

Optimum progress in understanding how plants suffer from disease will require much greater interaction of plant pathologists with scientists in other disciplines—especially plant physiology, biochemistry, biophysics, biomathematics, and ecology. We must learn to use the best analytical tools of these fields and to harness them systematically with an adequate conceptual theory of disease.

It will be exciting to watch and to participate in the progress of plant pathology in the years ahead. We believe that McNew's "Theory of Vital Functions" and Bateman's "Multiple Component Hypothesis of Disease" (see Chapter 3, this volume) will show us where the prospecting for new knowledge is most likely to pay off in improved understanding in the future.

As this volume goes to press, we pause once again to remember the prophet's prayer: "Lord, help our words to be gracious and tender today for tomorrow we may have to eat them."

V. SOME HIGHLIGHTS OF VOLUME III

This volume begins with two keystone chapters that set the stage for all the chapters that follow in Volumes III, IV, and V. Knowledge of "How Healthy Plants Grow" is essential to understanding how plants suffer from disease. In Chapter 2, Kozlowski outlines the array of physiological functions with which pathogens interfere. What are the links between these functions? How does the plant regulate its functions to cope with constantly changing stresses imposed by the environment as well as by pathogens?

In Chapter 3, Bateman has reviewed the strengths and weaknesses in the conceptual theories that have guided the progress of plant pathology since the time of Kühn. To guide future research he proposes that disease should be considered the end result of all, not just one or two, of the positive and negative influences of host on pathogen and pathogen on host. It makes a lot of sense.

With Chapter 4 we begin a series of chapters that expands McNew's list of six vital functions but retains its essential features. The first of these chapters is "The Capture and Use of Energy by Diseased Plants." Plants grow when the energy captured in photosynthesis exceeds the energy lost in respiration, including photorespiration. Many diseases interfere with photosynthesis, respiration, and photorespiration by the host. The pathogens also carry on respiration. How are these processes governed in diseased plants? Kosuge provides stimulus for our thinking about these processes in Chapter 4.

Many diseases cause dysfunction in the flow and storage of food. Any virus, mycoplasma, fungus, bacterium, or seed plant that disrupts the phloem disrupts the flow of food. In Chapter 5, Zimmerman and Mc-Donough tell how the phloem works and then go on to describe the suffering in plants that results when the phloem is disrupted.

Water is both an essential metabolite and a solvent for all the solutes in plants. Many diseases cause dysfunction of the water system. They interfere with uptake of water by roots, its movement into the xylem, its transport in the xylem, its movement in the lamina of leaves, and its loss through the stomates and hydathodes. In Chapter 6, Talboys describes the biophysics and physiology of these processes and tells where the prospecting for new knowledge will be productive in understanding "Dysfunction of the Water System."

Disturbed mineral nutrition is one of those fertile border zones between the fields of plant pathology and physiology, on the one hand, and the fields of agronomy, horticulture, and forestry, on the other. In Chapter 7, Huber tells how plants suffer from deficiencies of any one or more of 15 essential elements and excesses of toxic elements in soil.

Alteration of growth is a common dysfunction due to disease caused by many bacteria, fungi, viruses, and some pathogenic seed plants. Hormone imbalance often is involved. In Chapter 8, Helgeson has assessed our knowledge and research approaches in this field. Crown gall is an extreme case of altered growth. In Chapter 9, Merlo describes the physiological and biochemical basis of this special type of abnormal growth. Monstrosities of many types are induced by various teratogenic agents in plants—insect galls, tumors, burls, faciations, witches brooms, root knots, club roots, leaf enations, and phyllody. In Chapter 10, Wilson describes the common threads that connect these monstrosities.

Many plants need beneficial symbionts to stay healthy. Ectotrophic and endotrophic mycorrhizae and nitrogen-fixing bacteria on the roots of plants are the best known symbionts. They help obtain nutrients and protect against feeder root pathogens in return for photosynthate from the host. In Chapter 11, Bowen tells how plants suffer from dysfunction in their symbiotic associations.

Many plants suffer from what Gäumann called venereal disease—diseases that disrupt reproduction. How do diseases affect pollen production and function? How do they produce phyllody and transvestism? Many diseases are seed borne. How do pathogens affect the germinability of seeds and the growth of seedlings? Mathre lays this subject before us in Chapter 12.

One of the dramatic manifestations of disease is necrosis—tissues decay, tubers rot, wood rots, fruit rots, and leaves develop spots caused by pectinases, cellulases, ligninases, etc. In Chapter 13, Mount describes the state of our expanding knowledge of the disintegration of tissue.

Plants are cantilevered beams anchored in the soil. If the anchorage is weakened by rot in the roots, or the beam is weakened by cankers or decay in the stem, the plant may be toppled by wind or heavy loads of snow. Most plants that do not stand erect do not produce well or are not harvested. In Chapter 14, Cowling builds on the foundation in Chapter 13 to introduce, "The Engineering Mechanics of Pathogenesis."

Membranes are the structures that organize metabolites and organelles into functional units. When cells are diseased, the membranes are changed, the organelles are modified or disrupted, the flow of metabolites is altered, and the cells leak. In Chapter 15, Wheeler tells us what we know and how we can learn more about the "Alteration of Permeability and Membranes."

Healthy plants carry on their metabolism through a remarkable array of well-integrated synthetic and catabolic pathways. The products include a great diversity of organic compounds. Kuc tells us in Chapter 16 about these changes in intermediary metabolism during disease.

"Transcription and Translation in Diseased Plants" is the subject of Chapter 17. There Samborski, Rohringer, and Kim describe the effect of infection on synthesis of nucleic acids and proteins in plants infected by bacteriophages, viruses, bacteria, and fungi. This is a prospecting chapter with emphasis on promising approaches in future research.

Senescence and plant disease have much in common. This is the subject of Chapter 18 in which Farkas provides a comparative analysis of these systems with a view to testing the general theory that disease in many plants is really premature senescence.

Biological clocks have been developed by plants to help them cope with daily and seasonal rhythms in their environment. How do diseases affect these rhythms? In Chapter 19 Tibbitts has described biological rhythms in diseased plants—a new topic in plant pathology but an important aspect of plant disease.

VI. AN OVERVIEW OF THE TREATISE

Each volume of this treatise must stand alone. For this reason, we have included in this section a brief rationale for the treatise as a whole. This treatise is designed for advanced researchers in plant pathology, whatever their specialty and status may be, from applied mycologist to virologist and from graduate student to Nobel laureate. We hope it will broaden their view, stimulate their thinking, and help them to synthesize still newer ideas and to relate the previously unrelated. To do so this treatise must be comprehensive and timely, provocative and forward looking, practical and theoretical in outlook, and well balanced in its coverage. We hope you find it so.

We chose to call this treatise "Plant Disease," not "Plant Pathology." As discussed above, the term "plant pathology" means the study of suffering plants. Study is something man does. Man may suffer when disease hits his crops, but his suffering is secondhand. It is the plant that is sick, not the man. We seek to understand disease as plants experience it and thus to make this treatise plant- rather than man-centered.

To look at disease as plants do, plant pathologists must learn enough to predict the reactions of plants to pathogens. That is a tall order! It is hard enough to understand how healthy plants grow. It is even more difficult to understand how plants behave when they are sick. The acid test of our understanding of disease processes is our ability to predict the progress of disease both in individual plants and in whole populations in the field.

Many earlier books about disease in plants have been given titles with plural subjects like "Manual of Plant Diseases" (Sorauer, 1914; Heald, 1933); "Recent Advances in Plant Viruses" (Smith, 1933); "Plant Diseases and Their Control" (Simmonds, 1938), or "Pathology of Trees and Shrubs" (Peace, 1962). Use of the plural is understandable—the total number of diseases is as astronomical as the national debt.

It is impossible to learn all the diseases of plants. Even to learn all the diseases of one plant is a herculean task. So, we designed this treatise to emphasize the commonalities of disease in plants—the unifying principles and concepts that will integrate our thinking. This rationale was used in the design of the original treatise. It emphasizes the common features of disease rather than the diversifying factors that tend to fragment our thinking. Thus, as a symbol of our desire for synthesis and unification, we left the "s" off diseases and called this treatise just "Plant Disease: An Advanced Treatise." *

Given five volumes in which to set out the art and science of plant disease, we found the subject could be divided readily into the required five parts:

I. How Disease Is Managed
II. How Disease Develops in Populations

---

* After our decision was made we found that Russell Stevens (1974) had also used the title "Plant Disease" for his recent introductory text.

III. How Plants Suffer from Disease
IV. How Pathogens Induce Disease
V. How Plants Defend Themselves

When the original treatise was designed in the late 1950's, plant pathology was reaching for maturity as a science and basic research was coming into its own. For that reason the original treatise was organized around the scientific foundations of disease processes. But during the last decade, society has called on plant pathology to demonstrate its usefulness in a world of worsening hunger. This treatise was conceived in 1975, the year when the world passed another great milestone along the road to global starvation. In that year the world population reached 4 billion people, enough to form a column marching 30 wide and 1 m apart around the equator. It scared the wits out of us!

The growing urgency of the world food problem made us decide that it was timely, even urgent, that this treatise should begin with the arts of disease management and go on to the science of plant disease. In this way it could also relate the basics of our science more effectively to its usefulness to society.

After long and earnest debate we decided to move from the general to the specific—to put management in Volume I and epidemiology in Volume II. This sequence provides the strongest possible foundation for understanding both the theory of disease management and the dynamic nature of disease, in Volumes III through V.

Volume I is not a book about how to control specific diseases. Rather, it is a theoretical and philosophical treatment of the principles of managing disease—by altering the genes of plants, by changing the associated microbiota, by selecting or altering the environment, or by using chemicals. Since the first volume sets the stage for the others, Volume I also contains chapters on the profession of plant pathology, its sociology, and how it works to benefit society.

After management in Volume I comes epidemiology in Volume II— "How Disease Develops in Populations." Since 1960 explosive progress has been made in understanding epidemics of plant disease. The latest explosion has come in the mathematical analysis of factors that make epidemics wax and wane. This provides the foundation for the emerging new field of theoretical plant epidemiology (see Zadoks and Koster, 1976). Volume II also includes analyses of the genetic base of epidemics, the methodology and technology of epidemiological analysis and forecasting, the concepts of inoculum potential and dispersal, the climatology and geography of plant disease, agricultural and forest practices that favor epidemics, and the use of quarantines as a defense against epidemics of introduced disease.

In the present volume, we move from disease in populations to disease in individual plants—"How Plants Suffer from Disease." The early chapters set the stage for all the later chapters in Volume III plus those in Volumes IV and V. First, they describe how healthy plants grow. Next comes a modern conceptual theory of how disease develops in plants. Here disease is presented as the end result of all the positive and negative influences of hosts on pathogens and pathogens on hosts. The later chapters describe the many different kinds of impairments that can occur when the systems in the plant are diseased.

Having set out in Volume III the potential for dysfunction due to disease, Volume IV will consider how pathogens induce these various dysfunctions—"How Pathogens Induce Disease." This volume will describe the concepts of single-, multiple-, and sequential causality and their relationship to stress; the evolution and energetics of parasitism and pathogenism; the concepts of allelopathy and iatrogenic disease; the structure and function of toxins; and the role of pathogen enzymes and other regulatory metabolites in disease processes. Next we will compare and contrast the unique features of all the various pathogens of plants—fungi, bacteria, insects, mycoplasma, rickettsia, parasitic seed plants, nematodes, viruses, viroids, air pollutants, and so on. How are the effects of these pathogens similar? How are they unique? What offensive weapons does each type of pathogen use to be successful? Finally, we will consider the effects of diseased plants and pathogens on livestock and man.

Volume IV, in turn, sets the stage for Volume V—"How Plants Defend Themselves." The chapters of this final volume will be closely linked with the analogous chapters in Volumes III and IV. Volume IV deals with how pathogens thwart the defenses of the host. Volume V will describe how plants thwart the offenses of pathogens.

Plants have many natural enemies and they have evolved a magnificent array of armaments to keep their enemies out or to minimize the damage they cause once they get in. Some plants escape from disease. Others tolerate disease and grow well in spite of their sickness. Still others have evolved mechanisms to defend against disease, often with great success.

Volume V has been developed by analogy to the defense of a medieval castle. Those castles had defenses at the perimeter—the outer walls and gates. In case the gates did not hold, the castles also had internal defenses. There are dynamic defenses triggered by the invader and even defenses triggered by previous invaders. What did the previous invaders teach the occupants of the castle? The final chapter will describe how the metabolic resources of the plant are allocated to maintain and repair its defenses. The dynamic competition between the offensive weapons of the pathogen described in Volume IV and the defensive weapons of the host described in Volume V should read like a battle royal.

Since the treatise begins with management in Volume I, it is fitting that it end with defense in Volume V—a major goal of integrated management being the enhancement of natural defenses against disease.

References

Barrs, H. D. (1968). Determination of water deficits in plant tissues. *In* "Water Deficits and Plant Growth" (T. T. Kozlowski, ed.), Vol. 1, pp. 235–368. Academic Press, New York.

Bateman, D. F. (1970). Depletion of the galacturonic acid content in bean hypocotyl cell walls during pathogenesis by *Rhizoctonia solani* and *Sclerotium rolfsii*. *Phytopathology* **60**, 1846–47.

Brown, R. W., and van Haveren, B. P., eds. (1972). Psychrometry in water relations research. *Utah Agric. Exp. Stn.*, Utah State University, Logan, Utah.

Cook, R. J., and Papendick, R. I. (1972). Influence of water potential of soils and plants on root disease. *Annu. Rev. Phytopathol.* **10**, 349–374.

Cruikshank, I. A. M. (1963). Phytoalexins. *Annu. Rev. Phytopathol.* **1**, 351–374.

Daly, J. M. (1976). Some aspects of host-pathogen interactions. *In* "Physiological Plant Pathology" (R. Heitefuss and P. H. Williams, eds.), pp. 27–50. Springer-Verlag, Berlin and New York.

Duniway, J. M. (1976). Water status and imbalance. *In* "Physiological Plant Pathology" (R. Heitefuss and P. H. Williams, eds.), pp. 430–449. Springer-Verlag, Berlin and New York.

Friend, J., and Threlfall, D. R., eds. (1976). "Biochemical Aspects of Plant-Parasite Relationships." Academic Press, New York.

Gäumann, E. A. (1946). "Pflanzliche Infectionslehre." Birkhaeuser, Basel.

Goodman, R. N., Kiraly, Z., and Zaitlin, M. (1967). "The Biochemistry and Physiology of Infectious Plant Disease." Van Nostrand-Reinhold, Princeton, New Jersey.

Heald, F. (1933). "Manual of Plant Diseases," 2nd ed. McGraw-Hill, New York.

Heitefuss, R., and Williams, P. H., eds. (1976). "Physiological Plant Pathology." Springer-Verlag, Berlin and New York. 890 pp.

Horsfall, J. G., and Dimond, A. E., eds. (1959). "Plant Pathology: An Advanced Treatise," Vol. 1. Academic Press, New York.

Horsfall, J. G., and Dimond, A. E., eds. (1960). "Plant Pathology: An Advanced Treatise," Vol. 2. Academic Press, New York.

Kuć, J. (1976). Phytoalexins. *In* "Physiological Plant Pathology" (R. Heitefuss and P. H. Williams, eds.), pp. 632–752. Springer-Verlag, Berlin and New York.

Kurosawa, E. (1926). Experimental studies on the secretion of *Fusarium heterosporum* on rice plants. *J. Nat. Hist. Soc. Formosa* **16**, 213–227.

Link, K. P., and Walker, J. C. (1933). The isolation of catechol from pigmented onion scales and its significance in relation to disease resistance in onions. *J. Biol. Chem.* **100**, 379–383.

McNew, G. L. (1960). The nature, origin and evolution of parasitism. *In* "Plant Pathology: An Advanced Treatise" (J. G. Horsfall and A. E. Dimond, eds.), Vol. 2, pp. 19–69. Academic Press, New York.

Müller, K. O., and Börger, H. (1940). Experimentelle Untersuchungen über die Phytophthora-Resistenz der Kartoffel; zugleich ein Beitrag zum Problem der "erworbenen" Resistenz im Pflanzenreich. *Arb. Biol. Reichsanst. Land- Forst-wirtsch., Berlin-Dahlem* **23**, 189–231.

Peace, T. R. (1962). "Pathology of Trees and Shrubs." Oxford Univ. Press, London and New York.

Scheffer, T. C., and Cowling, E. B. (1966). Natural resistance of wood to microbial deterioration. *Ann. Rev. Phytopathol.* **4**, 147–170.

Schönbeck, F., and Schlösser, E. (1976). Preformed substances as potential protectants. *In* "Physiological Plant Pathology" (R. Heitefuss and P. H. Williams, eds.), pp. 653–673. Springer-Verlag, Berlin and New York.

Sequeira, L. (1963). Growth regulators in plant disease. *Annu. Rev. Phytopathol.* **1**, 5–30.

Sequeira, L. (1973). Hormone metabolism in diseased plants. *Annu. Rev. Plant Physiol.* **24**, 353–380.

Simmonds, J. H. (1938). "Plant Diseases and Their Control." David White, Brisbane, Australia.

Slovik, B. (1974). "Methods of Studying Plant Water Relations." Czech. Acad. Sci., Prague.

Smith, K. M. (1933). "Recent Advances in Plant Viruses." Mc-Graw-Hill (Blakiston), New York.

Sorauer, P. (1914). "Manual of Plant Diseases," Vol. I. Record Press, Wilkes-Barre, Pennsylvania.

Stakman, E. C. (1969). The need for intensified and integrated campaigns against pests and pathogens of economic plants. Unpublished memorandum to the Rockefeller Foundation, March 31, 1969. 47 pp.

Stevens, R. B. (1974). "Plant Disease." Ronald Press, New York.

Ten Houten, J. G. (1959). Scope and contributions of plant pathology. *In* "Plant Pathology: An Advanced Treatise" (J. G. Horsfall and A. E. Dimond, eds.), Vol. 1, pp. 19–60. Academic Press, New York.

Yabuta, T., and Sumiki, Y. J. (1938). Isolation of gibberellin. (In Japanese) *J. Agr. Chem. Soc. Japan* **14**, 1526.

Zadoks, J. C., and Koster, L. M. (1976). A historical survey of botanical epidemiology. A sketch of the development of ideas in ecological phytopathology. *Meded. Landbouwhogesch. Wageningen* **76**, 1–56.

Chapter 2

How Healthy Plants Grow

T. T. KOZLOWSKI

I. INTRODUCTION

Knowledge of how healthy plants grow is essential to understanding how plants suffer from disease. Since a plant is diseased when an alteration is induced in its vital functions by a pathogen, the more we understand about these vital processes in healthy plants the more adequately we will understand how they are altered during disease.

When the editors asked me, a plant physiologist, to write this chapter on how healthy plants grow, they said it was their hope that the chapter would not only provide a reference point for the chapters that follow but also that it would stimulate greater cooperation among plant pathologists and plant physiologists. A famous Dutch plant pathologist wrote some years ago (Ten Houten, 1959) that plant physiology was the science most closely related to plant pathology. Ten Houten's view is correct—not only in theory but also in practice. Plant pathologists discovered the so-called Bakanae effect caused by gibberellins and thus opened up a new dimension in the study of growth regulation in plant physiology. Similarly, plant physiologists first described the cyanogenic glucosides in plants and thus paved the way for analysis of their role in disease resistance.

Some years ago I was asked to prepare a review article entitled "Plant Physiology and Forest Pests" (Kozlowski, 1969). Writing that paper taught me a great deal about pathogens of forest trees and about the linkages and potential for mutually beneficial cooperation between our two fields.

With these ideas and experiences as background, this chapter was designed with two specific goals. The first is to outline in general terms the array of physiological functions that are essential in the growth and development of healthy plants. These are the vital processes with which pathogens interfere when they induce disease. They also provide the organizational framework for Chapters 4–18 in this volume. Many of these functions are linked together so that dysfunction in one system leads to dysfunction in other systems. The environment in which plants grow is constantly changing. Thus, we shall also discuss how plants regulate their functions so as to cope with the constantly changing stresses and strains induced by environmental as well as by pathological stresses.

The second major purpose of this chapter is to call attention to sources of some of the most reliable and useful methods that can be used to study dysfunction in the various vital processes of plants. These methods hold the key to advancing the frontiers of both of our fields. Using them to study healthy as well as diseased plants will enhance our understanding of both.

II. PHYSIOLOGY OF GROWTH

Growth of multicellular plants is the end result of a series of coordinated physiological processes that depend on the genetic characteristics of the plant, and a myriad of environmental factors. The various physiological and biochemical processes are linked together by regulatory processes which control photosynthesis, respiration, translocation, assimilation, etc.

The impact of disease on the growth of plants can be measured and analyzed by classical methods that have been applied to study genetic and environmental influences on growth. Single measurements of yield or plant size at the end of the growing season have often been used. Much more information can be gained, however, by repeated measurements at various intervals during a single or many growing seasons. In this way rates of growth can be determined quantitatively and growth effects partitioned among various genetic, pathological, and environmental components. At each experimental interval, various dimensions of harvested plants are measured—the height and girth of the whole plant; or the area, volume, fresh weight, or dry weight of the leaves, fruits, stems, roots, or other organs. These various components of growth are then assembled by regression or other methods of statistical analysis and the results expressed in such parameters as leaf area index or annual increase in biomass. These methods of growth analysis can be applied equally well to experimental plants in controlled environments, in field plots, or even to whole agricultural or natural ecosystems.

Increase in the size of plants is the result of changes in cell division and expansion. Cell enlargement is followed by differentiation and morphogenetic changes in which specialized organs are formed. Thus, from a developmental point of view, the sequential component processes of growth include cell division, cell enlargement, differentiation, and morphogenesis.

In the earliest stages of plant development, rates of increase in size or dry weight usually are approximately linear with respect to time. Eventually, however, various internal growth-controlling mechanisms induce departure from linearity so that, over a long period, growth can best be characterized by a sigmoid curve. Both seasonal and lifetime growth of shoots, roots, and reproductive structures conform to such a pattern.

Readers who wish to consider the general concepts and detailed procedures of growth analysis will find the book by Evans (1972) and the review by Erickson (1976) of special value.

A. Food Relations

Carbohydrates are the most important class of foods synthesized by plants. They are the direct products of photosynthesis and the raw materials from which proteins, fats, nucleic acids, and other cell constituents are synthesized. They also provide most of the energy required to drive metabolic processes (Kozlowski and Keller, 1966). Growing plants require a continuous supply of energy for the synthesis of protoplasm, maintenance of structure of organelles and membranes, and active transport of ions and molecules across membranes. Such energy is provided by respiration (oxidation of food resulting in the release of energy). Oxidation of glucose occurs in the cytoplasm by glycolysis (conversion of glucose to pyruvic acid), followed by the Krebs cycle (oxidation of pyruvic acid to CO_2 and water) in the mitochondria. In addition to producing a variety of intermediate compounds, the Krebs cycle produces large amounts of ATP which provides energy for various processes. Numerous enzymes are involved in glycolysis and the Krebs cycle (Meyer *et al.*, 1973).

When energy or substance is needed in other organs, carbohydrates are mobilized. This requires their conversion to soluble forms, loading into the phloem, flow under pressure, and assimilation into new tissues. Reproductive organs often place heavy demands on carbohydrate reserves. These demands are often met at the expense of vegetative tissues. Thus, when a heavy crop of fruit or seeds is being set, the reproductive structures become powerful sinks into which available carbohydrates are translocated rapidly (Dickmann and Kozlowski, 1968, 1970). Diseased tissues and organs often become similarly powerful metabolic sinks (Heitefuss and Williams, 1976).

Redistribution is the term plant physiologists use to refer to the process of mobilizing carbohydrates and mineral elements to meet metabolic needs in other parts of a plant. For excellent reviews of the general subject the reader is referred to the books by Crafts and Crisp (1971), Epstein (1972), Peel (1974), Zimmermann and Milburn (1975), and Wardlaw and Passioura (1976). In Chapter 4 of this volume Kosuge discusses dysfunction in the capture and utilization of energy and in Chapter 5, Zimmermann and McDonough discuss dysfunction in the flow of food.

1. Photosynthetic Pathways

The major sequential events in photosynthesis include trapping of light energy by chloroplasts, conversion of excitation energy to chemical energy, transfer of electrons leading to generation of NADPH and ATP in a phase requiring light, and use of the energy of NADPH and ATP to fix CO_2 and produce fructose.

In most higher plants, incorporation of carbon into organic molecules occurs in the stroma of the chloroplast and does not require light, even though it normally occurs in the light. With expenditure of additional energy of ATP more complex carbohydrates are produced from fructose.

Many intermediate products are formed in the three species-specific biochemical pathways of photosynthesis. These are listed below.

a. C_3 Plants. Most higher plants follow the Calvin cycle characterized by formation of stable three-carbon compounds (Zelitch, 1971; Whittingham, 1974). In these so-called C_3 plants, a five-carbon sugar, ribulose 1,5-bisphosphate (RuBP) is carboxylated to form two molecules of 3-phosphoglycerate (PGA) in a reaction catalyzed by RuBP carboxylase. At each turn of the Calvin cycle a molecule of CO_2 enters and is reduced while a molecule of RuBP is formed. After six revolutions one molecule of hexose sugar is formed and six molecules of RuBP are reformed to continue the cycle.

b. *C_4 Plants.* Many higher plants of the tropics and subtropics, including corn, sorghum, and sugar cane, follow the Hatch–Slack pathway in which the four-carbon acid, oxaloacetic acid, is formed when CO_2 is added to the three-carbon compound, phosphoenolpyruvate (PEP). The reaction is catalyzed by PEP carboxylase. In C_4 plants, malic and aspartic acids are formed and broken down to produce CO_2 and pyruvic acid. The CO_2 is transferred to RuBP of the Calvin cycle and pyruvic acid reacts with ATP to form more PEP. Hence photosynthesis of so-called C_4 plants differs from that of C_3 plants only in the initial steps.

The leaf structures of C_3 and C_4 plants are not the same. In C_3 plants, the chloroplasts are similar throughout the palisade parenchyma and the Calvin cycle occurs in each cell. In C_4 plants only certain specialized cells fix most of the CO_2. C_4 plants have a specialized "Kranz" anatomy with vascular bundles surrounded by large bundle sheath cells having prominent chloroplasts and abundant starch grains. By comparison, the chloroplasts of mesophyll cells usually lack starch. It is only the mesophyll cells that fix atmospheric CO_2 by the PEP system and they lack the Calvin cycle. Both malic and aspartic acids that form in the mesophyll cells are translocated to the bundle sheath cells where the Calvin cycle is operative and CO_2 is added.

C_4 plants appear to be unusually well adapted to regions undergoing stress conditions of drought and high temperatures. The optimum temperature and light intensity for photosynthesis are both much higher for C_4 than for C_3 plants. In fact, productivity of C_4 plants often is very high at temperatures that are lethal for C_3 plants. Atmospheric CO_2 is used more efficiently by C_4 than by C_3 plants. In C_3 plants the action of RuBP carboxylase is inhibited because O_2 competes with CO_2, causing a light-enhanced respiration (called photorespiration). In contrast C_4 plants do not show photorespiration. Methods of studying photorespiration are discussed by Jackson and Volk (1970). In C_4 plants the two separate carboxylation reactions lead to efficient CO_2 absorption even at low CO_2 concentrations followed by transfer to the Calvin cycle which produces the photosynthetic products needed in growth.

c. *Crassulacean Acid Metabolism (CAM) Plants.* In a number of succulent plants, stomata typically are open during the night and closed during the day. These plants also accumulate malic and citric acids at night and reconvert them to CO_2 during the day. The biochemical pathways of CAM plants are similar to those of C_4 plants. However, the PEP carboxylase reaction occurs primarily in the dark. The CO_2 that is then produced in the light from organic acids is subsequently fixed by RuBP carboxylase in a C_3 pathway.

Many plant diseases interfere with photosynthesis as Kosuge has described in Chapter 4, this volume. As his discussion shows, very few studies have been made of the physiological mechanisms by which this interference is achieved. Readers interested to better understand the biochemistry of photosynthesis will find the books by Zelitch (1971), Whittingham (1974), and Gregory (1977) of special interest. Methods useful in the analysis of dysfunction in photosynthesis are outlined by Sestak *et al.* (1971).

B. Water Relations

Water is essential for plants as the major constituent of physiologically active tissues. It is a reagent in photosynthesis and in hydrolytic processes as well as a solvent in which salts, sugars, and other solutes move from cell to cell. Water is also essential for maintenance of turgor in tissues and organs.

Most higher plants absorb water as a liquid from the soil. A few plants absorb water as vapor or as condensed water (dew) from the atmosphere. Thus, both roots and shoots of most plants are capable of absorbing moisture from the environment. Desert and other xerophytic plants have developed many specialized mechanisms for the absorption and conservation of water (Kozlowski, 1976b).

Most higher plants move large amounts of water from the soil to the atmosphere, but only a very small fraction of this water is used directly by plants in growth. Many plants require 250 to 1000 g of water to produce 1 g of dry matter. The remainder of the absorbed water merely goes through the plant and is lost to the atmosphere.

In many regions of the world, water supply limits the growth and survival of plants more than any other single factor. Soil moisture supplies rarely are optimal for growth. The difference in yield of crops with and without irrigation in arid regions shows the tremendous impact of water deficits in plants. But even in regions of adequate average rainfall, the staggering amounts of losses due to temporary water deficits often are not realized because data are not available to show how much more growth would occur if plants had favorable water supplies throughout the growing season (Kozlowski, 1968b). The book edited by Kozlowski (1978) and Chapter 6 in this volume by Talboys discuss how various pathogens induce dysfunction in the water system of plants.

The best measure of the water-status of plants is water potential (ψ)

—the difference between the free energy of water in a biological system and that of pure free water. The major forces affecting the water potential (ψ) of plant tissues are the solute (osmotic) potential (ψ_s), due to dissolved salts and sugars, and the pressure potential (ψ_p) caused by turgor pressure. Hence, the ψ of a cell can be characterized by:

$$\psi_{cell} = \psi_s + \psi_p$$

The value of ψ_s is negative; of ψ_p positive; and ψ_{cell} is negative except that in a fully turgid cell it is zero. The pressure bomb method and the highly reliable methods of thermocouple psychrometry are recommended for analysis of dysfunction in the water systems of plants. Barrs (1968) and Slavik (1974) give excellent discussions of the details and limitations of these and other methods for studying plant–water relations.

1. Development of Water Deficits

Soil–plant–atmosphere is a physical continuum through which water moves along a path of decreasing potential energy, that is, toward a more negative value of ψ. The ψ gradient is induced by transpirational water loss, thereby inducing a coupled water deficit in leaves, shoots, roots, and soil.

The path of water transport includes movement in soil toward roots, absorption by roots, movement across root tissues into xylem elements and upward to the leaves, evaporation into the intercellular spaces of the leaves, and diffusion as water vapor through the stomata into the atmosphere. Water encounters resistance to movement during its ascent. The resistance is greater in the soil than in the plant and is greatest in the transition from the leaves to the atmosphere where water changes from liquid to vapor (Hillel, 1970).

Upward water movement through the soil–plant–air continuum requires that the ψ_{plant} be more negative than the ψ_{soil}, with the lowest ψ being in the leaves. Evaporation and absorption from an initially wet soil progressively reduce both soil moisture content and ψ_{soil} (Fig. 1). Concomitantly ψ_{plant} is reduced, resulting in a plant water deficit. Hence, ψ of a plant growing in drying soil declines daily. Superimposed on this trend are diurnal variations in plant water balance. These are controlled by relative rates of absorption of water and of transpirational losses. During the day, when stomata are open, the rate of transpiration exceeds the rate of absorption, thus creating a water deficit in the plant. This deficit, however, is reduced or eliminated during the night, when both rates of absorption and transpiration are low, but the rate of absorption is greater, causing the plant to rehydrate (Kozlowski, 1968b).

Figure 1 shows the progressive changes in ψ_{leaf}, ψ_{root} (water potential

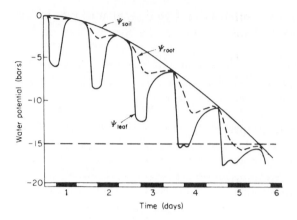

Fig. 1.   Changes over a 5-day period in leaf water potential (ψ_{leaf}), root water potential (ψ_{root}), and soil mass water potential (ψ_{soil}) as transpiration occurs from a plant rooted in originally wet soil. The horizontal dashed line indicates the value of ψ_{leaf} at which wilting occurs. For explanation see text. (From Slatyer, 1967.)

at the root surface), and ψ_{soil} as transpiration occurs over a 5-day period beginning when soil is at field capacity. During each day transpirational water loss reduces ψ in soil, roots, and leaves. Because absorption does not keep pace with transpiration, the deficit of water in the plant increases until absorption equals transpiration. When the soil is wet, water flow is maintained by only small differences in ($\psi_{soil} - \psi_{root}$). When transpiration declines late in the day and during the night, plant water content increases so $\psi_{leaf} = \psi_{root}$ and by early morning $\psi_{leaf} = \psi_{root} = \psi_{soil}$. However, as ψ_{soil} decreases (e.g., days 3 and 4) progressively higher values of ($\psi_{soil} - \psi_{root}$) are necessary to maintain water flow because the hydraulic conductivity of the soil rapidly decreases. By day 4, some stomatal closure decreases flow. Nevertheless,· equilibration of ψ_{leaf}, ψ_{root}, and ψ_{soil} occurs more slowly.

Leaf turgor declines during each successive day of a soil drying cycle and induces incipient wilting. This grades into temporary wilting, with drooping of leaves occurring during the day and recovery during the night. During sustained droughts, temporary wilting grades into permanent wilting, with leaves failing to recover turgidity at night. If it is assumed that wilting occurs when ψ_{leaf} reaches -15 bars (as shown in Fig. 1), then on day 4 the plants would be wilted for several hours during the day but would recover turgor during the night. However, by day 5, ψ_{soil} also decreases to at least -15 bars. Hence, the night of day 5, ψ_{leaf} would not recover sufficiently to induce turgor, since at the point when $\psi_{leaf} = \psi_{root} = \psi_{soil}$, the ψ would still be lower than -15 bars.

The plant would continue in a wilted state and become turgid only after soil water was replenished.

2. Effects of Water Deficits

Changes in physiological processes resulting from plant water deficits usually are reflected in reduced growth of vegetative and reproductive tissues. Shoot growth is inhibited by reduction in cell expansion of preformed shoot primordia as well as in formation of new primordia. In general, cell enlargement is more sensitive than cell division. Inhibition of shoot growth by water deficits varies among plant species and varieties; it also differs with the severity and timing of drought, and with different aspects of shoot growth such as bud formation, internode elongation, leaf expansion, and leaf abscission. In perennial plants various aspects of cambial growth, including number of cambial derivatives produced, seasonal duration of cambial growth, and early wood–late wood relations, are particularly responsive to plant water deficits. Water-stressed vegetative and reproductive tissues often shrink appreciably (Kozlowski, 1968e, 1972b; Chaney and Kozlowski, 1971; Pereira and Kozlowski, 1976). Water deficits in root meristems inhibit elongation, branching, and cambial growth in roots (Kozlowski, 1968a, 1971a,b). Severe water deficits may also inhibit various phases of reproductive growth, including floral initiation, fruit set, and fruit growth.

Sometimes, however, plant water deficits have some beneficial effects. For example, drought at a critical time in the reproductive cycle may stimulate formation of flower buds by inhibiting vegetative growth. Short periods of drought may also break the dormancy of the flower buds of some tropical species of woody plants (Alvim, 1977).

Internal water deficits affect a number of plant processes sequentially, and often concurrently. The first response to plant water deficit is likely to be a decrease in shoot and leaf growth caused by reduction in ψ. This is followed by reduced cell wall and protein synthesis in tissues with high growth potential. As tissue water deficits increase further, cell division may be inhibited and amounts of some enzymes may decline. Stomatal closure begins and rates of photosynthesis and transpiration consequently decrease. Abscisic acid (ABA) begins to accumulate (see Chapter 8 by Helgeson). By this time numerous secondary and tertiary changes occur. Further desiccation is accompanied by substantial decreases in respiration and translocation of carbohydrates and leaf cytokinins. Amounts of some hydrolytic enzymes may increase and ion transport decreases. Finally, proline accumulates and the rate of photosynthesis becomes low or negligible . Leaf senescence becomes apparent and old leaves are shed (see Chapter 18 by Farkas). These changes inevitably become associated with further reduction in plant growth (Hsiao, 1973). For more details on the effects of water deficits on various plant processes and on the growth of healthy and diseased plants the reader is referred to the five volumes on "Water Deficits and Plant Growth" (Kozlowski, 1968c,d, 1972a, 1976c, 1978).

Internal water deficits inhibit photosynthesis by inducing closure of stomata as well as changes in chloroplasts. The importance of stomatal control is shown by correlations between stomatal closing and decrease in photosynthesis during droughts and by midday decreases in photosynthesis during temporary stomatal closure. Photosynthesis also appears to be inhibited by changes in cell metabolism during drought. For example, in *Helianthus annuus* there was a general inhibition of the light reactions of photosynthesis when plant ψ dropped below -10 bars (Boyer, 1973). Decreased development of leaf area also contributes to reduction in photosynthetic activity of water-stressed plants. The combined loss in leaf area and photosynthetic activity represents a potentially large loss of photosynthate for crops.

3. Drought Resistance

Plants have developed two general mechanisms of tolerance to drought —drought avoidance and desiccation tolerance (capacity of protoplasm to survive when severely desiccated). Most higher plants depend on one or more of the drought-avoiding adaptations (Kozlowski, 1976a,b).

a. Leaf Shedding and Microphylly. In regions with pronounced dry seasons, many plants avoid desiccation by shedding leaves and twigs. For example, plants of the "caatinga" and "cerrado" ecosystems of Brazil shed leaves or branches during the dry season (see also Chapter 14 by Cowling, this volume). Many desert plants, such as species of *Artemisia, Noea, Haloxylon, Anabasis,* and *Zygophyllum* in Israel exhibit seasonal dimorphism by shedding and growing various types of branches at different seasons. Pathogens that inhibit leaf shedding would decrease the normal drought hardiness of these plants. Large winter leaves are replaced by small summer leaves. Orshansky (1954) considers the reduction of leaf surface to be the most important adaptation for drought avoidance of desert plants in the Near East. Pathogens that inhibit leaf shedding or alter normal patterns of seasonal dimorphism would decrease the normal drought hardiness of plants with these mechanisms of drought avoidance.

b. Control of Stomatal Aperture. Capacity to close stomata during droughts varies greatly both among species and within species of plants. Desiccation injury in abnormal diploid potato plants and wilt-prone mutants of tomato has been associated with their failure to close stomata under drought conditions (Waggoner and Simmonds, 1966; Tal, 1966). Xerophytic plants apparently have adapted to avoid drought by closing their stomata more firmly (show greater stomatal resistance than mesophytic plants) (Wuenscher and Kozlowski, 1971a,b). Plant pathogens may induce wilting and death of plants by interfering with normal mechanisms of stomatal closure (see Chapter 6 by Talboys, in this volume).

c. *Leaf Waxes*. Cuticular control of water loss is also extremely important in drought avoidance. Whereas many xerophytes can reduce transpiration by as much as 80–98% by closing stomata, mesophytes can reduce transpiration by only 50–80%; this emphasizes the importance of leaf waxes in drought avoidance. Wide variations occur in deposition of leaf waxes in different species of higher plants (Fig. 2). Superficial pathogens such as downy and powdery mildews may decrease drought resistance by increasing cuticular transpiration. Simulated acid rain has been shown to cause erosion of cuticular waxes (Shriner, 1976) and could decrease the drought resistance of plants.

d. *Stem Adaptations*. Many xerophytic plants avoid desiccation by producing few leaves and depending on green shoots for photosynthesis. Others produce living wood fibers that contain food reserves, develop a wide cortex that protects vascular tissues of young plants from desiccation before they develop periderm tissues, and develop interxylary cork (Kozlowski, 1976b).

e. *Root Adaptations*. A high ratio of root to shoot and foliar surface is conducive to drought avoidance. On dry sites, for example, *Eucalyptus socialis* is able to compete successfully with *E. incrassata* because of a higher root–shoot ratio (Parsons, 1969), seedlings of *Eucalyptus camaldulensis*, with stomata on both leaf surfaces, have higher rates of water loss than *E. globulus*, with stomata on the abaxial leaf surface only. Nevertheless, *E. camaldulensis* does not develop greater shoot water deficits than *E. globulus*, largely because the former produces a deeper and more branched root system that is very efficient in replacing transpirational loss (Pereira and Kozlowski, 1977). Feeder-root pathogens such as *Phytophthora cinnamomi* and structural root pathogens such as *Armillaria mellea* decrease the absorption and water-transport efficiency of plants.

C. Mineral Relations

Eleven mineral elements are essential for plant growth (K, P, S, Ca, Mg, Fe, Cu, Zn, B, Mn, and Cu). Each element plays a characteristic role in the structure and metabolism of plants. Many serve as coenzymes, buffers, regulators of osmotic and permeability phenomena, and anti-

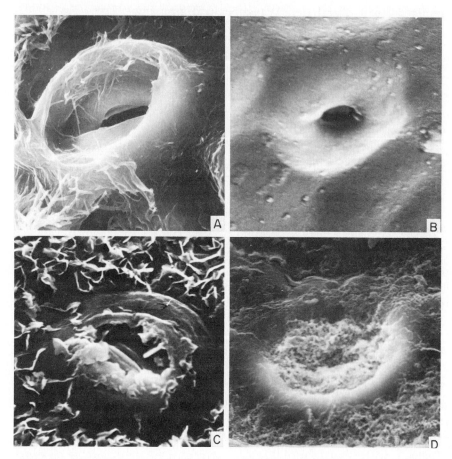

Fig. 2. Electron micrographs showing variations in structure and amount of wax deposited near stomata on leaves on (A) *Eucalyptus globulus,* (B) *Citrus mitis,* (C) *Acer saccharum,* and (D) *Pinus resinosa* (note occlusion of antestomatal chamber with wax). Photos by W. J. Davies and J. S. Pereira.

microbial substances. All of these essential nutrients are absorbed by roots from the soil solution. Dissolved substances in precipitation also are absorbed through foliar organs (Wittwer and Teubner, 1959). Foliar feeding is especially important in epiphytic plants and trees with a very large canopy of foliage.

Deficiencies of one or more elements result in both general and specific symptoms of mineral deficiency disease. These symptoms vary with the plant species, specific elements, and degree of deficiencies involved (Rains, 1976). In addition to inducing chlorosis and poor plant growth, symptoms of mineral deficiencies include necrosis, dieback of shoots, rosetting, gummosis, decreases in amounts of vascular tissue, retarded differentiation of tissues, injury to reproductive structures, and increased susceptibility to disease (Kozlowski, 1971a,b). The books by Hambidge (1941), Walker (1969), Epstein (1972), and Hewitt and Smith (1975) are excellent general references on effects of mineral deficiencies on plants. Disturbances of mineral relations induced by biotic pathogens are discussed by Huber, Chapter 7, this volume.

Excesses of mineral elements also cause disease in plants. Such excesses may decrease the soil ψ and thus induce water stress in plants through osmotic effects, and decreased absorption of water; they may also induce closure of stomata and thus inhibit transpiration and/or photosynthesis. Adding fertilizers to the soil late in the growing season sometimes maintains growth too long and does not allow enough time for perennial plants to develop frost resistance.

Mineral elements in seeds, together with carbohydrates, proteins, and fats, provide young seedlings with an initial supply of nutrients. Within a few days or weeks, however, these endogenous nutrients are depleted and additional supplies must be obtained from the soil or from precipitation. The elements in soil minerals vary from those relatively unavailable (for example, potassium ions bound in primary mineral lattices of feldspars) to readily available ones (such as ions in exchangeable, complexed, or surface forms that are more or less in equilibrium with ions dissolved in the soil solution).

Mineral requirements of plants have been studied by a variety of techniques including water cultures, sand cultures, soil cultures, plant injection, soil analysis, and foliar analysis. The limitations and advantages of these methods are discussed by Kramer and Kozlowski (1960), Smith (1962), Walsh (1971), and Van den Driessche (1974).

1. Absorption of Mineral Elements

Uptake of mineral elements from the rhizosphere involves a sequential series of steps: movement of ions to root surfaces, ion accumulation in root cells, radial transport of ions into xylem, and ion translocation from roots to shoots. Each step in this series is amenable to experimental analysis by procedures that are reviewed by Epstein (1972).

It often is assumed that absorption of ions occurs chiefly through the unsuberized region of the root just behind the root tip. However, considerable amounts of minerals also enter through suberized portions of roots several centimeters behind the tips. Salts enter suberized roots primarily by diffusion and by mass flow, through lenticels and openings caused by death of lateral roots.

As discussed more fully by Bowen in Chapter 11, this volume, and in the book by Marks and Kozlowski (1973), both mycorrhizae and rhizopheric microorganisms play important roles in increasing the availability of mineral elements in infertile soils. Although deficiency of an element can be partly overcome by mycorrhizal infection, the degree of plant response is also influenced by the concentration of nutrients in the soil (Bowen, 1973).

The distribution of mineral elements in mycorrhizal roots has been characterized most extensively with regard to phosphorus. Harley (1969) reported that 90% of the absorbed phosphate remained in the fungal sheath, did not mix with incoming phosphate, and was gradually released to the plant. In some plants movement of phosphate from the fungus to the higher plant occurs at the endodermis rather than to adjoining cells. In other plants with a restricted Hartig net, the movement of phosphate occurs first to cortical cells and then to the stele (Bowen, 1973).

Wilde (1968a,b) has emphasized that a considerable portion of the rhizosphere consists of fungi that are as important in symbiosis as are mycorrhizal fungi. In forest soils the rhizospheric mycelia, with an extensive surface area, derive energy from root exudates and from the decomposition of sloughed roots. Spyridakis et al. (1967) showed that chelating compounds formed in decomposition of tree roots were involved in transformation of biotite to kaolinite and subsequent availability of nutrients to trees.

2. Pathways of Mineral Transport

Ions move through the roots into the xylem and through leaves into the phloem. Once ions traverse the endodermis and enter the xylem of roots they are translocated upward in the sap stream, primarily by mass flow. Eventually ions reach the leaf veins where they enter the free (cell wall) spaces of mesophyll cells. These cells accumulate ions from the surrounding solution by active transport. Hence, before most ions reach protoplasts of leaf cells they are absorbed by cellular transport mechanisms at least twice—once by root cells and again by leaf cells. Ions in the cytoplasm of leaf cells move through the symplast (Epstein, 1972).

The rate of upward transport of ions in the xylem is influenced by the rate of transpiration and by utilization of minerals in metabolic processes. The rate of ascent of individual ions varies because of differences in adsorption of ions on xylem cell walls. Certain ions also tend to be precipitated in the xylem, causing deficiencies in the leaves even though appreciable amounts are absorbed. Iron, for example, is commonly precipitated as ferric phosphate in the vascular system.

Because leaf cells commonly receive more minerals than they can store or use, excess minerals are often retranslocated through the phloem. Redistribution of ions from old to young leaves, to reproductive structures, and to metabolically active tissues is well known. Hence, most elements become concentrated in meristematically active regions such as root tips, expanding shoots, and the cambial sheath. Certain ions are more readily redistributed than others. Some are highly mobile in the phloem (K, Rb, Na, Mg); others intermediately mobile (Fe, Mn, Zn), and still others relatively immobile (Ca, Li, Sr) (Bukovac and Wittwer, 1957). Much export of ions occurs from senescent leaves to twigs just before leaf abscission occurs. The amount retranslocated varies widely but may be as high as 90% of the maximum amount in leaves during the growing season (Biddulph, 1959). Loosely held mineral elements often are leached out of the cell wall spaces of leaves by rain or dew (Tukey, 1970). Some ions that are not used by shoots return to the roots by the phloem and leach out or are excreted again into the rhizosphere.

a. Apoplastic and Symplastic Movement. The unlignified cellulose walls of epidermal and cortical cells of roots have a system of microcapillaries that offers little resistance to movement of water and solutes. This "free space," comprising some 10–15% of the root volume, is freely accessible to solutes without the restraints of membrane permeability barriers. The free space apparently does not extend beyond the endodermis because the Casparian strips are a barrier to ion diffusion. If ions diffused freely into the stele, their concentration in the stelar and cortical cells would be similar. However, the concentration of ions in the stele often greatly exceeds that in the cortex. Movement of ions in free space is nonselective, reversible, and independent of metabolism.

Several investigators have shown that ions accumulate in the cytoplasm and move from cell to cell through plasmodesmata (symplastic movement). Methods for studying apoplastic and symplastic movement of ions as well as evidence indicating that the symplast comprises the most important pathway for ion movement may be found in the book by Clarkson (1974).

b. Active Transport. Active transport across cell membranes occurs against concentration gradients and requires metabolic energy. It is also a selective and irreversible process. Accumulation ratios (concentration in cells divided by the concentration in the surrounding solution) for some ions commonly are of the order of 1000:1 (Steward and Sutcliffe, 1960).

The most widely accepted explanation for movement of ions across membranes impermeable to them involves a carrier hypothesis. This envisages membrane subunits (carriers) binding an ion at the outer surface of the membrane. Utilizing the energy of ATP, the resulting complex then moves across the membrane, which the ion itself cannot penetrate. At the far end of the membrane the ion separates from the carrier and moves to the vacuole or some other part of the cytoplasm. Ion carriers have been postulated on the basis of kinetic evidence, but have not been chemically identified or isolated. It appears likely that they are proteins or protein derivatives (peptides). The rate of carrier-mediated transport depends on (1) a capacity factor, which sets the maximum rate of movement when all available carrier sites are loaded; and (2) an intensity factor which reflects the proportion of the carrier occupied at a specific concentration of the ion (Epstein, 1972).

There is considerable selectivity in carrier-mediated ion transport. For example, active sites of a carrier which bind and move potassium have little affinity for sodium. Similarly, absorption of chloride is not influenced by the presence of fluoride and iodide (Elzam and Epstein, 1965). Specificity among less closely related ions also occurs. For example, absorption of halides is not affected by the presence of sulfate or phosphate, or the reverse. On the other hand, certain ions that are closely related chemically do compete in absorption. Examples are potassium–rubidium, calcium–strontium, and chloride–bromide, indicating that both ions of a related pair fit into a common carrier site (Epstein, 1962).

In Chapter 15, this volume, Wheeler has discussed the alteration of membranes and permeability by biotic pathogens.

D. Hormone Relations

Plants are equipped with hormonal chemical messengers which direct cells to carry out the functions of growth and differentiation. The various organs and tissues of plants operate as parts of an organismic whole largely because of the integrative functions of the hormonal system, although nutritive interrelations among organs also are involved. At present, five groups of chemical growth regulators are well known: (1) auxins and (2) gibberellins which mainly stimulate cell expansion, (3) cytokinins which mainly stimulate cell division, (4) ethylene which stimulates growth of stems and roots and induces fruit ripening, and (5) inhibitors such as abscisic acid (ABA) which depresses cell enlargement. Each of these growth regulators is chemically distinctive and, although each has been associated with certain specific plant processes, each can influence most aspects of plant growth including cell division, increase in cell size, and differentiation. Thus, most changes in plant growth involve interactive influences of two or more growth regulators, rather than the influence of one alone. Hormonal interactions may be synergistic (with one hormone reinforcing the influence of another), or antagonistic (with one hormone counteracting the influence of another). This section will briefly describe some effects of hormones on vegetative and reproductive growth. For a more detailed discussion of mechanisms of hormone action the reader is referred to Chapter 8 by Helgeson in this volume.

1. Seed Dormancy

Mature seeds of many temperate-zone plants exhibit some degree of embryo dormancy and do not germinate soon after they are shed, even in the most favorable environment. Rather they must undergo prolonged chilling to break the dormant state. This requirement increases the survival of plants because it prevents seed germination until after killing frost in the spring. Individual seeds from the same plant vary widely in the degree of dormancy. Therefore, the period of germination often is spread over a very long time. This also is advantageous because it increases the survival of a species even though some plants may be killed by early frost or drought. Following release from dormancy by chilling, much higher temperatures are needed to induce germination. As discussed in this volume by Mathre (Chapter 12) many pathogens are known to inhibit germination of seeds. It is possible that some pathogens also affect seed dormancy.

2. Bud Dormancy

Buds of most woody plants alternate from active growth during the warm season to inactivity during the cold season. When seasonal shoot expansion stops, plants gradually enter a predormant, quiescent state during which they retain the capacity for growth but under a progressively narrowing range of environmental conditions. Gradually they lapse into a state of deep, physiological dormancy, and cannot grow even under ostensibly favorable environmental conditions. This protects the plants from damage by frost after early spring or midwinter warm spells. By spring the condition of deep dormancy has gradually dissipated until tissues are completely released from dormancy. Dormancy of buds is variously induced by low temperatures, short days, water deficits, or mineral deficiencies.

Dormancy of buds usually is broken by prolonged exposure to low temperatures. Although the duration of chilling required to break dormancy varies with plant species and varieties, it usually involves many weeks of low temperatures. For example, about 1000 hours of chilling were required to break the dormancy of *Pinus sylvestris* buds (Nagata, 1967) and about 2000 hours for *Acer saccharum* buds (Taylor and Dumbroff, 1975). After release from dormancy by chilling, buds require an exposure to high temperature for a critical period before normal growth can occur. Both induction and breaking of bud dormancy are mediated by delicate balances between endogenous growth promoters and inhibitors (Eagles and Wareing, 1963, 1964; Domanski and Kozlowski, 1968). Since many biotic pathogens produce auxins, gibberellins, cytokinins, and ethylene, it seems very likely that some may influence bud dormancy.

3. Leaf Senescence and Abscission

As temperatures decrease and the days shorten in late summer the leaves of many plants begin to senesce. The culmination of senescence is leaf abscission but before this occurs several changes occur in leaves. These include decrease in photosynthetic and respiratory capacity, changes in pigments, enzymatic changes, and decreases in capacity to synthesize nucleic acids and proteins.

Marked changes in color of leaves occur during senescence. Some plants (*Alnus, Salix*) show little change. Leaves of *Populus, Ginkgo, Fagus* and *Betula* change to yellow of different shades. Another group (*Acer, Rhus*) develop variegated leaf colors, primarily crimson or orange. The red colors are traceable to formation of anthocyanin pigments; the yellow colors to unmasking of residual carotenoid pigments by degeneration of the green chlorophylls. The most important environmental factors controlling autumn coloration are temperature, light, and water supply. The most vivid autumn colors occur under conditions of clear, dry, and cool weather during the autumn. Severe early frosts and rainy days near the peak time of autumn coloration decrease the intensity of colors (Kramer and Kozlowski, 1960).

Prior to leaf abscission some minerals are redistributed from leaves to stem tissues. Finally cellulase and pectinases are formed in the abscission zone and they begin to digest one or more layers of cell walls. Leaf shedding follows when the cells of an abscission layer are weakened sufficiently. In Chapter 14 (this volume) Cowling discusses this aspect of the engineering mechanics of pathogenesis.

According to Osborne (1973) abscission involves three sequential stages: stimulus, signal, and response (Fig. 3). As long as a leaf is exporting auxin, abscission is prevented. When a leaf senesces, however, auxin production and translocation across the abscission zone are reduced. Concurrently, ethylene production is increased and this initiates the processes of abscission (Leopold and Kriedemann, 1975). Normal leaf abscission can be delayed by exogenous auxin or gibberellin and

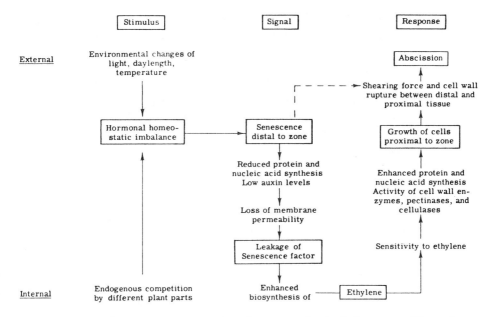

Fig. 3. Model of stimulus, signal, and response for leaf abscission. (From Osborne, 1973.)

greatly accelerated by concentrations of ethylene in the air as low as 0.1 μl/liter or by applications of Ethrel (2-chloroethylphosphonic acid) which is converted to ethylene. Although cytokinins, gibberellins, and inhibitors also are variously involved in abscission, their role may be that of modulating the primary auxin–ethylene regulatory system. Abscisic acid, for example, stimulates production of ethylene (Osborne, 1973).

4. Cambial Growth

Initiation of seasonal cambial growth depends on hormones translocated basipetally from active buds (Kozlowski, 1971b). Both auxins and gibberellins are needed for cell division and differentiation. The effects of either one, when applied alone, are due to the natural presence of the other (Thimann, 1972). When auxins and gibberellins are applied alone and in mixtures to disbudded shoots of woody plants, normal xylem is produced only by a synergistic effect of both. Inhibitors also are involved in cambial growth because they modify the auxin stimulation of cambial activity and xylem differentiation.

In leaning stems the distribution pattern and nature of cambial growth are altered. In gymnosperms xylem production is increased on the lower side of a leaning stem and decreased on the upper side. In addition, reaction wood, called compression wood, forms on the lower side. Such wood is characterized by abnormally rounded, thick-walled tracheids, with abundant intercellular spaces. In angiosperms, xylem production usually is stimulated and reaction wood, called tension wood, forms on the upper side of leaning stems. Tension wood has fewer and smaller vessels than normal wood and has an abundance of thick-walled, gelatinous fibers.

Reaction wood forms in response to geotropic redistribution of auxins. Compression wood in gymnosperms forms in response to a high auxin gradient causing mobilization of food. By contrast, the tension wood of angiosperms appears to be a response to auxin deficiency. This is shown by low amounts of auxin on the upper sides of leaning stems, reduction in wood formation following application of auxin to the upper side of tilted stems, and induction of tension wood formation by application of auxin antagonists (Kozlowski, 1971b). The auxin effects may be mediated, at least in part, by stimulating ethylene production.

5. Root Growth

Root growth requires a continuous supply of hormonal growth regulators. Defoliation by insects and leaf disease or stem girdling by cankers decreases the basipetal flow of hormones and rapidly inhibits root growth. Cambial growth of roots appears to be initiated and maintained by auxin from the stem if cytokinins and gibberellins are present. Hormones translocated basipetally play a prominent role in the complex of factors regulating formation of adventitious roots on cuttings. Formation of root initials depends on an auxin synergist as well as auxin. When both are present, RNA synthesis, necessary for formation of root primordia, is stimulated (Haissig, 1971).

6. Reproductive Growth

A good harvest of fruits or seeds implies success in each of the sequential processes of reproductive growth, including floral initiation and development, pollination, fertilization, growth and development of the embryo, growth of the fruit and seed, and ripening of fruits. A number of hormones are produced in surges during each phase of fruit development. Although the regulatory effects of hormones on reproductive growth are synergistic, specific growth regulators have prominent roles at different stages of the reproductive cycle. For example, auxin stimulates growth of flowers. Pollen contains auxin and growth of pollen tubes stimulates synthesis of additional auxin. Increase in fruit size is regulated by auxin and gibberellin supply. According to Crane and van Overbeek (1965), the primary physiological role of fertilized ovules or seeds in fruit growth is synthesis of hormones which sustain a gradient along which metabolites are transported to the growing fruit. Ethylene plays a predominant role in fruit ripening but other hormones also are involved. Abscisic acid also stimulates fruit ripening whereas auxins, gibberellins, and cytokinins tend to inhibit ripening (Sacher, 1973).

Synthetic growth regulators have been used to alter the course of reproductive growth. For example, certain applied gibberellins induce precocious flowering in the Cupressaceae and Taxodiaceae. Fruits of fig, grape, avocado, and pear can be set by applying auxins, but this cannot be done in stone fruits (except for apricot) and many pomaceous fruits. In some species fruits that can be set with auxins can be set as well or better with applied gibberellins, as in citrus, fig, pear, and apple.

III. PLANT RESPONSES TO ENVIRONMENTAL CHANGES

The constantly changing environment requires both diurnal and seasonal changes in how plants function. This section will provide some examples of how plants adapt to cope with environmental changes and stresses. In Chapter 19 (this volume), Tibbitts has discussed how disease alters the biological rhythms plants have developed to cope with daily and seasonal changes in the environment in which they grow.

A. Diurnal Changes

Photosynthesis follows a daily cycle that often is reasonably well correlated with stomatal aperture. Thus the resistance offered by stomata to CO_2 uptake often comprises a major limitation for photosynthesis. Early in the morning, as light intensity and temperature increase, the rate of photosynthesis may accelerate rapidly and reach a maximum before noon. In the late afternoon, when light intensity and temperature decrease, the rate of photosynthesis declines markedly. Often there are small or large midday decreases, followed by increases, in photosynthesis. The importance of stomatal diffusion resistance is shown by high correlations between midday dips in photosynthesis and stomatal aperture.

The rate of photosynthesis and accumulation of photosynthetic products may vary appreciably from an idealized day-to-day pattern. These deviations are caused by the innumerable combinations of light and temperature throughout the day and the differential effects of high temperatures on photosynthesis and respiration. For example, Gates (1965) compared diurnal changes in the responses of leaves in full sun and others in shade on a relatively warm day (temperature range of 29°–40°C) and on a

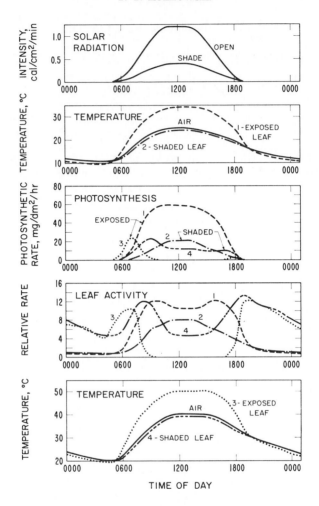

Fig. 4. Influence of a cool summer day and a warm summer day (with solar radiation as shown) for sunlit and shaded leaves on the photosynthetic rate and biochemical activity of a leaf at various times of day. Curves 1 and 2 refer to an exposed and shaded leaf, respectively, on the cool day and curves 3 and 4 to the exposed and shaded leaf on the warm day. (From Gates, 1965.)

cool day (temperature range of 12°–25°C). As shown in Fig. 4, on a cool day a high rate of photosynthesis was observed in sunlit leaves (curve 1) and a low rate in shaded leaves (curve 2). By contrast, on a warm day the rate of photosynthesis in shaded leaves (curve 4) exceeded that in sunlit leaves (curve 3). This difference is due to the inhibitory effect of high temperature on photosynthesis. Photosynthesis was negligible during midday and recovered only slightly during late afternoon. On the warm day, total accumulation of photosynthetic products in the sunlit leaves was less than 10% of that in sunlit leaves on the cooler day.

Superimposed on the diurnal changes in physiological processes that are controlled by changes in environmental factors are diurnal circadian rhythms which occur in plants independently of environmental fluctuations. Such oscillations can result from a wide variety of feedback systems. Circadian rhythms in stomatal aperture, photosynthesis, and respiratory gas exchange have been demonstrated in many plants (Queiroz, 1974).

B. Seasonal Changes

Temperate zone plants synchronize their physiological activities with the changing seasons. This facilitates their adaptation to periods of environmental stress.

In perennial woody plants progressively decreasing temperatures and shortening days lead to cessation of growth, development of buds, redistribution of nutrients from leaves to stems, abscission of leaves, development of a metabolically dormant state, and induction of frost hardiness. These adaptations facilitate the survival of plants during the winter and additional metabolic changes occur that are necessary for the resumption of growth in the spring. We have dealt with all of these except frost hardiness, at least in a general way. Hence, we will now discuss frost hardiness.

Temperate zone plants that cannot survive temperatures even slightly below freezing in midsummer often can survive superlow temperatures during the winter. Frost hardiness typically develops in stages (Fig. 5). The first stage involves cessation of growth and formation in the leaves of a translocatable factor that promotes hardiness; it begins before the first frost of the season. The second stage is triggered by the first frost, and the third stage by very low temperatures, usually below $-30°C$ (Weiser, 1970). During development of frost hardiness, changes occur in sugars, proteins, amino acids, nucleic acids, and lipids. Tissue hydration is reduced and membrane permeability increases. The frost-hardened state has been causally linked to some of these changes. In most species of woody plants, for example, sugars increase in the autumn as frost hardiness develops and then decrease in the spring as hardiness dissipates. Sugars may increase hardiness by decreasing intracellular ice formation and thus also reducing freeze-induced cell dehydration. The sugars may also be metabolized and produce as yet not understood

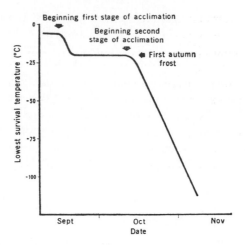

Fig. 5. Typical seasonal pattern of cold resistance in the living bark of *Cornus stolonifera* stems in Minnesota. Acclimation of this hardy shrub proceeds in distinct stages. (From Weiser, 1970.)

protective changes (Levitt, 1972). Some investigators have emphasized a strong relationship between increase in water-soluble proteins and development of frost hardiness. Changes in soluble proteins may be associated with a decrease in free cellular water, resulting in a free energy gradient that would tend to reduce formation of intracellular ice during late development of frost hardiness (Brown and Bixby, 1975).

Bacteria on leaf surfaces act as ice nuclei and materially increase the sensitivity of unhardened plants to frost. For example, *Zea mays* plants with *Pseudomonas syringae* on their leaves were killed when the temperature dropped to −4°C, whereas those lacking surface populations of *P. syringae* were not substantially injured until the temperature decreased to −8°C (Arny *et al.*, 1976).

IV. THE PHYSIOLOGICAL IMPACT OF ENVIRONMENTAL AND PATHOLOGICAL STRESSES

Environmental stresses and biotic pathogens adversely influence plant growth by interfering with rates and balances among internal physiological processes, especially food, hormone, and water relations. Pathogens often set in motion or accelerate a sequential and very complicated series of metabolic disturbances, rather than a simple change in only one process, such as photosynthesis, as is sometimes supposed. Eventually disease symptoms may be variously expressed as color changes, necrosis, wilting, leaf spots, atrophy, overgrowth, dieback, and abscission of plant parts. Some of these localized responses promote changes in physiological activity which affect other processes in more distant tissues and organs. For example, chlorosis results in a depression of photosynthesis and a decrease in the basipetal transport of growth hormones from shoots to the lower stem and roots. These changes adversely affect cambial growth and, sequentially, root growth. Root diseases at first may decrease absorption of water and minerals. Decreased absorption of water subsequently leads to stomatal closure, then to reduced carbohydrate production and translocation, and finally to reduced supply of growth regulators. This sequence of events eventually further decreases root growth, which decreases absorption of water even more, and so on. Similarly, defoliation affects processes other than carbohydrate production. It is important to reemphasize that the physiological impact of a localized attack by a pathogen in one part of the plant usually is transmitted to distant organs and tissues in a complex manner and eventually may affect the entire plant.

The effects of an environmental stress or of a given disease may or may not be drastic, depending on the extent to which various processes are disrupted and on the phenological stage and physiological status of the plant at the time of stress or fungus invasion (Kozlowski, 1969). Growth of annual and herbaceous perennial plants generally is altered rapidly after an environmental stress or after attack by a pathogen. In contrast, growth inhibition of some woody perennial plants may be postponed for a long time.

In the temperate zone several different groups of species of woody plants are recognized with respect to shoot growth characteristics. In one group the shoots are fully predetermined in the winter bud—that is, the shoots exhibit "fixed" growth. Shoot formation involves bud differentiation during one year and extension of the preformed parts within the bud into a shoot during the next year. Northern pines and some maples and ashes are of this type. Another group of species exhibits "free" growth. This involves elongation of a shoot by simultaneous initiation and elongation of new stem units as well as expansion of preformed parts (Pollard and Logan, 1974). Shoots of species that exhibit free growth often have two sets of leaves: (1) early leaves that were relatively well developed in the winter bud, and (2) late leaves which continued to form and elongate while the shoot internode was expanding. Such free growth resembles that in a number of herbaceous plants. Woody plants that exhibit free growth include poplar, apple, birch, *Eucalyptus*, larch, and some tropical pines. In still another group, such as the southern pines and many species of tropical angiosperms, shoot elongation consists of a series of recurrent growth flushes from the opening of a series of buds produced during the same growing season. When recurrently flushing species are moved to the hot, wet tropics their shoots may grow continuously throughout the year rather than in periodic surges (Kozlowski and Greathouse, 1970).

In woody plants with fixed growth, potential shoot growth is largely governed by the number of stem units present in the unopened bud. Hence the phase of bud formation determines ultimate shoot length. No matter how favorable the environment is during the year of bud expansion, shoots of species with fixed growth expand for only a few weeks in the early part of the frost-free season. Thus, environmental conditions during the previous year control the ultimate shoot length of such species more than those during the year of bud expansion into a shoot. A favorable environment during the year of bud formation causes formation of large buds, which produce long shoots in the subsequent growing season. Bud size can thus be used as an index of shoot growth potential of certain species and their provenances and of different shoots on the same tree. Kozlowski et al. (1973) found strong correlations between bud size and shoot length of Pinus resinosa trees (Table I). The amount and rate of shoot growth was greater in upper whorls, which have the largest buds, than in lower whorls.

Several investigators found strong correlations between shoot growth of species with fixed growth and weather conditions of the previous year. For example, when late-summer temperatures were low, the period of bud differentiation of Picea abies plants was short and this was reflected in production of short internodes during the following year (Heide, 1974). Mikola (1962) emphasized that the annual height growth of Pinus silvestris in Finland was determined largely by the air tempera-

TABLE I

Relation of Bud Size (March 20, 1970) to Final Shoot Length (August 19, 1970) of 8-Year-Old *Pinus resinosa* Trees [a]

	Bud diameter (mm)	Bud length (mm)	Shoot length (mm)
Terminal leader	8.2 ± 0.7	38.0 ± 2.8	742.0 ± 26.7
Whorl 1 shoots	5.9 ± 0.1	27.3 ± 0.7	484.8 ± 11.0
Whorl 2 shoots	5.5 ± 0.1	22.9 ± 0.8	403.2 ± 13.0
Whorl 3 shoots	4.5 ± 0.2	16.6 ± 0.9	271.4 ± 19.1
Whorl 4 shoots	3.8 ± 0.3	12.5 ± 1.0	132.1 ± 20.6
Whorl 5 shoots	3.7 ± 0.3	9.9 ± 0.8	65.2 ± 16.0
Whorl 6 shoots	3.3 ± 0.4	8.6 ± 1.4	74.4 ± 31.5

[a] From Kozlowski et al. (1973).

ture of the previous summer. Clements (1970) noted that irrigation of *Pinus resinosa* trees in late summer caused formation of large buds which produced long shoots in the subsequent growing season. By comparison, late summer droughts induced small buds which produced shoots with short internodes. Irrigation in the spring did not appreciably influence shoot length during the same year, emphasizing the paramount importance of shoot predetermination on shoot length. It should be remembered, however, that the dependence of shoot length on prior-year weather applies primarily to species with preformed shoots whose internodes complete expansion during the very early part of the frost-free season. Thus, whereas environmental stresses late in the summer often do not limit current-year shoot expansion of such species, they will inhibit the expansion of shoots of heterophyllous and recurrently flushing species, which do not have fully preformed shoots in the winter bud and continue to expand their shoots late into the summer (Kozlowski and Clausen, 1966).

It should be obvious that defoliation by fungi in midsummer will not inhibit the amount of current-year growth of shoot internodes by species with fixed growth, which by midsummer generally have completed all internode elongation. By comparison, the same degree of defoliation will variously inhibit current-year shoot elongation of species that exhibit free growth, recurrently flushing species, or continuously growing tropical trees, and even those species that exhibit fixed growth but also have a tendency to produce abnormal late-season lammas shoots.

Because of differences among species of woody plants in seasonal distribution of radial growth, blockage of the phloem in midsummer will prevent most cambial growth below the blocked area of the stem, but some species may already have completed much or most of their cambial growth and others only a small amount. Ladefoged (1952) noted that *Fraxinus excelsior* formed its annual xylem ring rapidly. Half of it was formed in May and June, and after July only a small amount of xylem was deposited. In contrast, *Fagus sylvatica* formed most of its annual ring late in the season, with one-third to one-fourth produced in August and early September. These variations emphasize that the impact of phloem blockage by disease will not be the same for all species of woody plants.

V. CONCLUSIONS

This chapter has strongly emphasized the importance of the effects of pathogenic diseases on plants through the intermediation of physiological processes. For this reason more cooperation between plant pathologists and physiologists is needed in evaluating the impact of disease on plant dysfunction. Some of the most difficult disease problems probably can best be investigated with a team research philosophy and pooling of talents and facilities.

It should be recognized at the outset that the state of knowledge of some physiological processes and their controls is imperfect. There also are wide differences in our capacity to quantify dysfunction in various physiological processes. For example, disease-induced reductions in transpirational water loss or CO_2 absorption by leaves are more easily quantified than are changes in hormone or enzyme balances. The growth of plants is complex. It is influenced by a constantly changing environmental nexus and a complicated biochemistry involving many partial processes and synergisms among internal growth requirements. These considerations argue for input by well-educated and experienced specialists to studies of disease-induced plant dysfunction. Increasingly sophisticated experiments in well-equipped laboratories are needed. Small isolated field laboratories without access to good equipment and other specialists will be at a disadvantage.

Experienced investigators are steadily improving techniques for quantifying and characterizing physiological processes in plants. For example, radioactive isotopes, remote sensing techniques, scanning electron microscopes, computers in modeling, and controlled environment facilities have become important research tools.

There are many problems in studying the impact of disease in the field or greenhouse because it is difficult to define conditions in a reproducible way and to separate the effects of pathogens and of environmental factors on plant dysfunction. Light, temperature, humidity, and other factors are so interdependent that a change in one alters the others. Thus, it is extremely difficult to evaluate the effects of individual factors, or interactions among them, on the impact of disease on plant processes. Furthermore it is virtually impossible in the field to reproduce precisely the environment of one day on succeeding days. Phytotrons and biotrons can be used to dissect or construct a given environment and study concurrently the effects of disease over a wide range of environmental factors, or in simulated daily and seasonal environmental regimes for various regions from the Arctic to the tropics. The great value of biotrons is that they decrease variability and increase reproducibility of research results.

Although the costs of biotron research are high, they often are overestimated when compared with costs of field or greenhouse research. Actually results obtained in biotrons often are much more useful and cost less than those obtained in the field because of the smaller number of plants necessary in the biotron and the high reproducibility of data.

These considerations make short-term experiments useful and result in reduced costs of laboratory analyses. Furthermore, in comparing costs, consideration often is not given to the expenses of maintaining experimental areas, field travel, damage to field experiments by weather and insects (requiring repetition of experiments), and delays in waiting for specific weather conditions.

References

Alvim, P. de T. (1977). Cacao. In "Ecophysiology of Tropical Crops" (P. de T. Alvim and T. T. Kozlowski, eds.), pp. 279–313. Academic Press, New York.

Arny, D. C., Lindow, S. E., and Upper, C. D. (1976). Frost sensitivity of *Zea mays* increased by application of *Pseudomonas syringae*. *Nature* (*London*) **262**, 282–284.

Barrs, H. D. (1968). Determination of water deficits in plant tissues. In "Water Deficits and Plant Growth" (T. T. Kozlowski, ed.), Vol. 1, pp. 235–368. Academic Press, New York.

Biddulph, O. (1959). Translocation of inorganic solutes. In "Plant Physiology" (F. C. Steward, ed.), Vol. 2, pp. 553–603. Academic Press, New York.

Bowen, G. D. (1973). Mineral nutrition of ectomycorrhizae. In "Ectomycorrhizae" (G. C. Marks and T. T. Kozlowski, eds.), pp. 151–205. Academic Press, New York.

Boyer, J. S. (1973). Response of metabolism to low water potentials in plants. *Phytopathology* **63**, 466–472.

Brown, G. N., and Bixby, J. A. (1975). Soluble and insoluble protein patterns during induction of freezing tolerance in black locust seedlings. *Physiol. Plant.* **34**, 187–191.

Bukovac, M. J., and Wittwer, S. H. (1957). Absorption and mobility of foliar applied nutrients. *Plant Physiol.* **32**, 428–435.

Chaney, W. R., and Kozlowski, T. T. (1971). Water transport in relation to expansion and contraction of leaves and fruits of Calamondin orange. *J. Hortic. Sci.* **46**, 71–81.

Clarkson, D. T. (1974). "Ion Transport and Cell Structure in Plants." Wiley, New York.

Clements, J. R. (1970). Shoot responses of young red pine to watering applied over two seasons. *Can. J. Bot.* **48**, 75–80.

Crafts, A. S., and Crisp, C. E. (1971). "Phloem Transport in Plants." Freeman, San Francisco, California.

Crane, J. C., and van Overbeek, J. (1965). Kinin-induced parthenocarpy in the fig, *Ficus carica* L. *Science* **147**, 1468–1469.

Dickmann, D. I., and Kozlowski, T. T. (1968). Mobilization by *Pinus resinosa* cones and shoots of ^{14}C-photosynthate from needles of different ages. *Am. J. Bot.* **55**, 900–906.

Dickmann, D. I., and Kozlowski, T. T. (1970). Mobilization and incorporation of photoassimilated ^{14}C by growing vegetative and reproductive tissues of adult *Pinus resinosa* Ait. trees. *Plant Physiol.* **45**, 284–288.

Domanski, R., and Kozlowski, T. T. (1968). Variations in kinetin-like activity in buds of *Betula* and *Populus* during release from dormancy. *Can. J. Bot.* **46**, 397–403.

Eagles, C. F., and Wareing, P. F. (1963). Dormancy regulators in woody plants. Experimental induction of dormancy in *Betula pubescens*. *Nature* (*London*) **199**, 874–875.

Eagles, C. F., and Wareing, P. F. (1964). The role of growth substances in the regulation of bud dormancy. *Physiol. Plant.* **17**, 697–709.

Elzam, O. E., and Epstein, E. (1965). Absorption of chloride by barley roots: Kinetics and selectivity. *Plant Physiol.* **40**, 620–624.

Epstein, E. (1962). Mutual effects of ions in their absorption by plants. *Agrochimica* **6**, 293–322.

Epstein, E. (1972). "Mineral Nutrition of Plants: Principles and Perspectives." Wiley, New York.

Erickson, R. O. (1976). Modeling of plant growth. *Annu. Rev. Plant Physiol.* **27**, 407–434.

Evans, G. C. (1972). "The Quantitative Analysis of Plant Growth." Univ. of California Press, Berkeley.

Gates, D. M. (1965). Energy, plants, and ecology. *Ecology* **46**, 1–13.

Gregory, P. F. (1977). "Biochemistry of Photosynthesis." Wiley, New York.

Haissig, B. E. (1971). Influence of indole-3-acetic acid on incorporation of ^{14}C-uridine by adventitious root primordia in brittle willow. *Bot. Gaz.* (*Chicago*) **132**, 263–267.

Hambidge, G., ed. (1941). "Hunger Signs in Crops." Am. Soc. Agron. and Natl. Fertilizer Assoc., Washington, D.C.

Harley, J. L. (1969). "The Biology of Mycorrhiza." Leonard Hill, London.

Heide, O. M. (1974). Growth and dormancy in Norway spruce ecotypes (*Picea abies*). I. Interaction of photoperiod and temperature. *Physiol. Plant.* **30**, 1–12.

Heitefuss, R., and Williams, P. H., eds. (1976). "Physiological Plant Pathology." Springer-Verlag, Berlin and New York.

Hewitt, E. J., and Smith, T. A. (1975). "Plant Mineral Nutrition." Wiley, New York.

Hillel, D. (1970). "Soil and Water: Physical Principles and Processes." Academic Press, New York.

Hsiao, T. C. (1973). Plant responses to water stress. *Annu. Rev. Plant Physiol.* **24**, 519–570.

Jackson, W. A., and Volk, R. J. (1970). Photorespiration. *Annu. Rev. Plant Physiol.* **21**, 385–432.

Kozlowski, T. T. (1968a). Soil water and tree growth. *In* "The Ecology of Southern Forests" (N. E. Linnartz, ed.), pp. 30–57. Louisiana State Univ. Press, Baton Rouge.

Kozlowski, T. T .(1968b). Introduction. *In* "Water Deficits and Plant Growth" (T. T. Kozlowski, ed.), Vol. 1, pp. 1–21. Academic Press, New York.

Kozlowski, T. T., ed. (1968c). "Water Deficits and Plant Growth," Vol. 1. Academic Press, New York.

Kozlowski, T. T., ed. (1968d). "Water Deficits and Plant Growth," Vol. 2. Academic Press, New York.

Kozlowski, T. T. (1968e). Diurnal changes in diameters of fruits and tree stems of Montmorency cherry. *J. Hortic. Sci.* **43**, 1–15.

Kozlowski, T. T. (1969). Tree physiology and forest pests. *J. For.* **69**, 118–122.

Kozlowski, T. T. (1971a). "Growth and Development of Trees," Vol. 1. Academic Press, New York.

Kozlowski, T. T. (1971b). "Growth and Development of Trees," Vol. 2. Academic Press, New York.

Kozlowski, T. T., ed. (1972a). "Water Deficits and Plant Growth," Vol. 3. Academic Press, New York.

Kozlowski, T. T. (1972b). Shrinking and swelling of plant tissues. In "Water Deficits and Plant Growth" (T. T. Kozlowski, ed.), Vol. 3, pp. 1–64. Academic Press, New York.

Kozlowski, T. T. (1976a). Drought and transplantability of trees. In "Better Trees for Metropolitan Landscapes" (F. S. Santamour, Jr., H. Gerhold, and S. Little, eds.), Gen. Tech. Rep. NE–22, pp. 77–90. U.S. For. Serv., Upper Darby, Pennsylvania.

Kozlowski, T. T. (1976b). Water relations and tree improvement. In "Tree Physiology and Yield Improvement" (M. Cannell and F. T. Last, eds.), pp. 307–327. Academic Press, New York.

Kozlowski, T. T., ed. (1976c). "Water Deficits and Plant Growth," Vol. 4. Academic Press, New York.

Kozlowski, T. T., ed. (1978). "Water Deficits and Plant Growth," Vol. 5. Academic Press, New York.

Kozlowski, T. T., and Clausen, J. J. (1966). Shoot growth characteristics of heterophyllous woody plants. Can. J. Bot. 44, 827–843.

Kozlowski, T. T., and Greathouse, T. E. (1970). Shoot growth and form of tropical pines. Unasylva 24, 6–14.

Kozlowski, T. T., and Keller, T. (1966). Food relations of woody plants. Bot. Rev. 32, 293–382.

Kozlowski, T. T., Torrie, J. H., and Marshall, P. E. (1973). Predictability of shoot length from bud size in Pinus resinosa. Ait. Can. J. For. Res. 3, 34–38.

Kramer, P. J., and Kozlowski, T. T. (1960). "Physiology of Trees." McGraw-Hill, New York.

Ladefoged, K. (1952). The periodicity of wood formation. Dan. Biol. Skr. 7, 1–98.

Leopold, A. C., and Kriedemann, P. E. (1975). "Plant Growth and Development." McGraw-Hill, New York.

Levitt, J. (1972). "Responses of Plants to Environmental Stresses." Academic Press, New York.

Marks, G. C., and Kozlowski, T. T., eds. (1973). "Ectomycorrhizae." Academic Press, New York.

Meyer, B. S., Anderson, D. B., Bohning, R. H., and Fratianne, D. G. (1973). "Introduction to Plant Physiology." Van Nostrand-Reinhold, Princeton, New Jersey.

Mikola, P. (1962). Temperature and tree growth near the northern timber line. In "Tree Growth" (T. T. Kozlowski, ed.), pp. 265–274. Ronald Press, New York.

Nagata, H. (1967). Studies on the photoperiodism in the dormant bud of Pinus densiflora Sieb. et Zucc. (II). Effects of temperature and photoperiod on the breaking of winter dormancy of first-year seedlings. J. Jpn. For. Soc. 49, 415–420.

Orshansky, G. (1954). Surface reduction and its significance as a hydroecological factor. J. Ecol. 42, 442–444.

Osborne, D. J. (1973). Internal factors regulating abscission. In "Shedding of Plant Parts" (T. T. Kozlowski, ed.), pp. 125–147. Academic Press, New York.

Parsons, R. F. (1969). Physiological and ecological tolerances of Eucalyptus incrassata and E. socialis to edaphic factors. Ecology 50, 386–390.

Peel, A. J. (1974). "Transport and Nutrients in Plants." Butterworth, London.

Pereira, J. S., and Kozlowski, T. T. (1976). Diurnal and seasonal changes in water balance of Abies balsamea and Pinus resinosa. Oecol. Plant. 11, 413–428.

Pereira, J. S., and Kozlowski, T. T. (1977). Leaf anatomy and water relations of Eucalyptus camaldulensis and E. globulus seedlings. Can. J. Bot. 7, 145–153.

Pollard, D. F. W., and Logan, K. T. (1974). The role of free growth in the differentiation of provenances of black spruce *Picea mariana* (Mill.) B. S. P. *Can. J. For. Res.* **4**, 308–311.

Queiroz, O. (1974). Circadian rhythms and metabolic patterns. *Annu. Rev. Plant Physiol.* **25**, 115–134.

Rains, D. W. (1976). Mineral metabolism. *In* "Plant Biochemistry" (J. Bonner and J. E. Varner, eds.), 3rd ed., pp. 561–597. Academic Press, New York.

Sacher, J. A. (1973). Senescence and post harvest physiology. *Annu. Rev. Plant Physiol.* **24**, 197–224.

Sestak, Z., Catsky, J., and Jarvis, P. G. (1971). "Plant Photosynthetic Production; Manual of Methods." Junk, The Hague.

Shriner, D. S. (1976). Effects of simulated rain acidified with sulfuric acid on host-parasite interactions. *In* "Acid Precipitation and the Forest Ecosystem" (L. S. Dochinger and T. A. Seliga, eds.), Gen. Tech. Rep. NE-23, pp. 919–925. U.S. For. Serv., Upper Darby, Pennsylvania.

Slatyer, R. O. (1967). "Plant-Water Relationships." Academic Press, New York.

Slavik, B. (1974). "Methods of Studying Plant Water Relations." Springer-Verlag, Berlin and New York.

Smith, P. F. (1962). Mineral analysis of plant tissue. *Annu. Rev. Plant. Physiol.* **13**, 81–108.

Spyridakis, D. E., Chesters, G., and Wilde, S. A. (1967). Kaolinization of biotite as a result of coniferous and deciduous seedling growth. *Soil Sci. Soc. Am., Proc.* **31**, 203–210.

Steward, F. C., and Sutcliffe, J. F. (1960). Plants in relation to inorganic salts. *In* "Plant Physiology—A Treatise" (F. C. Steward, ed.), Vol. 2, pp. 253–478. Academic Press, New York.

Tal, M. (1966). Abnormal stomatal behavior in wilty mutants of tomato. *Plant Physiol.* **41**, 1387–1391.

Taylor, J. S., and Dumbroff, E. B. (1975). Bud, root, and growth-regulator activity in *Acer saccharum* during the dormant season. *Can. J. Bot.* **53**, 321–331.

Ten Houten, J. G. (1959). Scope and contributions of plant pathology. *In* "Plant Pathology: An Advanced Treatise" (J. G. Horsfall and A. E. Dimond, eds.), Vol. 1, pp. 19–60. Academic Press, New York.

Thimann, K. V. (1972). The natural plant hormones. *In* "Plant Physiology" (F. C. Steward, ed.), Vol. 6B, pp. 3–145. Academic Press, New York.

Tukey, H. B., Jr. (1970). The leaching of substances from plants. *Annu. Rev. Plant Physiol.* **21**, 305–324.

Van den Driessche, R. (1974). Prediction of mineral nutrient status of trees by foliar analysis. *Bot. Rev.* **40**, 347–394.

Waggoner, P. E., and Simmonds, N. W. (1966). Stomata and transpiration of droopy potatoes. *Plant Physiol.* **41**, 1268–1271.

Walker, J. C. (1969). "Plant Pathology." McGraw-Hill, New York.

Walsh, L., ed. (1971). "Instrumental Methods for Analysis of Soils and Plant Tissue." Soil. Sci. Soc. Am., Madison, Wisconsin.

Wardlaw, I. F., and Passioura, J. B., eds. (1976). "Transport and Transfer Processes in Plants." Academic Press, New York.

Weiser, C. J. (1970). Cold resistance and injury in woody plants. *Science* **169**, 1269–1278.

Whittingham, C. P. (1974). "The Mechanism of Photosynthesis." Am. Elsevier, New York.

Wilde, S. A. (1968a). Mycorrhizae and tree nutrition. *BioScience* **18**, 482–484.

Wilde, S. A. (1968b). Mycorrhizae: Their role in tree nutrition and timber production. *Univ. Wis., Coll. Agric. Life Sci., Res. Div., Bull.* **272**.

Wittwer, S. H., and Teubner, F. G. (1959). Foliar absorption of mineral nutrients. *Annu. Rev. Plant Physiol.* **10**, 13–32.

Wuenscher, J. E., and Kozlowski, T. T. (1971a). The response of transpiration resistance to leaf temperature as a desiccation resistance mechanism in tree seedlings. *Physiol. Plant.* **24**, 254–259.

Wuenscher, J. E., and Kozlowski, T. T. (1971b). Relationship of gas-exchange resistance to tree seedling ecology. *Ecology* **52**, 1016–1023.

Zelitch, C. P. (1971). "Photosynthesis, Photorespiration, and Plant Productivity." Academic Press, New York.

Zimmermann, M. H., and Milburn, J. A., eds. (1975). "Phloem Transport," Vol. 1. Springer-Verlag, Berlin and New York.

Chapter 3

The Dynamic Nature of Disease

DURWARD F. BATEMAN

I. INTRODUCTION

The term "disease" represents a phenomenon exclusive to life forms. The study of life, biology, because of its diversity and complexity, has necessarily evolved with foci on divisions or aspects of the whole (Arber, 1954) as opposed to a comprehensive consideration of the totality of organic evolution. The disciplinary approach, which man has imposed on the study of the biological realm, coupled with the methods of science, has enabled development of insights into and an appreciation of nature not possible before.

During the early phases of modern biology the taxonomic approach arose and spawned many disciplines such as mycology, bacteriology, and entomology. The study of life from other perspectives led to development of other disciplines such as physiology, pathology, cytology, and genetics that transverse taxonomic divisions. Yet another approach, namely the study of the interactions between and among individuals, species and communities, and the abiotic environment, designated the science of ecology, provides other insights into the living world. The various biological disciplines relate to portions or aspects of a multifaceted continuum, and the knowledge generated by each provides the greatest insights when viewed and examined in the context of the whole. Disease is the central theme of pathology; it pertains to all levels of biological organization and thus should be examined in this broad context.

A. Pathology—A Biological Discipline

Pathology is a biological discipline in the same sense as is physiology (Link, 1932). Physiology and pathology are biological sciences that cover aspects of the same reality, i.e., they are both concerned with the functions and vital processes of life forms, only their emphases differ. The focus of physiology is the elucidation of those vital processes and functions of developing and mature organisms which enable them to adapt to their environment, maintain themselves, and reproduce. Physiology thus seeks understanding of the mechanisms that lead to full functional competency in living systems. The science of pathology embodies these same objectives, but the focus is on those mechanisms which operate when living systems cannot maintain functional competency. The objective of pathology, then, is to elucidate the basis of functional incompetency in life forms; the basis of functional incompetency should be embodied in the concept of disease.

The disease concept thus becomes the cornerstone of the science of pathology; other concepts and terms of this science should be based on, and related to, the concept of disease. For the purposes of economy of effort and clarity of communication, all fundamental concepts and terms originating in the various biological disciplines should find universal applicability; at worst, they should not exhibit incongruence. As we consider the disease concept, we should also recognize its desired utility as a central concept in an experimental science.

The disease concept should be applicable at all levels of biological organization, and both phytopathology and zoopathology should share a common group of concepts and theory. The modern applied branches of pathology—plant pathology, veterinary medicine, and human medicine —have derived much common benefit from the discoveries that dispelled spontaneous generation and established the role of microbes in disease. On the other hand, they have developed independently as applied disciplines and, in general, have failed to benefit from a science of comparative pathology. In this chapter I wish to consider the disease concept and the discipline of pathology from the point of view of biology as a whole without injecting the confusion that can arise by taking more limited views of these subjects. The applied branches of pathology tend to diminish the view of pathology as a basic science. The basis for this can be seen in the forces at play in the field of plant pathology.

B. Plant Pathology—An Applied Science

When man has sought to control a portion of his ecosystem through an applied science like plant pathology, the bringing together of relevant biological disciplines coupled with an empirical experimental approach has proved to be quite successful. Plant pathology represents one of the more complex branches of applied science. A working knowledge in this field requires a broad comprehension of diverse disciplines and an ability to integrate and apply them to an "artificial" ecosystem, the agroecosystem. In the area of application much is to be gained from the science of pathology. The theory of pathology as viewed in a biological context should meld into that of disease control strategies and serve to aid the choices and directions taken in the development and application of disease control strategies.

The activities within plant pathology are diverse. They range from the purely theoretical, concerned with the nature and cause of disease, to the practical arts of disease management. This diversity is both a strength and a weakness. Strength is derived from insights gained from the many areas of activity; the field is constantly infused with new ideas from diverse sources. But scientists contributing to various facets of plant pathology, because of its diversity, often tend to become isolated from some segments of the science, and thus fail to benefit from a comprehensive view of pathology as a biological discipline.

Other factors also contribute to this problem. Since plant pathology is directly relevant to man's need to grow enough food and fiber to sustain civilization, society has demanded rapid solutions to the control of plant diseases. This pressure has fostered the initially more rapid empirical approach to problem solving which has tended to decrease emphasis on the development of the theory and science of pathology. More than 45 years ago Link (1932) noted that plant pathology, through neglect of its theoretical aspects and preoccupation with its empirical and applied phases, has permitted the development of theory and systematization of knowledge to lag behind practice, and that plant pathology had failed to build a formally expressed system of relations. This situation still exists today, and it is leading to an increased inefficiency in the applied phases; too many researchers are continually "reinventing the wheel" and our journals are becoming crowded with isolated facts of seemingly little import. A more comprehensive body of theory is needed to help coordinate these facts into a more workable system of knowledge and to aid in helping to reveal their utility in furnishing new ideas to fuel the empirical and theoretical approaches for controlling plant diseases. This same need appears to be evident in the other applied fields of pathology (Cohen, 1977).

In the United States most plant pathology research has been organized along commodity lines. This, coupled with short-term funding, has dictated a preoccupation with short-range solutions to immediate disease problems. By its design our system results in relatively little input into the basic areas of pathology and presses toward a technology with only tenuous connections to pure science. This drift, if not checked, is likely to prove detrimental both to the theoretical and applied phases of plant pathology and will place upon society an unnecessary burden because of inefficiency in developing new, more effective disease control strategies. Plant pathology must embrace much more than man's efforts to control plant diseases if it is to be of maximum benefit to the society which pays its bills.

Another factor contributing to the lack of a coherent philosophy of pathology in plant pathology is related to the diverse training and experience of scientists and technologists in this field. Many plant pathologists view their work from the perspective of an agronomist or horticulturist, or from the perspective of a mycologist, bacteriologist, virologist, etc., without focusing on their activities from the perspective of pathology as a biological discipline. Many of the insights to be gained from the science of pathology are not likely to be realized when the problems of pathology are approached with a distinct bias from the perspective of the agent of disease or the suscept (plant or host) without a focus on the nature of disease.

Plant pathology is a branch of the biological discipline of pathology. The development and expansion of the theory and principles of pathology can greatly aid all of the applied branches of the discipline as well as contribute to basic biological knowledge. Also, an appreciation of disease and its significance in a biological context provides a greater insight into nature and, at the same time, points the way for man's greater utilization of biological knowledge. The theory, concepts, terminology, and propositions of plant pathology should be compatible with an understanding of the larger discipline of pathology.

C. Goals of the Chapter

The primary objective of this chapter is to present a concept of disease, compatible with current biological knowledge, that can function as a central theme for the science of pathology. Other major objectives are to examine the nature of the relationships between organisms which result in disease and to present a hypothesis to aid in the elucidation of host–parasite and suscept–pathogen relationships. The propositions to be advanced will apply to various levels of biological organization—cells, organisms, species, communities, etc.; only the tools and methods used in elucidating a given problem would differ substantially. The focus of this presentation will be directed toward the cellular level of organization. Subsequent chapters in this volume will be directed at specific physiological and biochemical factors.

II. THE NATURE OF DISEASE

As biologists what do we mean by the term "disease"? Is disease a condition? Is disease a state? Is disease tangible? Is disease injurious? Is disease a process? Can disease be analyzed? Can disease be expressed in quantitative terms? What is the most useful definition or concept of disease? The term "disease" should embrace the concepts and theory on which the science of pathology is based.

The Webster's Third New International Dictionary defines disease as

> an impairment of the normal state of the living animal or plant body or of any of its components that interrupts or modifies the performance of the vital functions, being a response to environmental factors, to specific infective agents, to inherent defects of the organism, or to combinations of these factors.

From this we gain the view that disease is an impairment of the vital functions in a living organization. Also, disease may be induced by factors that are internal or external to living systems. All of this is compatible with the commonly held notion that disease is responsible for a departure from health, and when health is impaired the living system is said to be diseased. In this context the terms disease and health encompass polar concepts that speak to a common reality, i.e., the state of a living system at a given point in time. Thus no clear-cut separation between disease and health is possible except by definition. The dictionary definition of disease is informative, but it fails to identify the essence of disease, i.e., to pinpoint a tangible reality that can serve as the focal point for the science of pathology.

Many plant pathologists have thought seriously about the nature of disease and others have not. Space does not permit an extensive consideration of recorded definitions or statements about disease by renowned pathologists, but a few examples could be helpful.

Julius Kühn (1858)   "The diseases of plants are to be attributed to abnormal changes in their physiological processes; they are disturbances in the normal activity of their organs."

H. Marshall Ward (1896)   disease is "a condition in which the functions of the organism are improperly discharged; or in other words, it is a state which is physiologically abnormal, and threatens the life of the being or organ."

Robert Hartig (1900)   "I do not regard the investigation of mere sickness as a task of pathology. It is only when the sickly condition leads to the death of some part of the plant that we speak of actual disease."

H. Morstatt (1923)   disease "is the sum of the abnormal life processes going on in the body."

H. H. Whetzel (1935)   "Disease in plants is injurious physiological activity, caused by the continued irritation of a primary causal factor, exhibited through abnormal cellular activity, and expressed in characteristic pathological conditions called symptoms."

J. G. Horsfall
and A. E. Dimond (1959)   . . . "Disease Is Not a Condition . . . Disease Is Not the Pathogen . . . Disease Is Not the Same as Injury . . . Disease Results from Continuous Irritation . . . Disease Is a Malfunctioning Process . . ."

G. C. Kent (1973)   "Disease is the diminution or loss of the ability to correlate the utilization of energy within an individual as a result of the continued irritations of a pathogen."

Most of these definitions or statements deal in one way or another with the theme of abnormal physiology. Other definitions are focused on the theme of economics and deal with loss of economic value of plants and plant products. Since the latter do not provide a biological perspective or aid in a mechanistic analysis of disease or its causation, they will not be considered here.

Julius Kühn, who is considered to be the father of modern plant pathology, attributed disease to changes in the physiological processes in plants. According to Hartig's statement, not just any change in physiology can be considered as disease, but rather the change must be of such magnitude that it is injurious to the living system. Link (1932), in addressing this point, notes that "pathic events" (disease) may be spoken of when the change in vital processes falls outside the range of easy tolerance or adaptiveness within a living system. H. Marshall Ward, perhaps the greatest of the British plant pathologists, suggested that disease arises from a condition in which the functions of an organism are improperly discharged. Morstatt's statement implies that all the functions of the living system in question have to be considered, but the key element is life processes.

Whetzel's statement provides us with a specific view of the nature of disease—"Disease . . . is injurious physiological activity." This statement provides considerable insight, and just as important, a basis for experimental analysis. As noted by Horsfall and Dimond, disease is not a condition, it is not the same as injury, it is not the pathogen, but rather it is a malfunctioning process. Thus, disease is not static but rather dynamic, ever changing. It is useful to distinguish disease from injury; as noted by Whetzel, disease is fostered by the continued irritation of a causal factor (or factors) as opposed to injury which results from an instantaneous event. The precept that disease is injurious is essential to an adequate concept of disease.

Kent's statement about disease adds an element that is very helpful in considering the nature of disease, for it links disease with the loss of ability to correlate the utilization of energy and vital functions within an organism. The words of Tansley (1929) add clarity to this view.

> Life indeed, may almost be said to consist in the continual ordered discharge (utilization) of energy by the organism. When this discharge is at a maximum, life is being lived at its highest intensity: When it is at a minimum, life is at a low ebb: When the coordination of discharge is impaired the organism is unhealthy; and when the organism is no longer capable of ordered discharge of energy it is dead.

The linking of coordinated energy utilization to the concept of disease facilitates application of the disease concept to all levels of organization in biology—cells, organisms, communities, ecosystems. In a recent treatment of metabolic regulation in cells, Kosuge and Gilchrist (1976) noted "The most fundamental requirement of metabolic regulation is the insured maintenance of a steady state between the energy-generating catabolic processes and the myriad of energy-requiring synthetic reactions proceeding concurrently in the cell." They go on to say "that the result of any unrelieved disruption of the metabolic state, regardless of the cause, results in the diseased state." The maintenance of an appropriate "energy balance" is fundamental at all levels of biological organization.

A concept of disease in modern terms may be stated as follows: Disease is the injurious alteration of one or more ordered processes of energy utilization in a living system, caused by the continued irritation of a primary causal factor or factors. When a living system is altered beyond its range of easy tolerance in the sense of Link (1932) it is diseased as opposed to healthy; if the system is pushed beyond its limits of absolute tolerance it will die.

The thesis that disease is injurious, uncoordinated dissipation of energy essential for the vital functions of a living system can serve as a base on which to build an understanding of disease and the experimental science of pathology. This concept of disease is compatible with the widely accepted views that disease may be triggered or initiated by biotic or abiotic factors internal or external to living systems. This concept is applicable to living systems over a spectrum of organizational levels: ecosystems, populations, individuals, organs, tissues, and cells. The consideration of energy and its utilization make this broad application possible.

In pathology the focus usually is on the individual or population, but in a consideration of the molecular aspects of disease the cell becomes a more logical unit of organization for study. Pathological events occur and require evaluation at each level in the hierarchy of biological organization, but the same principles and similar concepts can be usefully applied at each level. Only the plane on which the problem is considered and, of course, the experimental tools employed need differ. For example, resistance in wheat to *Puccinia graminis* var. *tritici* may be achieved through the hypersensitive reaction of the invaded cell. The plant in a statistical sense remains healthy and yet the invaded cell is diseased. If our focus is the plant we are dealing with a mechanism of disease resistance. If we are interested in cellular pathology we are dealing with a case of extreme susceptibility to disease. An improved understanding is generated through analysis from both perspectives.

Our theory and terms in pathology should be related to our concept of disease and our concept of disease needs to be compatible with other biological concepts. There is a great tendency to equate the pathogen with the disease it induces. An example of the confused terminology this can generate is seen in the unwise extension of the valuable concept of the life cycle to the often-used term "disease cycle." This term is not compatible with the concept of disease outlined above since disease is not a cyclic process. Disease is a couplet of the duality health and disease. The concept of the cycle in biology is an extremely useful tool, particularly when applied in the context of the duality life and death, but even here the cycle is best viewed as one turn of a helix extending through time and space. The concept of the cycle applied to disease merely adds confusion. Disease may occur when the life cycle of a suscept and the life cycle of a biotic pathogen impinge upon each other at the proper times and places in a suitable environment, but clarity is not achieved if either health or disease is viewed as being cyclic.

III. DISEASE ETIOLOGY

The examination of the cause of disease is not a simple matter because of the complexities involved with factors both external and internal to the suscept that effect disease causation. There has been a great tendency to focus on the pathogen as the cause of disease. Link (1933), in his excellent treatment of etiological phytopathology, notes that this "arose as a consequence of the quest for single external causes, which dominated the atomistically oriented natural sciences of the last half of the 19th century and manifested itself in part in the naive version of the germ theory of disease that characterized the biology and pathology of that period." The adoption of this pseudoetiology, i.e., the pathogen being the cause of disease, inhibits a consideration of the nature of disease and its mechanisms of causation; further, it fails to engender the type of experimental analysis essential for the understanding of disease as a biological phenomenon. The affliction, propagated by a pseudoetiology, has carried over into our consideration of the basis for disease resistance, susceptibility, and pathogenicity; thought and effort is too often focused on elucidating the cause for a phenomenon as opposed to the complex of determining elements. A "thoroughgoing etiology" of disease or susceptibility, resistance, and pathogenicity must, because of their nature, deal with causal complexities, i.e., the totality of the antecedents of the phenomenon (Link, 1933).

Plant pathologists have utilized the disease triangle and more recently the disease pyramid (Browning et al., 1977) to help visualize the complex of factors that relate to disease causation. A "thoroughgoing etiology" of disease must deal with the component nature of the elements of the base of the disease pyramid—suscept (host), pathogen (parasite), and environment in relation to the other element, time. Understanding of disease causation requires that causal components be dissected conceptually and experimentally and the determining factors be reassembled in terms of their spatial and temporal occurrence and interaction in disease causation; this approach to disease etiology holds the key to greater understanding of disease causation. Such a "thoroughgoing etiology" also forms the basis of comparative pathology and will serve to integrate pathology with other experimental biological sciences.

The use of the term "cause" for the pathogen and linking of disease with pathogens that are often also parasites has given rise to the widely held view that parasites are necessarily harmful to their hosts (Odum, 1959; Vines and Rees, 1964). This point of view as we shall see can be quite misleading if we wish to examine the total spectrum of heterotrophic food relationships among organisms. Link (1933) suggested that the term "incitant or some better equivalent" be used to designate the causal agent of disease; his intent was to recognize the many factors involved in the complex web of causality.

Plant pathology tends to focus on diseases induced by biotic pathogens. Since many pathogens are parasitic, a consideration of parasitism and how it relates to disease will precede consideration of hypotheses aimed at helping to elucidate the nature and causes of disease and/or parasitism.

IV. PARASITISM AND DISEASE

The term "parasitism" is generally used to designate the relationship or association between organisms, usually belonging to different species, in which one party, the parasite, benefits from the other, the host. It is often assumed, sometimes incorrectly, that the parasitic relationship is necessarily harmful to the host. Parasitism and disease are distinct biological phenomena that should not be confused. This discussion follows the thesis of Smith (1934) which associates parasitism with a relationship in which the parasite obtains nourishment from its host. The struggle for existence among living forms has led to two major types of heterotrophic food relationships characterized by the predatory and parasitic habits. The spectrum of heterotrophic food relationships characterized by predation and parasitism involves most species of life, including both plant and animal forms.

The distinction between predators and parasites is easily made when the extremes are considered, but as might be expected, there exist relationships where elements of both habits may be present. Generally predators are larger than their prey, while parasites are normally smaller than their hosts. Predators ingest their food while parasites absorb theirs. Predators inflict injury but not disease; parasites, if injurious, normally induce disease. Our interest is in parasitism and its relationship to disease. An examination of the diverse biological relationships involving parasitism will reveal that parasitism is a characteristic feature of most life forms and that parasitism does not necessarily indicate a pathological relationship, but rather, it is a normal condition having its origin in the interdependence of living organisms (Smith, 1934).

A. Heterotrophic Food Relationships Involving Parasitism

The spectrum of relationships and the conditions that theoretically govern parasitic associations are diagrammed in Fig. 1. Note that the four elements of the disease pyramid (Browning et al., 1977) are applicable to all parasitic relationships; all we need to do is substitute the parasite for the pathogen. Numerous types of relationships fall within the spectrum of parasitism. These may involve associations in which the coactions of hosts and parasites result in a relationship of mutual benefit to the associates (mutualistic); or relationships in which only one of the

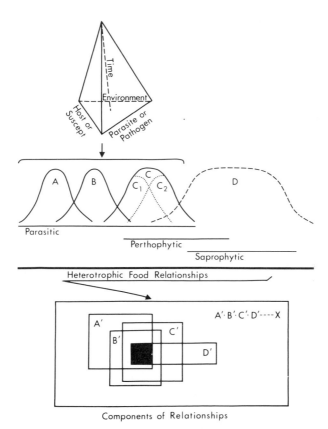

Fig. 1. Outline of heterotrophic food relationships. The types of relationships (A) mutualism, (B) commensalism, (C) pathosism, and (D) saprophytism are shown in relation to modes of existence—parasitic, perthophytic, or saprophytic. Of the total components (X) in a given relationship those that determine the type of the relationship are designated by A', B', C', and D', respectively, for mutualism, commensalism, pathosism, and saprophytism. It is conceived that relationships A, B, and C are determined by the components within the elements embodied in the host–parasite–environment–time pyramid.

associates, usually the parasitic partner, benefits without a harmful effect upon the other associate (commensalistic); or relationships in which only the host or both the host and parasite are injured or even killed (pathosistic). Although the spectrum of possible associations can be divided into a large number of specific categories (Burkholder, 1952), it will serve our purpose to divide parasitism into three overlapping groups characterized by mutualism, commensalism, and pathosism.

The fungal and algal symbionts in lichens or nitrogen-fixing bacteria

(*Rhizobium* sp.) associated with root nodules of legumes may be regarded as examples of parasitic associations exhibiting mutualism. Commensalism is seen in the associations of many rhizosphere and phyllo-
sphere microbes associated with higher plants. Where parasitic associations result in disease we can speak of pathosism. The connotations of
the terms mutualism, commensalism, and pathosism in biology transcend
parasitic associations. These terms are applicable to biological relationships other than those involved with food or nourishment.

It should be clear that not all parasites are pathogens and not all
pathogens are parasites. Pathogens may be biotic or abiotic. Biotic pathogens may be parasitic or nonparasitic. An organism that is susceptible to
disease is designated a suscept and may or may not be a host. The term
"suscept" is useful since the pathogen may not be a parasite. We thus have
the couplets host–parasite and suscept–pathogen. The couplet host–
pathogen is correct only if the pathogen is a parasite. In all cases where
disease is involved the suscept–pathogen couplet is correct and has specific meaning. For the sake of brevity this discussion is limited to biotic
pathogens and the diseases they induce, but the basic terms and concepts
elaborated have universal application in pathology.

The spectrum of food relationships presented in Fig. 1 shows that
parasitism grades into saprophytism. Saprophytes colonize and obtain
their food from dead organisms. To bridge the area between parasitism
and saprophytism one can include an intermediate type of food relationship in which the "parasite," most commonly a pathogen, kills host cells
before they are colonized. This mode of attack is characteristic of perthophytes (Münch, 1929) and the association is termed "perthophytic."
Perthophytism is characteristic of plant pathogens such as *Sclerotium
rolfsii* and certain strains of *Rhizoctonia*. Thus, it seems desirable to
divide pathosistic associations into two overlapping groups (Fig. 1, C_1
and C_2) of relationships related to mode of existence—parasitic and perthophytic. This division is useful since the mechanisms of pathogenesis
by parasites and perthophytes may differ substantially, although they
may be expected to involve many common components.

B. Parasitism and Pathogenesis

Pathology includes those parasitic associations that result in disease. An
understanding of disease and its causation may or may not encompass
an understanding of the mechanisms of parasitism, but an understanding
of pathogenesis is critical to an understanding of pathology. An analysis
of host–parasite relationships that result in disease usually reveals that the
damage inflicted upon the suscept results more from the "attacking

mechanisms" of the pathogen (i.e., degradative enzymes, toxins) and adverse metabolic reactions induced in the suscept (i.e., decompartmentalization of enzymes, release of phytoncides) than can be attributed to utilization of suscept nutrients by the pathogen. In other words, the mechanisms evolved by pathogens (both parasites and perthophytes) that render nutrients in living cells of suscepts readily available to pathogens, and the impact of these mechanisms on suscept cells in terms of disruption of metabolic regulation as well as release of toxic substances of suscept origin (phytoncides), and the synthesis of antimicrobial substances (phytoalexins) by suscepts, are likely to be more injurious than the mere loss of nutrients per se. This point of view is supported by the fact that obligate parasites that are pathogens generally cause less physical disruption of their hosts; a more compatible existence has evolved. In parasitic relationships involving mutualism the components of cooperation between the symbionts outweigh the components of conflict. Such relationships have evolved to the point where the mechanisms essential for establishing the parasitic habit must be subject to control and regulation by the coactions of the mutualistic partners.

Historically, attempts to understand the parasitic habit have focused on elucidation of parasite nutrition, availability of appropriate nutrition for the parasite in the host, and the nature of the antagonistic environment (resistance) in the host to the parasite (Lewis, 1953; Garber, 1956). Because of the often close association between parasitism and pathogenicity the possible role of parasite nutrition has assumed a major role in attempts to understand both phenomena. Based on current knowledge of suscept–pathogen relationships (Heitefuss and Williams, 1976), the role of nutrition per se may not be important for understanding pathogenicity and disease. The following brief review of several hypotheses that seek to explain the basis of parasitism and pathogenicity, and the new hypothesis outlined below, may aid in our understanding of disease processes and stimulate improved approaches to the solution of pathological problems.

V. HYPOTHESES RELATING PARASITISM AND PATHOGENICITY

Biotic pathogens are usually parasitic or perthophytic. Generally, distinction has not been made between these two groups of disease agents, although the physiological and biochemical mechanisms for establishing the two habits are very different. Earlier attempts to formulate hypotheses to explain diseases induced by biotic pathogens have tended to center

on the theme of parasitism. Historically, a feeling has evolved that elucidation of the nutritional requirements of pathogens and the ability of
suscepts to provide these requirements would provide insights into disease and explanations of it. This vein of thought led to development of
the nutrition hypothesis of parasitism which attempts to explain susceptibility and resistance of suscepts (hosts) to disease in terms of the nutritional requirements of the pathogen (parasite) and the biochemical
composition of the suscept.

A. The Nutrition Hypothesis of Parasitism

The nutrition hypothesis states that a parasite grows in a particular host or
organ because in that host or organ there is available to the parasite the
kinds and quantities of foods that are necessary for its life, and in resistant hosts or organs the necessary kinds and/or quantities are not present.
The possible combinations of foods present plus the capacity of the parasite to respond differently to these combinations accounts for degrees of
resistance and susceptibility not explainable in other terms. (Lewis, 1953.)

Although this hypothesis embraces concepts that are straightforward
and reasonable, the hypothesis is deficient from a number of perspectives,
the primary one being testability. As Lewis pointed out in 1953 and as
is true today, there is no case on record where the nutrition hypothesis
has been thoroughly tested by examining the definitive nutrition of a
parasite and the relevant composition of its host in the detail that our
biochemical knowledge will permit. Some reasons for lack of pursuit of
the approaches suggested by the nutrition hypothesis include the following: (1) the apparent difficulty of determining the complex nutrition
of some parasites, (2) the failure to correlate parasitism with nutritional
complexity, (3) the similarity in the *in vitro* nutritional requirements of
some parasites that have very different host ranges, (4) the view that
analysis of a host will not provide a true picture of the nutrients that may
be available to a parasite, (5) the knowledge of active defense mechanisms in both animals and plants, and (6) the impact of the concept of
the biochemical unity of life. For these several reasons the nutrition
hypothesis of parasitism has not been widely accepted as a profitable
avenue for elucidating the mechanisms of parasitism or pathogenesis.

The nutrition hypothesis of parasitism is not without experimental
support, though it be quite limited. This support has been derived for
the most part through the use of biochemical mutants of pathogens and
alteration of host nutrition. For example, Bacon *èt al.* (1951) have shown
that mutants of *Bacterium typhosum* deficient for *p*-aminobenzoic acid
or adenine were much less virulent to mice than the parent strain Ty 22.

The virulence of either mutant was restored when *p*-aminobenzoic acid or adenine was injected into the host with the appropriate mutant. The full virulence of the mutant deficient for *p*-aminobenzoic acid was also restored when additional *p*-aminobenzoic acid was incorporated in the diet of the host. The studies of Boone (1971) and associates with nutritional mutants of *Venturia inaequalis*, the apple scab pathogen, have shown that the ability of the host, apple, to supply adequate nutrition for the parasite may be a critical factor in determining success or failure in establishment of a parasitic relationship. Induced nutritional mutants of *V. inaequalis* requiring choline, riboflavin, or certain amino acids for growth were avirulent on apple leaves unless the nutrient for which the mutant was deficient was added to the inoculum during the incubation period. These mutants were able to grow on apple leaf extract, indicating that the host tissue contained adequate nutrition, but a sufficiency of the required nutrients was not available to meet the requirements of the mutants during the critical phase of establishment. Nonidentical nutritional mutants were able to grow syntrophically when paired in a minimum medium. Also, infection occurred and disease developed in apple leaves coinoculated with nonidentical mutants. Mutants of *V. inaequalis* deficient for biotin, inositol, nicotinic acid, pantothenic acid, or reduced sulfur maintained virulence to apple, indicating that the host (suscept) furnished adequate quantities of these nutrients for the parasite in the infection court.

Nutritional mutants of *V. inaequalis* failed to exhibit host specificity as is observed for wild-type strains. Loss in virulence due to a nutritional deficiency resulted in loss in virulence on all host varieties. Wild-type isolates of *V. inaequalis* often exhibited differential growth on apple tissue extracts of different varieties, but virulence of these isolates was not necessarily correlated with their growth responses on host tissue extracts (Boone, 1971). It therefore appears that factors aside from nutrition are of paramount importance in controlling host–parasite specificity.

B. The Balance Hypothesis of Parasitism

Lewis (1953) enunciated the balance hypothesis of parasitism which holds that the environment in which some nutrient compounds exist will influence their effects on organisms. Some nutrients at metabolic concentrations in appropriate environments may actually inhibit growth. Nutrients in a given nutritional environment have been classified as belonging to one of four groups: (1) essential, (2) stimulatory, (3) inhibitory, or (4) apparent (Snell, 1949). Growth of an organism will occur only in the presence of essential foods. Stimulatory foods are those which stimulate growth but are not essential. Inhibitory foods are those that may serve as a food for one organism, but inhibit another or serve as food in one environment and as an inhibitor in a different environment. Apparent foods are those that are essential in one environment but not in another. The *in vitro* nutrition of a number of plant pathogenic parasites tends to confirm the above classification of nutrients.

Studies by Lewis (1953) of the *in vitro* nutrition of *Alternaria solani*, a pathogen of tomato, and *Claviceps purpurea*, the ergot fungus, illustrate the differential effects of specific nutrients or foods in different environments on a given parasite (pathogen). *Alternaria solani* grew on a medium containing only glucose and inorganic salts, but the lag and linear growth phases were long. Upon addition of biotin, choline, and thiamine to the minimum medium, the length of the lag phase was shortened and the growth rate was markedly increased. Addition of amino acids individually and in combinations to the enriched medium yielded a number of interesting effects. Aspartic and glutamic acids increased growth while β-alanine and cystine inhibited growth. Furthermore, the inhibitory effect of β-alanine was overcome by 15 of 22 amino acids tested. Comparable studies with *C. purpurea* revealed that biotin and glutamic acid were required in addition to inorganic salts and glucose for growth. In this medium β-alanine enhanced growth of *C. purpurea* ·when used at concentrations which inhibited growth of *A. solani*.

Based on these and other comparable studies of the *in vitro* nutrition of other parasites, Lewis (1953) formulated the balance hypothesis of parasitism. The basic tenets of this hypothesis are the following:

Among the host's metabolites available to a potential parasite there are some substances that may favor and some substances that may hinder the growth of the parasite. These substances are vitamins, amino acids, organic acids, and others; they are the substances common to the metabolism of many different species. The kinds and concentrations of metabolites in a host vary as conditions change. This variation, in substances capable of affecting the growth of a parasite, may account for the changes in resistance and susceptibility which sometimes occur within a single host. In different environments the individuals of a host variety may likewise vary in the concentrations of metabolites present and thus the variety may be susceptible in one set of conditions and resistant in another. Among different hosts the kinds and concentrations of metabolites vary from variety to variety and from species to species. These variations plus the different responses of different parasites can be expected to account for the complex patterns of varietal and species resistance and susceptibility. It is conceived that, in general, innate immunity is determined by the normal host metabolites that are present and unfavorable to the growth of the parasite, or by the partial or complete absence from the host metabolites of substances that are required by the parasite, or by a combination of these. Immunity of this sort is not determined by special substances or mechanisms. The metabolites of a susceptible host do not inhibit, but rather favor the growth of the parasite. One can imagine combinations of metabolites that would permit any intermediate condition between complete resistance and complete susceptibility. There is, in a sense, a sort of balance in the host between the substances that inhibit and the substances that promote the growth of a parasite. From host to host or from one host condition to another, the balance may tip toward resistance on one hand, toward susceptibility on the other, or remain between the two possible extremes.

The balance hypothesis seeks to explain selective parasitism and/or resistance and susceptibility of hosts without recourse to phytoalexins, phytoncides, mechanical barriers, enzymes, toxins, etc. The differing effects on parasites of nutrients in different environments greatly complicates any possible analysis of the role of specific nutritional constituents in host–parasite systems. This hypothesis appears to provide avenues for understanding problems of obligate parasitism and species and varietal susceptibility in complexes where delicately balanced host–parasite relationships are involved. Upon careful scrutiny it must be concluded that the major tenets of the balance hypothesis are not testable in a definitive way, no matter how appealing the general concept is.

C. The Nutrition–Inhibition Hypothesis of Pathogenicity

Garber (1956) has formulated "the nutrition–inhibition hypothesis of pathogenicity." This hypothesis embodies elements of the balance hypothesis of parasitism but adds a significant component, the concept of host defense. The parasite is visualized to face two possible "environments" within the host, a "favorable environment" and an "unfavorable environment," the latter encompassing more than the presence or absence of adequate nutrition.

> The nutrition–inhibition hypothesis of pathogenicity involves the two environments in the host which directly affect the fate of an invading parasite: a nutritional environment and an inhibitory environment. If the nutritional environment is inadequate for the parasite, the latter can neither proliferate nor metabolize extensively and will not be virulent; if the nutritional environment is adequate for the parasite, the latter will be able to proliferate or to metabolize extensively but may or may not be virulent. An effective inhibitory environment may or may not also be present. In the absence of an effective inhibitory environment, a parasite in an adequate environment generally will be virulent; in the presence of an effective inhibitory environment, the parasite will not be virulent even though the nutritional environment is adequate. Of the four possible combinations of host environments, only the adequate nutrition-ineffective inhibition environment results in virulence; the other environments result in avirulence.

Garber (1956) cites a number of plant and animal studies in which biochemical mutants of pathogens have been used to demonstrate that the nutrition available to a parasite or pathogen can influence its virulence. But the question remains, do experiments with nutritional mutants of pathogenic parasites merely represent interesting laboratory studies of secondary importance in the real world? It is well known that both plants and animals in a state of low vigor, whether due to malnutrition or something else, are often more susceptible to attack by certain pathogens; but is this due to availability of nutrients to the parasite in the host environment or to a reduction of the active and passive defense mechanisms of the host? Also, the toxigenic, antigenic, morphological, and enzymatic characteristics of parasites complicate attempts to understand and examine pathogenic host–parasite relationships in terms of the nutritional "environment" in the host.

All three of these hypotheses—the nutrition and balance hypotheses of parasitism, and the nutrition–inhibition hypothesis of pathogenicity—represent significant efforts aimed at developing a theoretical structure to help explain host–parasite relationships, and at the same time suggest profitable avenues of research involving host–parasite systems. These hypotheses apparently have not had a major impact upon research in pathology for the reasons enumerated above. A more adequate conceptual scheme is needed to materially enhance understanding, organize our current knowledge, and aid in formulating more profitable research approaches. The following hypothesis is offered to aid in these processes.

D. The Multiple-Component Hypothesis of Pathogenesis and Parasitism

Pathogenicity, susceptibility, resistance, and pathogenesis are all processes or attributes governed by the genetic characteristics of living systems and the physical environment in which they operate. The multiple-component hypothesis focuses attention on the nature of the interacting components of host (suscept) and parasite (pathogen) as determining factors in parasitic or pathogenic relationships. This hypothesis is predicated on a consideration of the multiple-component nature of life, the existence of both physical barriers and physiological barriers in living systems, and the assumption that both hosts and parasites contain certain factors that are potentially favorable as well as those that are potentially unfavorable to each other.

1. The Multiple-Component Nature of Living Systems

Living systems are multiple-component systems. As biologists, regardless of discipline, we draw our conclusions directly or indirectly from measurements of the rates of processes ultimately traceable to chemical reactions subject to the controls in the simplest unit of life, the cell

(Thimann, 1956). It is from the acceleration or inhibition of these rates that we make deductions about our subject and thereby advance our science. This principle holds whether our focus is on an individual biochemical process or the behavior of a species within an ecoysystem. All biological processes are subject to regulation through factors which stimulate or inhibit them. In some cases, a single given factor may serve as an inhibitor or as a stimulator of a process, depending upon its concentration and the "environment" in which it operates. In living systems many components influence the rate of a given process, and it is the sum of the inhibitory and the stimulatory components that determines the rate of the process. Most biological processes are subject to multiple controls which provide for enormous flexibility in living systems. Thimann (1956) notes, "Perhaps indeed it is the interaction and balance between multiple systems that is the real distinguishing characteristic of living systems."

As discussed earlier, plant pathology has perhaps focused too much on studies of "the cause" of disease, i.e., seeking to attribute disease to the agent or cause without due emphasis to what may be termed the "causal complex." This attitude no doubt stems from the great impact the germ theory of disease had on the science of pathology during its formative years. The view that disease is due to a cause or that resistance of a suscept is due to the presence or absence of a factor persists even today. In view of our current understanding of living systems, this thesis would appear to be untenable.

2. Physical Barriers and Physiological Barriers in Host–Parasite Systems

The establishment of a host–parasite relationship involves the bridging of two groups of barriers, physical barriers and physiological barriers. The physical barriers separate the host and parasite and prevent intimate contact or ramification of the host tissues by the parasite. Parasites may enter the host passively through wounds or natural openings or they may possess an active means of ingress. For a parasitic relationship to be established in a functional sense, both groups of barriers must be successfully breached.

Some organisms become parasitic only after they find their way into their host through a natural opening or wound. They have evolved a means to overcome the physiological barriers, but lack the ability to actively traverse the physical barriers. Parasites of this type are generally pathogenic when introduced into a host. Other parasites have evolved means of direct penetration of host tissues and are able to actively gain ingress. Once physical barriers are bridged, the success or failure of the

potential parasite depends upon its ability to cope with the physiological
barriers present in the potential host. For example, many plant pathogenic fungi have evolved specialized structures (i.e., appressoria) for
effecting ingress into plants. Such organisms are capable of breaching
both the physical and physiological barriers of their suscepts. Many of
these same organisms are quite capable of gaining ingress into nonhosts
but fail to breach the physiological barriers of these potential hosts.

The bridging of physical and physiological barriers between hosts and
parasites is accomplished in various ways. Multiple factors in both the
host and the parasite determine the manner in which a given barrier is
bridged or not bridged. These factors are regulated by the genetic potentials of the host and the parasite, subject to different degrees of expression due to the particular physical environment present and to modifications by the interaction of the existing components. How does one
examine such a complex system? What are the significant components
in a given host–parasite complex that govern the outcome in a given
environment?

Obviously, studies centered on the parasite alone are not likely to yield
satisfactory explanations for disease, virulence, resistance, susceptibility
or mutualism, commensalism or pathosism. Even if a "primary determinant" of the relationship is revealed, it must operate in a background of
more subtle factors in order to be functionally operative in the host–
parasite complex. The explanation of host–parasite relationships requires
a consideration of all salient phenomena.

3. The Concept of Pathogen (or Parasite) and Suscept (or Host) Environments

It is conceived that a potential parasite and a potential host are each
theoretically capable of presenting each other with two possible "environments," an "unfavorable environment" and a "favorable environment."
A given environment may consist of multiple components and include
both morphological and biochemical components. Thus, four potentially
interacting environments may be postulated to interact and exist in a
given host–parasite (or suscept–pathogen) complex. In a given host–
parasite relationship at least one of the postulated environments has to
be operative, but all four are likely to be. If the host–parasite complex
is mutualistic, the favorable environments offered by the participants predominate. If the relationship is commensalistic then the favorable environment offered by one member of the complex benefits the other
member, and the potential unfavorable environments of the two and the
favorable environment of the other member are not sufficiently operative
to be of significance. When the parasitic relationship is pathosistic the

unfavorable environment of one or both members of the complex predominates. If it is the unfavorable environment offered by the host (suscept) which predominates, then the host will be resistant. On the other hand, if the unfavorable environment of the parasite is dominant then the host will be susceptible to disease.

The multiple-component hypothesis of pathogenesis and parasitism seeks to provide a theoretical framework to embrace the many facets of host–parasite interaction that contribute to the outcome of host–parasite encounters whether they be mutualistic, commensalistic, or pathosistic (see Fig. 1). The parasitic habit of a parasite and the susceptibility of a host to a given parasite are considered to result from the sum of the total interactions between the two living systems as they are influenced by the physical environment and time. The parasitic ability of the parasite and the resistance or susceptibility of a host are considered to be determined ultimately by the genetic attributes of the participants. These attributes of the participants encompass both their biophysical and biochemical characteristics.

4. The Multiple-Component Hypothesis—A Basis for Dissecting Biological Relationships and Pathogenesis

The multiple component hypothesis is applicable to all of the relationships of life associations diagrammed in Fig. 1. It facilitates consideration of the factors pertinent to heterotrophic food relationships and provides the mechanism for conceptually dissecting the two environments, favorable and unfavorable, that each host or parasite may bring to a given relationship. The elements of the multiple-component hypothesis are diagrammed in Fig. 2.

In most host–parasite associations it is visualized that all four potential environments are operative; and thus it is the balance or sum of these environments, including their interactions, which determines the success or failure of parasitism and whether or not the relationship between host and parasite is mutualistic, commensalistic, or pathosistic. It is considered that some host–parasite relationships can exist in more than one type of parasitic association (mutualism, commensalism, or pathosism), depending upon the physiological condition of the participants and the physical environment in which they exist.

The favorable environment that a parasite (Fig. 2) may provide a host might include nutrients for which the host is deficient (e.g., minerals, growth factors, utilizable nitrogen). The dominance of this type of environment can be seen in relationships between rhizobia and legumes, between mycorrhizal fungi and roots of many plant species, and so on. Components of an unfavorable environment a parasite (Fig. 2) may offer

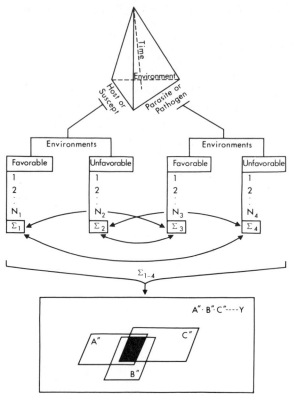

Host & Parasite Components of Relationships

Fig. 2. Outline of the multiple-component hypothesis of parasitism. The host and parasite are each conceived to contain a favorable and an unfavorable environment for each other. Each of these four environments is of a multiple-component nature. It is the sum of the components of the four environments and their interaction (Σ_{1-4}) that determines the type of host–parasite relationship—mutualistic, commensalistic, or pathosistic. Of the total components (Y) in the parasite and host elements of the host–parasite–environment–time pyramid, those that determine the type of relationship are indicated by A″, B″, and C″ for mutualism, commensalism, and pathosism, respectively.

a host might include toxins and enzymes that degrade host constituents, excess production of growth regulators, etc. The favorable environment of the host (Fig. 2) includes the availability of those nutrients and other factors needed by the parasite for its growth and reproduction. The unfavorable environment of the host (Fig. 2) would include physical barriers, phytoncides, and phytoalexins. Thus, the individual theoretical environments of the host–parasite complex are viewed as being multicomponent in nature. In terms of the impact of a given environment one can visualize that some of the components will be of more importance than others.

Components of a given environment may be constitutive or inductive. Phytoncides are constitutive components and phytoalexins are inductive components of the unfavorable environment a given plant may present to a potential parasite; also, certain toxins may be regarded as constitutive and certain enzymes as inductive components of the potential unfavorable environment a parasite may present to a potential host. In the case of inductive components of the different environments, the component in question is elicited after exchange of a chemical message(s) between the host and the parasite.

In host–parasite combinations that result in disease, in the host element of the complex, the unfavorable environment of the parasite and the favorable environment of the host normally will be dominant. It is visualized that in many instances the favorable environment of the host to the parasite may be created, at least in part, by components of the unfavorable environment of the parasite, designated "attacking mechanisms" by Brown (1955). These attacking mechanisms aid the parasite (or perthophyte) in gaining ingress by altering host structure, cell permeability, and in some cases, facilitating the colonization of host tissues. A strong battery of attacking mechanisms is often characteristic of facultative parasites and perthophytes, and they may be operative throughout the period of host–parasite interaction. It is considered that the environments of one party in a host–parasite complex influence the environments in the other, as depicted in Fig. 2. In cases where the host–parasite interaction is of a more compatible nature, the battery of attacking mechanisms of the unfavorable environment of the parasite may be expected to be less intense or subject to control or regulation by the host–parasite complex. The controlled production of plant cell wall-degrading enzymes by mycorrhizal fungi may be regarded as an example of such control.

In most diseases, unless the host is killed rapidly by a potent component(s) of the unfavorable environment of the parasite, the disease agent becomes limited to a given host tissue or area of the plant by the active defense mechanisms called into play by the suscept. These active defense mechanisms are components of the unfavorable environment of the host (or suscept). Even in cases where we regard the host reaction as being susceptible, the parasite is usually limited to lesions or specific areas of the host and fails to ramify throughout all or even the greater portion of the suscept unless the parasite becomes systemic. The active defense mechanisms called into play in resistant and susceptible suscepts or hosts of a given species are generally similar, the primary difference being the speed and/or intensity of host reaction.

In a number of respects the susceptible plant which is not rapidly killed by a given pathogen may represent a better system for elucidating components of the unfavorable environment of the suscept than a system involving a highly resistant suscept. Similarly, a system involving a susceptible host and a virulent parasite represents the suscept–pathogen system best suited for elucidation of the unfavorable environment of the pathogen. Because of the number of factors involved and their potential interactions, the multiple-component hypothesis of parasitism and pathogenesis suggests that attempts to elucidate the components of the unfavorable environment of the host (resistance mechanisms) by comparative studies of the reaction of resistant and susceptible plants to a pathogen would not necessarily represent the most expedient approach. This is true even in cases where isogenic lines of a suscept differing by only genes controlling disease reaction may be available, since each host–parasite interaction is predicted to be unique and characteristic for the particular suscept–pathogen combination in a given environment.

Most published studies on the nature of resistance in plants to pathogens involve comparative studies of resistant and susceptible plant varieties with the apparent aim of distinguishing the feature or character which accounts for host resistance. Despite enormous investment of resources and time, this approach has not resulted. in an adequate explanation at the biochemical level for disease resistance in a single host–parasite combination. The failure of this approach to provide an understanding of disease resistance is understandable, logical, and even predictable in the context of the multiple-component hypothesis.

As noted above, for disease to occur in a plant interacting with a biotic agent, the unfavorable environment of the biotic agent must be functional. Likewise, if the host range of the biotic agent is restricted, then the unfavorable environment of the nonhost members of the species must be operative. The simplest case of disease induction might then be envisioned where the unfavorable environment of the disease agent contains one component of primary significance and this component creates in the susceptible host a favorable environment due to a lack of a means on the part of the suscept to negate the action of this component of the pathogen's unfavorable environment. The host-specific toxins could conceivably correspond to such components in the unfavorable environment of certain pathogens.

No such simple cases are yet known. For example, Yoder and Scheffer (1969) have shown that the host-specific toxin, victorin, produced by *Helminthosporium victoriae* serves to aid the pathogen during ingress and establishment. They further demonstrated that the ability of *H. victoriae* to form appressoria, attach to the host surface, and penetrate oat leaves are characteristics essential for successful parasitism. The latter abilities may be regarded as components of the unfavorable environment of *H. victoriae*. They represent components shown to be lacking in a true saprophyte, *Neurospora tetrasperma*. Yoder and Scheffer also showed that *H. carbonum*, a pathogen of corn but not of oats, and a victorinless isolate of *H. victoriae* both formed appressoria, attached to the plant surface, and penetrated oak leaves of a *H. victoriae*-susceptible variety but they were not virulent or parasitic. When victorin was added to the infection court, both *H. carbonum* and the victorinless *H. victoriae* isolate invaded and parasitized oat tissue. The victorin merely created in the host a favorable environment for growth of the potential parasites by destroying the permeability of host cells and negating the plant's defense mechanisms. Although victorin appears to be a key factor in parasitism and disease, the ability of the pathogen to form appressoria, attach to the host surface, and penetrate also are important considerations in nature.

In terms of susceptibility of susceptible oats to *H. victoriae* it also needs to be stressed that the receptor site in the suscept for victorin is just as important to disease development as is victorin itself. Thus, the occurrence of an attacking mechanism such as a potent host-specific toxin and its involvement in disease is best considered in the context of the four theoretical environments of the host–parasite complex envisioned in the multiple component hypothesis of parasitism and pathogenesis.

Elucidation of host–parasite systems in terms of mechanisms at the cellular or biochemical level is a complex task. It is postulated that a disease such as victoria blight of oats may represent an extreme in terms of simplicity with respect to unravelling the salient features related to pathogenesis, and that diseases of this type may be atypical and represent chance occurrences in nature. In most host–parasite combinations it is postulated that host–parasite compatibility or incompatibility rests mainly on the occurrence and interactions of multiple components in the four theoretical environments postulated to exist in host–parasite systems. Thus, in any given host–parasite system it is not unexpected that toxins, enzymes, ability to form infection structures, etc., may act simultaneously as components of the unfavorable environment of a parasite, or that phytoncides, phytoalexins, morphological barriers, etc., will act simultaneously as components of the unfavorable environment of a host, or that the favorable environments of both the host and parasite are composed of multiple factors acting simultaneously. In such a system it may be extremely difficult to demonstrate the relevance of a given component to the host parasite interaction because of the inherent flexibility likely to be operative in many host–parasite combinations.

For example, rigid proof of the role of phytoalexins in disease resistance is still lacking. Does this mean we should ignore these compounds in a consideration of host–parasite relationships? Not at all! We merely need to change our approach to solving the problem relating to phytoalexins. The same may be said of other components of host parasite systems suspected of being related to virulence, pathogenesis, resistance, susceptibility, etc.

Cellular pathology is still in its descriptive phase. This is true even though the functional aspects of certain components in host–parasite interactions are generally appreciated, e.g., the relationship of certain toxins to the alteration and injury of host cell membranes (Wood *et al.*, 1972) and the relationship of pectic enzymes to tissue maceration and plant cell death (Bateman and Basham, 1976). The multiple-component hypothesis of parasitism and pathogenesis suggests the need for further description and characterization of elements of host–parasite systems that may be assigned to the four theoretical environments of host–parasite complexes. We must know and appreciate the key elements of these environments before we can integrate these key elements to explain the basis of host–parasite interaction in a systematic, mechanistic manner.

All the facts need not be known before attempting a rational analysis of host–parasite systems. The multiple-component hypothesis predicts that the group of elements or mechanisms involved in a given parasitic relationship will be unique for a given parasite and host species or variety. This does not mean that common elements will not be found in many host–parasite combinations. The hope and expectation with respect to pathosistic relationships is that there are many elements in common, and that they merely need to be identified and analyzed in the context in which they operate. The contribution of a given element to the host–parasite relationship has to be analyzed in the context of the total host–parasite environment in which it operates.

Thus, it can be envisioned that parasites a and b are both capable of producing enzyme x. Enzyme x may be a key element of the "attacking mechanisms" of parasite a and of relatively little consequence to parasite b, if the host of parasite b contains a repressor of enzyme x or if parasite b produces a toxin or some other metabolite which negates the need for enzyme x, etc.

When we narrow our focus to host–parasite relationships that result in disease, it is reasonable to focus on the unfavorable environment of the parasite and the favorable environment and unfavorable environment of the host. In most cases it appears that the favorable environment in the host is created at least in part by the action of components of the unfavorable environment of the parasite. Since all living systems contain many constituents in common, it would appear that nutrition per se as a factor in determining host–parasite specificity in nature may be legitimately removed from serious consideration in most instances. Thus, the focus centers on the unfavorable environment of the parasite and the unfavorable environment of the host for seeking an explanation of host–parasite specificity, mechanisms of pathogenesis, and a basis for resistance or susceptibility of hosts or suscepts. This type of rationalization is needed for the sake of expediency, but it must be remembered that the multicomponent hypothesis warns that it may not be feasible to provide adequate explanations for host–parasite systems apart from adequate consideration of the four theoretical environments because of possible interactions of components in the total system.

VI. UTILITY OF THE MULTIPLE-COMPONENT HYPOTHESIS IN PATHOLOGY

Disease is the theme of pathology and is defined as the injurious alteration of one or more processes of ordered energy utilization in a living system caused by the continued irritation of a primary causal factor or factors. Thus, the focus of the science of pathology is disease etiology, i.e., the elucidation of the factors involved and the systematization of the knowledge of their relationships to disease causation. The science of pathology must utilize theory to organize and integrate the facts relevant to disease into an ordered scheme of relationships. This is essential if our knowledge of pathology is to advance in an efficient manner. Advancements in the science of pathology are needed to provide stability and permanence to knowledge (Wigglesworth, 1955) that can give direction to the applied phases of pathology concerned with disease control strategies.

The multiple-component hypothesis of pathogenesis and parasitism represents a scheme for compartmentalizing components of the complex of factors involved in establishment of a spectrum of heterotrophic food relationships, including those that result in disease. When applied to a given suscept–pathogen relationship, this hypothesis may serve as a useful guide in considering the factors involved and their possible roles in disease. The current methods of experimental pathology permit the identification of many potential components of the four theoretical environments embodied in the multiple-component hypothesis of pathogenesis and parasitism. Although the complexities that could be involved in determining a given host–parasite relationship may be beyond our capabilities to solve at present, the hypothesis suggests a way of looking at host–parasite relationships that may be useful in identifying components of the system and in developing procedures by which they can be experimentally isolated and verified.

Current literature in the area of cellular pathology deals primarily with components of the unfavorable environments of parasites and the unfavorable environments of hosts or suscepts (Heitefuss and Williams, 1976). Considerable factual information exists that can be assigned to these environments. This literature has been generated primarily by investigators seeking to determine the basis of pathogenicity or the basis of plant resistance, without, in many cases, due consideration to the likely involvement of multiple components and their interactions governing attributes of pathogens and suscepts. Consequently, the information available represents mostly isolated facts pertaining to many different diseases or host–parasite systems. Prior to seeking a mechanistic explanation of a given disease, the multicomponent hypothesis dictates that pertinent components of the four theoretical environments in the given suscept–pathogen complex be catalogued or known in sufficient detail to permit construction of a model that can be analyzed and tested. To construct a model to explain a given disease will not necessarily require that all of the components or elements be known. This approach may in fact be used as a tool to direct the research needed for elucidating some of the more subtle components involved in suscept–pathogen complexes.

Current information is not adequate for the construction of a mechanistic model of a parasitic or pathosistic relationship. Also, many existing facts may be of doubtful value since their involvement or association with suscept–pathogen complexes has been demonstrated only after the critical periods in establishment of a parasitic or perthophytic relationship. Too often investigators have examined or studied the biochemical or other elements of suscept–pathogen interactions after the critical events of the interactions have passed. The presence of toxins, enzymes, phytoalexins, etc., in tissues with already well-defined lesions may represent discarded or residual armament or products of necrobiosis which have little relevance to the suscept–pathogen encounter.

The multiple-component hypothesis of pathogenesis and parasitism indicates the need for coordinated investigation of a limited number of host–parasite systems. The need exists for detailed characterization of pertinent components in the four theoretical environments of a given host–parasite complex. This is no simple task. Adequate characterization of a specific component of a given environment may in some instances require years of dedicated effort. Only after the nature of the given component is known and its potential in the context of the suscept–pathogen complex has been defined will the component become useful in the formulation of a mechanistic model of the suscept–pathogen complex.

The multicomponent hypothesis suggests abandonment of the current, often favored approach in which the factor is sought to explain parasitism or pathogenesis or resistance. According to the multiple-component hypothesis such a factor does not exist, except perhaps in the most extreme cases that would be atypical of most host–parasite systems. The investigator who uses the multiple–component hypothesis as a guide in formulating an approach to cellular pathology must be content, at least for the present, to dedicate himself to the elucidation of the individual components of the theoretical environments of a host–parasite system, without knowing the real contribution of these components being investigated to the success or failure of parasitism or pathosism. Satisfaction would be derived from adding another element of knowledge to our understanding of nature and potentially aiding the ultimate elucidation of a host–parasite system at the molecular level. The choice of components to study is by no means a random process, but the selection of a given component for investigation based on current knowledge does not assure one that the component selected will be a primary determinant in a given host–parasite system. Once sufficient data are available for the development of a mechanistic model for a given suscept–pathogen relationship it can be evaluated, refined, and used as a predictive tool to guide future work.

Based on current knowledge, efforts aimed at the understanding of disease causation might best be focused on elucidation of the components of the unfavorable environment of the pathogen and the unfavorable environment of the suscept. For elucidating a given pathosistic association it may also be necessary to clearly elucidate components of the favorable environments of the parasite and the suscept, but logic dictates that emphasis be focused on the unfavorable environments.

In conclusion, the multiple-component hypothesis of pathogenesis and parasitism takes into consideration the dynamic nature of disease and the dynamics of its causation. It represents a systematic approach to the solution of the problems of pathogenesis and parasitism. Research with host–parasite systems over the past century indicates that there are no simple solutions to these problems. The multiple-component hypothesis may enable us to rethink where we are in the study of host–parasite systems and pathogenesis as well as enable us to develop approaches to research in pathology that will enhance our understanding of the causal complexes responsible for disease.

References

Arber, A. (1954). "The Mind and the Eye." Cambridge Univ. Press, London and New York.

Bacon, G. A., Burrows, T. W., and Yates, M. (1951). The effects of biochemical mutation on the virulence of *Bacterium typhosum*: The loss of virulence of certain mutants. *Br. J. Exp. Pathol.* 32, 85–96.

Bateman, D. F., and Basham, H. G. (1976). Degradation of plant cell walls and membranes of microbial enzymes. *In* "Physiological Plant Pathology," Encyclopedia of Plant Physiology (n.s.) (R. Heitefuss and P. H. Williams, eds.), Vol. 4, pp. 316–355. Springer-Verlag, Berlin.

Boone, D. M. (1971). Genetics of *Venturia inaequalis*. *Ann. Rev. Phytopathol.* 9, 297–318.

Brown, W. (1955). On the physiology of parasitism in plants. *Ann. Appl. Biol.* 43, 325–341.

Browning, J. A., Simons, M. D., and Torres, E. (1977). Managing host genes: Epidemiologic and genetic concepts. *In* "Plant Disease: An Advanced Treatise" (J. G. Horsfall and E. B. Cowling, eds.), Vol. 1, pp. 191–212. Academic Press, New York.

Burkholder, P. R. (1952). Cooperation and conflict among primitive organisms. *Am. Sci.* 40, 601–631.

Cohen, S. S. (1977). A strategy for the chemotherapy of infectious disease. *Science* 197, 431–432.

Garber, E. D. (1956). A nutrition-inhibition hypothesis of pathogenicity. *Am. Nat.* 90, 183–194.

Hartig, R. (1900). "Lehrbuch der Pflanzenkrankheiten." Springer-Verlag, Berlin and New York.

Heitefuss, R., and Williams, P. H., eds. (1976). "Physiological Plant Pathology," Encyclopedia of Plant Physiology (n.s.), Vol. 4. Springer-Verlag, Berlin and New York.

Horsfall, J. G., and Dimond, A. E. (1959). The diseased plant. *In* "Plant Pathology: An Advanced Treatise" (J. G. Horsfall and A. E. Dimond, eds.), Vol. 1, pp. 1–17. Academic Press, New York.

Kent, G. C. (1973). Lecture notes, Cornell University, Ithaca, New York (unpublished).

Kosuge, T., and Gilchrist, D. G. (1976). Metabolic regulation in host–parasite interactions. *In* "Physiological Plant Pathology," Encyclopedia of Plant Physiology (n.s.) (R. Heitefuss and P. H. Williams, eds.), Vol. 4, pp. 679–702. Springer-Verlag, Berlin.

Kühn, J. (1858). "Die Krankheiten der Kulturgewächse, ihre Ursachen und ihre Verhütung." G. Bosselmann, Berlin.

Lewis, R. W. (1953). An outline of the balance hypothesis of parasitism. *Am. Nat.* 87, 273–281.

Link, G. K. K. (1932). The role of genetics in etiological pathology. *Q. Rev. Biol.* 7, 127–171.

Link, G. K. K. (1933). Etiological phytopathology. *Phytopathology* 23, 843–862.

Morstatt, H. (1923). "Einführung in die Pflanzenpathologie." Borntraeger, Berlin.

Münch, E. (1929). Ueber einige Grundbegriffe der Phytopathologie. *Z. Pflanzenkr. Gallenkd.* 39, 276–286.

Odum, E. P. (1959). Principles and concepts pertaining to organization at the inter-species population level. *In* "Fundamentals of Ecology," Chapter 7, pp. 225–244. Saunders, Philadelphia, Pennsylvania.

Smith, T. (1934). "Parasitism and Disease." Princeton Univ. Press, Princeton, New Jersey.

Snell, E. E. (1949). Nutrition of microorganisms. *Annu. Rev. Microbiol.* 3, 97–120.

Tansley, A. G. (1929). "The New Psychology and Its Relation to Life." Dodd, Mead, New York.

Thimann, K. V. (1956). Promotion and inhibition: Twin themes of physiology. *Am. Nat.* 90, 145–162.

Vines, A. E., and Rees, N. (1964). "Plant and Animal Biology," Vol. 1, Chapter 30, pp. 1103–1115. Pitman, London.

Ward, H. M. (1896). "Diseases of Plants." Society for Promoting Christian Knowledge, London.

Whetzel, H. H. (1935). The nature of disease in plants. *In* "Elementary Plant Pathology," Lecture Notes. Cornell University, Ithaca, New York.

Wigglesworth, V. B. (1955). The contribution of pure science to applied biology. *Ann. Appl. Biol.* 42, 34–44.

Wood, R. K. S., Ballio, A., and Graniti, A. (1972). "Phytotoxins in Plant Diseases." Academic Press, New York.

Yoder, O. C., and Scheffer, R. P. (1969). Role of toxin in early interactions of *Helminthosporium victoriae* with susceptible and resistant oat tissue. *Phytopathology* 59, 1954–1959.

Chapter 4

The Capture and Use
of Energy by Diseased Plants

TSUNE KOSUGE

I. INTRODUCTION

The capture of energy from the sun by photosynthesis is one of the most important processes carried out by living organisms. In green plants, solar energy is converted to chemical energy by photophosphorylation. The discovery of this process must be recognized as one of the most important of our time. All other energy yielding reactions of the plant, such as oxidative phosphorylation, utilize substrates initially produced by photoassimilation and thereby deduct from the overall yield of the process. Thus the overall development of the plant is a function of the efficiency of capture of the sun's energy.

Several methods are used to assess the efficiency of energy capture in photosynthesis. The most useful are direct measurements of the accumulation of dry matter by plants. Farmers measure the efficiency of energy capture as yield of grain, or potatoes, or alfalfa hay per hectare. Plant physiologists measure efficiency as net photosynthesis—the difference between gross photosynthesis and photorespiration. Plant pathologists recognize photosynthesis as a major function with which pathogens can interfere. It is our intention to understand how plant pathogens reduce the yield and quality of crop plants and how these effects can be minimized.

This chapter is designed to show how disease affects the capture, transfer, and utilization of energy by plants. It concerns the various photochemical and biochemical processes associated with energy capture and transfer, their contribution to yield, and their disruption by disease.

II. AN OVERVIEW OF ENERGY CAPTURE AND UTILIZATION IN PLANTS

A. A Diagrammatic View of Energy Capture, Production, and Utilization

Figure 1 shows the various components of photosynthesis and photorespiration in plants. The apparatus for photophosphorylation resides in the chloroplast lamellae and provides energy and reducing power for fixation and assimilation of CO_2. The light reaction systems which carry out the transport of electrons from water to ferredoxin are called Photosystems I and II. Photosystem I leads to oxidation of an intermediate chain of electron carriers and concomitant reduction of ferredoxin. Photosystem II leads to oxidation of water and the accompanying reduction of an intermediate chain of electron carriers (Fig. 2). Through these two photosystems, light energy is transformed to chemical energy which then

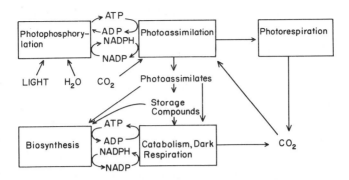

Fig. 1. Diagrammatic presentation of the biochemical machinery of a plant. Photophosphorylation by Photosystems I and II (see Fig. 2) captures light and produces energy (ATP) and reducing power (NADPH) which are used for the photoassimilation of CO_2. Certain photoassimilates, intermediates, and products of photoassimilation via the reductive pentose phosphate cycle have several fates. In C_3 plants, photorespiration accounts for large losses in assimilated carbon; part of the CO_2 evolved by photorespiration can be reassimilated. Catabolism of various compounds via "dark respiration" will provide energy when the photochemical apparatus is absent (as in nonchlorophyll-containing tissue) or nonfunctional (as in green leaves in the dark). See text for further discussion. (I am indebted to Dr. D. E. Atkinson for the idea of presenting the plant biochemical components in this fashion.)

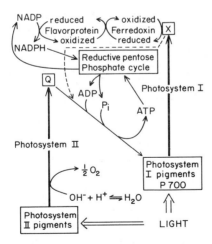

Fig. 2. Dependence of the reductive pentose phosphate cycle on energy, and reducing power generated by Photosystems I and II. Energy (ATP) and reducing power (NADPH) produced by the noncyclic and cyclic photophosphorylation components of the photosystems drive the reductive pentose phosphate cycle. (– – –) Cyclic flow of electrons. See text for further discussion.

is made available in plants in the form of ATP and NADPH. This is accomplished via cyclic photophosphorylation:

$$\text{light} + \text{ADP} + \text{Pi} \rightarrow \text{ATP} + \text{H}_2\text{O}$$

and via noncyclic photophosphorylation:

$$\text{light} + \text{NADP} + \text{ADP} + \text{Pi} + 2\text{H}_2\text{O} \rightarrow \text{NADPH} + \text{H}^+ + \frac{1}{2}\text{O}_2 + \text{ATP} + \text{H}_2\text{O}$$

The photophosphorylation systems provide ATP and NADPH for the photoassimilation system which resides in the stroma or embedding matrix of chloroplasts. Photosystem II and the noncyclic photophosphorylation system may occur only in the grana, whereas photosystem I and the cyclic photophosphorylation system may reside in both stroma lamella and grana.

Although the plant has an ingenious mechanism for capturing light and converting it to energy, it is nevertheless inefficient in this process. Only about 10–12% of available light energy is converted to chemical energy in photosynthesis (Zelitch, 1971). Photorespiration is estimated to reduce net photosynthetic efficiency by as much as 50%. Thus, photorespiration is a major barrier to yield (Evans, 1975; Zelitch, 1971). Its relationship to infection will be discussed later.

Fixation of CO_2 by the reductive pentose phosphate cycle initiates a process which on a global scale results in the production of 140×10^7 metric tons of edible dry matter per year. The flow of CO_2 from the atmosphere into the chloroplasts of leaves is an uncomplicated process of diffusion that is regulated by the several factors discussed below (Zelitch, 1971). Since the cuticle of leaves is a substantial barrier to diffusion of CO_2, control of CO_2 uptake is achieved by the opening and closing of stomata. Once past the stomatal opening, CO_2 must diffuse into the stomatal cavity, across the cell membrane, into the cytoplasm, across the chloroplast membrane, into the chloroplast, and finally into the stroma where it is captured in a reaction catalyzed by ribulose diphosphate carboxylase. In a later discussion, it will be seen that any one of the above barriers to diffusion of CO_2 may increase or decrease in infected plants.

In plants with the C_3 pathway of photosynthesis, assimilation of CO_2 occurs in a reaction catalyzed by ribulose diphosphate carboxylase in the reductive pentose phosphate cycle (Fig. 3). In plants with the C_4 pathway of photosynthesis, atmospheric CO_2 is first fixed into C_4 acids via phosphoenolpyruvate (PEP) carboxylase and then "refixed" by the reductive pentose phosphate cycle. This cycle is driven by NADPH and ATP produced during photophosphorylation. Light directly or indirectly activates several enzymes of this cycle and assures that it will function only in the light. Once formed, the products of the reductive pentose

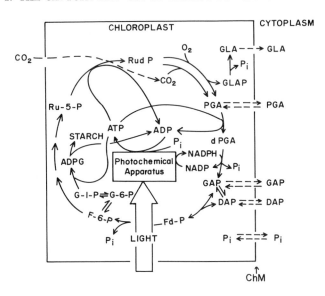

Fig. 3. Schematic diagram of the reductive pentose phosphate (Calvin) cycle in C_3 plants. Diffusive and transport processes (‒ ‒ ‒); enzyme-catalyzed reactions (———). Light-generated ATP and NADPH drive the cycle by involvement in the reactions indicated. Phosphoglycerate (PGA), glyceraldehyde 3-phosphate (GAP), dihydroxyacetone phosphate (DAP), and inorganic phosphate (P_i) are transported across the chloroplast membrane (ChM); the other intermediates have restricted movement. Diffusion of CO_2 from the atmosphere to the site of assimilation is regulated by diffusive resistances described in Fig. 5. Oxygenase and carboxylase activities of ribulose diphosphate carboxylase initiate photorespiration and photoassimilation, respectively. Phosphoglycolate (GLAP), a product of oxygenase activity, is hydrolyzed to glycolate (GLA) which moves to the cytoplasm and participates in photorespiration reactions as described in the text. RudP, ribulose diphosphate; ADPG, adenosine diphosphate glucose; dPGA, diphosphoglyceric acid; Fd-P, fructose diphosphate; G-1-P, glucose 1-phosphate; G-6-P, glucose 6-phosphate; F-6-P, fructose 6-phosphate; Ru-5-P, ribulose 5-phosphate.

phosphate cycle may remain in the cycle for regeneration of the CO_2 acceptor. Alternatively, they may be transported from the chloroplast or diverted into other metabolic pathways in the chloroplast such as the reactions for starch synthesis.

For maximum plant growth and development, photoassimilates must be translocated efficiently from the chloroplast to other parts of the cell and to other parts of the plant where they are used for synthesis of essential cell constituents or reutilized to regenerate energy via the so-called dark respiration processes. Movement of photoassimilates from the chloroplast is regulated by the chloroplast membrane; only a few key intermediates including the triose phosphates are transported into the cytoplasm (Heber, 1974). Here they are converted to other metabolites for synthesis of essential cell components or to compounds such as sucrose for transport to other parts of the leaf and plant (Kelly *et al.*, 1976).

Transport of photoassimilates throughout the plant is energy dependent and occurs largely through the phloem (see Chapter 5, this volume). Net movement of photoassimilates into nonphotosynthetic tissues is particularly important for the development of the plant. The transport assimilates are used in sink or recipient tissues for synthesis of needed cell constituents, deposited as storage metabolites, or used as sources of energy. Photoassimilates also are used for energy sources in green tissue when the output of the photochemical system is inadequate as is the case in darkness or in diseased tissue.

Each plant cell is a carefully designed unit with systems for synthesis and catabolism (Kosuge and Gilchrist, 1976). Even in its normal healthy state, cells contain enzymes and metabolites that would be deleterious to the cell as a whole were it not for compartmentation of the different metabolic systems and various metabolic controls. The diagrammatic presentation of the biochemical components of a plant in Fig. 1 makes it possible to visualize how plant disease affects each metabolic component and how interference with the functioning of any one component can affect another component and ultimately the overall development of the plant. Disease affects the control mechanisms in plants in various ways. The collective disruptive effects of disease on these controls help determine the severity of disease.

Although presentation of metabolic schemes in simplifying diagrams such as Fig. 1 has certain advantages, it also has limitations. For example, it is likely that all the components in the boxes shown in Fig. 1 may be affected simultaneously to varying degrees in a given host–pathogen system. By focusing attention on individual components, the obvious, that is, the effect on the whole plant, may be overlooked. Readers should be mindful that diagrams such as those presented in Fig. 1 tend to provide a static view of what truly is a dynamic, ever changing situation and that each time an experimental sampling is done the investigator catches a view of a living system only in one fleeting moment of its dynamic existence.

B. Components of Yield

1. Source–Sink Relationships

In considering the effects of disease on the capture and utilization of energy in plants, it will be useful to think in terms of the physiological components of yield: source, translocation, and sink. These terms will be used in their broadest sense and identified according to their major biochemical roles.

a. Source. Source includes those mechanisms concerned with the capture of light energy and the assimilation of CO_2 into energy-yielding and cell-building compounds that are available for transport to sinks. Processes such as photorespiration reduce the efficiency of the source in capturing and transferring energy (Zelitch, 1971).

b. Translocation. The capacity of the plant to transport assimilates from source to sink is an essential component of yield. The processes include movement of certain photoassimilates from the chloroplast into the cytoplasm, synthesis of the photoassimilates into transport forms such as sucrose, loading of the transport forms into the phloem, and translocation in the phloem to the sinks.

c. Sinks. In the broadest sense, metabolic sinks refer to those plant parts or tissues that are the recipients of translocated metabolites, generally photoassimilates. Some crop physiologists narrow the definition to include those organs concerned with storage of assimilates, specifically those that concern yield. In this discussion, sinks will be used in the broader context and will include those tissues or parts that cannot produce sufficient energy and/or metabolites for their own needs and therefore depend upon surrounding tissues for those supplies. Sinks are centers of high metabolic activity with high rates of respiration and/or biosynthesis.

In developing plants, photoassimilates are partitioned among several sinks such as storage, root, and meristemic tissues as well as young leaves (Fig. 4). Under stress conditions abnormal partitioning of photoassimilates may occur. In some diseased plants, particularly those infected with a virus or a biotrophic fungus, one or a few sinks usually become dominant. Thus, the dominant sinks prosper at the expense of other sinks and normal development of the plant cannot occur.

2. Crop Respiration

The division of crop respiration into maintenance and growth respiration is useful and will be employed here. "Maintenance respiration" refers to the respiratory activity required to maintain and replace structures in the plant undergoing turnover (McCree, 1970). The expenditure of energy for maintenance respiration will vary with age of tissue. Relative to the total weight of the plant, it will decrease substantially as the plant approaches maturity (Evans, 1975). This component of crop respiration is susceptible to considerable change in infected plants.

"Development respiration" refers to energy diverted directly to the development of the plant. It can be expressed as the metabolic energy cost of converting translocated products of photosynthesis into structural, cytoplasmic, and storage compounds. This component is directly con-

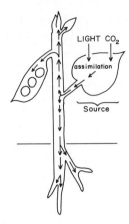

Fig. 4. Schematic diagram of photoassimilate source–sink relationships. Photoassimilates are produced in actively photosynthesizing leaves (source) and are translocated and partitioned into sinks such as young leaves, apical meristem, developing seeds, and roots. Certain pathogens cause source leaves to become sinks, reduce translocation of photoassimilates to other sinks, and produce uneven partitioning of photoassimilates. For other details see text.

nected with crop yield and is particularly vulnerable to disruption in diseased plants.

III. EFFECTS OF DISEASE ON THE EFFICIENCY OF ENERGY CAPTURE BY PLANTS

A. Reduction of Plant Density

Full coverage of land by foliage is essential for maximum capture of solar energy. On a global basis maximum capture of energy per unit of land area is required to meet the food and fiber needs of the world. Limits on production are imposed by such environmental factors as heat, cold, and deficiency of water or nutrients. Production also is limited by disease which injures or kills plants and thus reduces the number of plants per unit area of land. In theory it is possible to calculate losses due to reduction in stand density. In practice, however, most diseases that reduce plant density also reduce plant size, leaf size, and canopy density as noted below.

B. Alteration of the Plant Canopy

The ability of a given plant to capture light is affected by the leaf area and the orientation of its leaves with respect to the sun. Leaf orientation for maximum interception of light has been studied in crops such as maize and barley. For barley, plants with erect leaves intercept more light than plants with horizontally oriented leaves. Leaf area index is one measure of the potential efficiency of utilization of solar energy by plants in a stand. The main leaf angle from the horizontal also regulates the degree of penetration of incident light. The penetration of incident light into a crop canopy has been defined as follows:

$$\log_e (I/I_0) = -\alpha_L (\text{LAI})$$

where I and I_0 are the same terms used in Beer's law, α_L is the crop extinction coefficient, and LAI is the leaf area index (Saeki, 1963). It can be calculated that in a crop canopy with a leaf area index of 1 and an α_L value of 1, 37% of the incident radiation would pass through to the leaves below the top of the canopy. Calculated α_L values for stands of dicotyledonous field crops are between 0.7 and 1.0 and for grasses 0.3 to 0.5 (Zelitch, 1971).

Disease symptoms, such as leaf epinasty, reduction in leaf size, and change in leaf orientation can alter the leaf canopy and thus reduce the plant's capacity to intercept light. Leaf area of some plants infected with certain viruses is reduced by the formation of spindly and otherwise malformed leaves. Obviously, the most devastating effect of disease on leaf canopy would be defoliation at a stage of development when plant growth is directly correlated with photosynthesis. In cotton plants infected with *Verticillium dahliae*, among other diseases, defoliation and the accompanying loss in photosynthetic capacity during early stages of plant development have significant effects on yield. Yield is not affected, however, if the components of boll yield have been deposited prior to the time of defoliation (J. E. DeVay, unpublished observations). In contrast, defoliation at any stage of development would affect yield and quality of alfalfa since the entire upper part of the plant is used for animal feed.

Effects of disease on the ability of the whole plant to capture solar energy may escape notice if a single alfalfa plant dies from diseases in an otherwise dense stand. Along a tree-lined boulevard or in an orchard, however, death of a single tree may attract considerable attention.

C. Disruption of Photochemical Systems within the Plant

The most drastic effects of disease on the photochemical machinery of plants may be expected in cases such as the hypersensitive destruction of leaf tissue and the accompanying disruption of chloroplast structure (Carroll and Kosuge, 1969). The process appears to be initiated by a loss of membrane integrity leading to the loss of compartmentation and disruption of the photochemical apparatus. This would cause a rapid decline in capacity to trap light energy, in output of ATP, and in reducing power. The diseased plant would become increasingly dependent upon utilization of reserve foods and dark respiration for its sources of energy and reducing power. These changes from dependence on photosynthesis and current photosynthate to metabolism of reserve foods obviously will be detrimental to yield.

Virus multiplication requires host energy. Thus it is logical to assume that the process could be linked to the production of energy via photophosphorylation. This appears to be true in the case of turnip yellows mosaic virus (TYMV)—photophosphorylation increases coincidentally with the period of rapid synthesis of the virus and then declines (Goffeau and Bove, 1965). The virus multiplies in chloroplasts where a convenient supply of ATP is available for synthetic purposes. In other virus infections, photosynthesis is needed for virus synthesis and photosynthesis is impaired only after a period of rapid virus synthesis (Magyarosy et al., 1973, Zaitlin and Jagendorf, 1960).

Photosynthetic systems escape significant alteration in plants during the early stages of infection by biotrophic fungi such as rusts or powdery mildews (Montalbini and Buchanan, 1974). In rust-infected plants, noncyclic photophosphorylation was inhibited 20–30% but cyclic photophosphorylation was reduced no more than 10%. Similarly noncyclic photophosphorylation was inhibited 50% but cyclic photophosphorylation was unaffected in sugar beets infected with powdery mildew (Magyarosy et al., 1976). Thus, some photosynthesis seems necessary for development of biotrophic fungi. Photoassimilation may help maintain host tissue in a condition conducive for development of the parasite.

In diseases in which the pathogen can exist saprophytically, such as bacterial and necrotrophic fungal infections, the pathogen is not dependent upon the host energy-generating system; it possesses its own systems for generating energy by utilizing plant components. In such cases it is common to observe significant reductions in the photochemical activities of plants during early stages of infection.

IV. THE EFFECT ON CARBON ASSIMILATION

A. Alteration in Diffusion of CO_2

Barriers to movement of gases into leaf tissue may be depicted as diffusive resistances (Fig. 5). Since water vapor and CO_2 have a common path of movement in plants, stomatal resistance to water movement will likewise provide a barrier to CO_2 uptake and ultimately photosynthesis

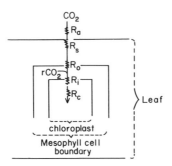

Fig. 5. Diagrammatic view of factors controlling CO_2 uptake by the plant leaf. Movement of CO_2 from the atmosphere to site of fixation in chloroplast is regulated by resistances: R_a, air boundary layer; R_s, the stomatal aperture; R_0, the mesophyll cell boundary; R_i, internal physical barrier separating incoming CO_2 from the carboxylation system; R_c, limitations of the carboxylation system. Respiratory CO_2 (rCO_2) generated by dark and photorespiration joins the main flow of atmospheric CO_2 mostly outside the chloroplast and provides an opposing flow of varying magnitude. (Adapted from Waggoner, 1969; and Zelitch, 1971.)

(Duniway, 1976). Equations for calculating the resistances shown in Fig. 5 have been published (Waggoner, 1969; Zelitch, 1971) and permit quantitation of the effect of disease on the various components of photosynthesis. Thus, various phenomena associated with infection could affect certain resistances as follows: stomatal closure, increased R_s; increased photorespiration or dark respiration, increased R_i; decreased ribulose diphosphate carboxylase activity, increased R_c. R_0 would be affected by changes in permeability of the mesophyll cell plasmalemma or the mesophyll cell wall. Of the factors controlling movement of gases into leaves of diseased plants, stomatal closure has been studied most. Water stress during infection may cause stomatal closure and indirectly cause reduced photosynthesis through reduced CO_2 uptake (Duniway and Slatyer, 1971). In sugar beets infected with the beet yellows virus, reduced stomatal aperture was qualitatively the most important factor limiting CO_2 uptake and one of the principal factors limiting photosynthesis in infected plants (Hall and Loomis, 1972). Infected leaves also had 16% fewer stomates than healthy leaves.

In tomatoes infected with *Fusarium*, resistance to CO_2 diffusion increased because stomates did not open normally (Duniway, 1976). Some toxins produced by plant pathogens, such as fusicoccin, decrease diffusive resistance by causing stomata to open in both light and darkness (Turner and Graniti, 1976). Abnormal stomatal opening also has been observed in plants infected with other diseases but the exact cause of the phenomenon has not been determined (Ayres and Jones, 1975).

Not all diseases affect CO_2 diffusion by interfering with normal stomatal opening and closure. Rupture of cuticle and epidermis of plant leaves occurs during the formation of uredosori on barley and movement of CO_2 into infected leaf tissue continues without interruption (Duniway and Durbin, 1971).

As yet no generalization can be made on the effects of disease on CO_2 diffusion—disease may impair or facilitate movement of CO_2 into the plant.

B. Disease-Induced Changes in the Reductive Pentose Phosphate Cycle

The reductive pentose phosphate cycle is carefully regulated both by light and by intermediates in the cycle itself (Bassham, 1971; Kelly and Latzko, 1977). Several enzymes of this cycle are light activated, including glyceraldehyde-3-phosphate dehydrogenase, ribulose-5-phosphate kinase, fructose-1,6-diphosphate phosphatase, and ribulose-1,5-diphosphate carboxylase. In addition, light-generated ATP and NADPH "drive" the cycle by providing substrates and reducing power at key points (see Fig. 3) (Bassham, 1971). The function of the reductive pentose phosphate cycle is photoassimilation of CO_2 and production of substrates for metabolic pathways that reside both within and outside the chloroplast. Since this cycle is the beginning point of all carbon metabolism in the plant, its disruption by disease obviously would deprive the plant of essential metabolites.

Much research has been done on the enzyme ribulose diphosphate carboxylase, which is the Fraction I protein first isolated by plant virologists (Kawashima and Wildman, 1970). In cells actively synthesizing virus, the amount of this enzyme decreases because enzyme synthesis ceases and enzyme turnover continues unchecked. Synthesis of ribulose diphosphate carboxylase is inhibited because virus infection results in general disruption of protein synthesis in chloroplasts (Hirai and Wildman, 1969). Depending on the severity of infection, ribulose diphosphate carboxylase activity may be reduced 20% or more. Other enzymes of the cycle surely would be affected in a similar manner if protein synthesis in the chloroplast is curtailed. Apart from direct effects on enzymes, reduced input of ATP and NADPH from the photochemical machinery of the plant likewise will interfere with the activity of the cycle. Various estimates of the extent of reduction in output of photophosphorylation have been given in a succeeding section and need not be repeated here. Suffice it to say that there will be a direct correlation between the amount of output from the photochemical machinery and the photo-assimilatory yields of the chloroplast.

C. Disease-Induced Changes in C_4 Metabolism

Plants with the C_4 pathway of photosynthesis fix atmospheric CO_2 in mesophyll cells by the phosphenol pyruvate (PEP) carboxylase reaction: $CO_2 + PEP \rightarrow$ oxalacetate $+ P_i$. However, net photosynthetic CO_2 assimilation is accomplished by the ribulose diphosphate carboxylase reaction (Fig. 6). The PEP carboxylase reaction initiates the C_3 acid $\rightarrow C_4$ acid $\rightarrow C_3$ acid transformation in the C_4 cycle that shuttles CO_2 from mesophyll cells to bundle sheath cells where CO_2 is assimilated by ribulose diphosphate carboxylase in the reductive pentose phosphate cycle. Movement of the C_3 acid back to mesophyll cells and regeneration of PEP complete the C_4 cycle (Fig. 6).

Although in all C_4 plants PEP carboxylase catalyzes the $C_3 \rightarrow C_4$ conversion and the reaction product oxalacetate is readily converted to malate and aspartate, variations exist among C_4 plants as to which of the two compounds participate in the CO_2 shuttle system. Plants that use malate as the CO_2 carrier convert oxalacetate to malate by the malate dehydrogenase reaction: oxalacetate $+ NADH \rightarrow$ malate $+ NAD$. Malate then is transported to bundle sheath cells where it is converted to pyruvate by a NADP-specific malic enzyme: malate $+ NADP \rightarrow$ pyruvate $+ CO_2 + NADPH$.

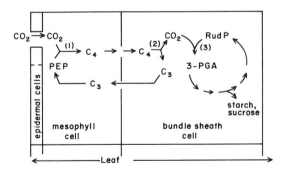

Fig. 6. Diagrammatic view of the C_4 pathway of photosynthesis. CO_2 is assimilated by PEP carboxylase in reaction (1), forming oxalacetate which is transformed to malate or aspartate, depending upon the plant species. The C_4 compound is transported from the mesophyll to the bundle sheath cell where it undergoes metabolism to a C_3 compound and CO_2 (reaction 2). CO_2 is assimilated via the ribulose diphosphate carboxylase reaction (3). The C_3 compound is returned to the mesophyll cell and converted to PEP to complete the cycle. Details of the individual reactions appear in the text. (After Hatch, 1976.)

Pyruvate then moves to the mesophyll cells and is converted to PEP by the pyruvate P_i dikinase reaction: pyruvate $+ ATP + P_i \rightarrow PEP + AMP + PP_i$.

In C_4 plants that use aspartate as the CO_2 carrier, aspartate is formed from oxalacetate by transamination in mesophyll cells, then moves to bundle sheath cells where it undergoes $C_4 \rightarrow C_3$ conversion by sequential reactions catalyzed by aminotransferase (1) and PEP carboxykinase (2):

$$\text{aspartate} + \alpha\text{-ketoglutarate} \rightarrow \text{oxalacetate} + \text{glutamate} \qquad (1)$$

$$\text{oxalacetate} + ATP \rightarrow PEP + CO_2 + ADP \qquad (2)$$

PEP is transported to the mesophyll cells to complete the cycle. Still other variations of the $C_4 \rightarrow C_3$ transformation occur but in all cases the principal intermediates are one or more of the C_4 and C_3 compounds described above (Hatch, 1976).

As in C_3 plants, light-generated ATP and NADPH drive photosynthetic CO_2 assimilation in C_4 plants. The energy requirement for carbon assimilation in C_4 plants is 5 ATP and 2 NADPH per mole of CO_2 fixed (2 ATP for the C_4 cycle, 3 ATP and 2 NADPH for the reductive pentose phosphate cycle). Although C_4 plants require more energy for photoassimilation, they have very low rates of photorespiration, do not show O_2 inhibition of photosynthesis, and consequently are photosynthetically more efficient than C_3 plants (Edwards et al., 1976).

Studies on the effects of disease on C_4 plants have focused on net photosynthesis and ribulose diphosphate carboxylase and have largely neglected components of the C_4 cycle. However, one observation may be pertinent here. Virus infection of some C_3 plants causes changes in early products of CO_2 fixation that resemble shifts from the C_3 to the C_4 type of metabolism. Thus, in leaf tissues of the C_3 plant, *Brassica pekinensis*, infection by TYMV causes a shift in the pattern of early products of CO_2 fixation from decreased incorporation into 3-phosphoglycerate to increased incorporation into malate and aspartate. Although reminiscent of a C_4 type of metabolism, that pattern of CO_2 assimilation is the consequence of reduced ribulose diphosphate carboxylase and increased PEP carboxylase activities (Bedbrook and Matthews, 1972).

Studies are needed on the effects of disease on energy capture and utilization in succulent plants which contain a unique pathway of CO_2 assimilation known as crassulacean acid metabolism (CAM). In darkness, these plants open their stomatal pores, fix atmospheric CO_2 via PEP carboxylase, and store CO_2 in C_4 acids such as malic acid. In daylight stomatal pores close, limited CO_2 and water vapor exchanges occur between leaves and the atmosphere, the dark-accumulated malic acid is decarboxylated, and the released CO_2 is photoassimilated by the reductive pentose phosphate cycle.

D. Alteration in the Compartmentation of the Photoassimilatory System

The photophosphorylation and photoassimilation systems of the plant reside in the lamellae of the chloroplast. They are organized in a manner conducive to the capture of photons and transfer of electrons from one photosystem to the next. In addition, the movement of key metabolites of the reductive pentose phosphate pathway through the chloroplast membrane is carefully and selectively regulated. Certain metabolites are confined within the chloroplast to assure maintenance of optimal concentrations of metabolites which are essential for optimal rates of photoassimilation. Such selectivity for metabolites provides a regulatory influence on cellular metabolism in photosynthetic tissue owing to the concentration of NADP–NADPH and intermediates of the reductive pentose phosphate cycle in the chloroplast.

Loss of chloroplast integrity has been described in many plant diseases, particularly those in which hypersensitive reactions occur in leaves. Chloroplast membranes are ruptured and general loss of structural integrity follows. This causes loss of both the energy and reducing power necessary to drive the reductive pentose phosphate cycle. Consequently, activity of ribulose diphosphate carboxylate would diminish. In tobacco leaves systematically infected with tobacco mosaic virus, synthesis of ribulose diphosphate carboxylase decreased in the chloroplast but active synthesis of viral protein and RNA occurred in the cytoplasm (Hirai and Wildman, 1969). Also, "fraction I protein" (ribulose diphosphate carboxylase) was released in the cytoplasm when chloroplast integrity was lost in rice leaves infected with the blast fungus (Akazawa and Ramakrishnan, 1967). In both cases, activity of the enzyme was lost and CO_2 assimilation in the infected plant was reduced.

Loss of integrity of the chloroplast membrane also would allow free movement of cofactors such as NADP and metabolites from the chloroplast into the cytoplasm and would negate any regulatory influence imposed on metabolism by compartmentation of these compounds. However, there are diseases in which more discrete changes in chloroplast compartmentation occur. The chlorophyll-deficient areas of tobacco leaves infected with tobacco mosaic virus contain smaller than normal plastids that are low in, or devoid of, chlorophyll. In these tissues, protein synthesis is inhibited in the chloroplast but continues rapidly in the cytoplasm (Hirai and Wildman, 1969). Thus, the normal division of activity between chloroplast and cytoplasm is lost. There are other virus diseases in which ribosomes and Fraction I protein in the chloroplast are un-affected until advanced stages of the disease (Mohamed and Randles, 1972).

E. Change in the Assimilation of Storage Carbohydrates

Starch is one of the major products of photoassimilation and a major form of stored energy in the plant. Adenosine diphosphate glucose syn-thase is the enzyme which regulates starch synthesis in green leaves. It is activated by 3-phosphoglycerate and other intermediates of the reduc-tive pentose phosphate cycle and inhibited by orthophosphate (Ghosh and Preiss, 1966). Thus, by controlling production of ADPG, the enzyme controls starch synthesis.

When CO_2 concentrations are not limiting, starch accumulates in chlo-roplasts during daylight periods when light intensity and ATP and NADPH production are high. Activity of the reductive pentose phosphate cycle would be high and concentrations of effectors of ADPG synthase would be optimal for full activation of that enzyme (Ghosh and Preiss, 1966). Starch accumulation is most active in plant leaves when demands for essential cell reactions are met and the cell is in a condition of energy excess. Starch is broken down during dark periods when energy capture and photoassimilation are low. It is converted to products that can move from the chloroplast into the cytoplasm and to other parts of the plant (Peavey et al., 1977).

Turnover of starch and its reutilization has been studied in a number of fungus–plant interactions. Starch breakdown occurs in the presence of pathogen enzymes that catalyze its hydrolysis to glucose which then can be utilized as a source of carbon. An unusual case occurs in the bean rust pathogen *Uromyces phaseoli* which in contact with host tissue ap-parently excretes a substance that activates the host amylase (Schipper and Mirocha, 1969). The latter then catalyzes the breakdown of host starch to fragments such as glucose and maltose that may be utilized as sources of carbon by the pathogen. In some host–parasite interactions, such as stripe rust of barley, an apparent transient accumulation of starch may occur (MacDonald and Strobel, 1970). Here the host's enzymes controlling starch synthesis are activated and starch accumulates. It has not been conclusively established if decreased turnover also helps ac-count for the increased accumulation of starch.

In plants affected with certain virus diseases, starch accumulated in chloroplasts during light periods does not break down during dark periods. Large starch grains become prominent intrachloroplast inclusions in plants infected with viruses such as TMV (Carroll and Kosuge, 1969). The prevention of starch turnover may be a consequence of factors that prevent transport of photoassimilates in virus-infected plants as will be described later.

V. ALTERATION IN THE EXPORT OF REDUCING POWER AND ENERGY

As stated earlier, normally the inner envelop of the chloroplast is essentially impermeable to NADP and NADPH. This prevents direct transport of reducing power from chloroplast to cytoplasm. However, indirect transport can occur by a shuttle system involving glyceraldehyde 3-phosphate, which moves from the chloroplast to the cytoplasm where it is oxidized to NADPH and 3-phosphoglycerate by a NADP-linked, irreversible glyceraldehyde 3-phosphate dehydrogenase. The 3-phosphoglycerate then returns to the chloroplast where it can again be converted back to glyceraldehyde 3-phosphate by the reductive pentose phosphate cycle. This and other NADPH shuttle systems involving dicarboxylic acids and orthophosphate (Heber, 1974) provide rapid export of reducing power from chloroplast to cytoplasm. Similarly, ATP does not readily move through the chloroplast membrane, and shuttle systems provide for rapid movement of photochemically generated energy in the form of ATP from chloroplast to cytoplasm. All photoassimilates are oxidizable substrates and thus are potential sources of stored energy and reducing power. For example, triose phosphates can be exported from chloroplasts into the cytoplasm where they can be converted to hexose phosphates. These in turn can be oxidized by appropriate reactions to yield ATP, NADH, NADPH, or converted to transport forms for movement to other parts of the plant.

Transport of assimilates occurs by at least two separate steps, namely phloem loading and translocation. Energy provides the driving force for solute transport and is associated with phloem loading. The latter involves movement of sugar from the mesophyll and its accumulation in the phloem (Giaquinta, 1977). Translocation in the phloem occurs by "pressure" flow arising from osmotically generated turgor gradients.

Inhibited transport of photoassimilates is common in virus-infected plants. For example, soluble carbohydrates increased 400% in wheat leaves infected with barley yellow dwarf virus even though the dry weight of the infected plant was only half that of noninfected plants (Jensen, 1972). This suggests that virus infection inhibited loading of phloem perhaps because available energy was not sufficient to drive the process or phloem transport was impaired. Disruption of phloem transport is a characteristic symptom of diseases in which the viral pathogen is confined to the phloem (Jensen, 1972; Panopoulus et al., 1972). Such infections cause aberrations in phloem structure and probably cause interference with normal translocation (Esau, 1941). Inhibited translocation may in turn prevent the normal breakdown of starch in virus-infected leaves and be responsible for the prominent starch granules that accumulate in chloroplasts (Carroll and Kosuge, 1969; Diener, 1963).

The transport of photoassimilates is reversed in wheat and bean leaves bearing rust pustules. Rust infection centers become metabolic sinks and draw photoassimilates from surrounding tissue and nearby healthy leaves (Livne and Daly, 1966). The sinks are created by concentration gradients of photoassimilates between host tissues depleted of solutes by the parasites and the surrounding healthy host tissue. It is possible that photoassimilates are retained by sink leaves by the same mechanism, whether the tissue is infected by a virus or by a fungus. Mobilization of energy sources to sites of infection at the expense of surrounding host tissue is an advantage to an invading pathogen which may need an abundant supply of energy if it is to multiply and sporulate. Cytokinins produced by the pathogen may be the stimulus for the establishment of a metabolic sink. These plant hormones are known to cause mobilization of nutrients when applied to healthy leaves (Thimann et al., 1974).

VI. EFFECT ON RESPIRATION

A. Photorespiration

Photorespiration appears to be a misdirection of photoassimilation since it reduces net photosynthesis by as much as 50% in some plants (Zelitch, 1975). It is curious that the major enzyme involved in photorespiration, ribulose diphosphate carboxylase, is the same enzyme concerned with photoassimilation (Ryan and Tolbert, 1975). The oxygenase activity catalyzed by the enzyme ribulose diphosphate carboxylase initiates the photorespiration process (Fig. 7).

The fate of phosphoglycolate is uncertain but it is proposed that 2 moles of phosphoglycolate ultimately give rise to one mole each of CO_2 and serine. The other product of ribulose diphosphate oxygenase activity, 3-phosphoglycerate, could be recycled through the reductive pentose phosphate cycle to give rise to ribulose diphosphate, a wasteful process that consumes both ATP and NADPH.

Since photoassimilation and photorespiration are mediated by the same

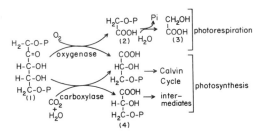

Fig. 7. Schematic presentation of the oxygenase/carboxylase activities of ribulose diphosphate carboxylase. Oxygenase activity initiates photorespiration by conversion of ribulose diphosphate (1) to phosphoglycolate (2) and phosphoglycerate (4). Phosphoglycolate is hydrolyzed to glycolate (3), which moves to the cytoplasm where it undergoes further metabolism in the photorespiration pathway. Carboxylase activity catalyzes the photoassimilation of CO_2, yielding two phosphoglycerates which remain in the reductive pentose phosphate cycle. Other details of the pathways are described in the text.

enzyme, decline in one activity in diseased tissue should parallel reduction in activity of the other process. This appears to be the case in oak leaves inoculated with the powdery mildew fungus. Gross photosynthesis remained constant but photorespiration decreased 1 day following inoculation, and caused an apparent, transient increase in net photosynthesis (Hewitt and Ayres, 1975). Thereafter both photosynthesis and photorespiration declined, as would be expected if both the oxygenase and the carboxylase activities of ribulose diphosphate carboxylase were lost.

Certain metabolites of the reductive pentose phosphate cycle such as ribose 5-phosphate, fructose 6-phosphate, or glucose 6-phosphate apparently can bring about simultaneous inhibition of oxygenase and activation of the carboxylase activities of ribulose diphosphate carboxylase. During the early stages of infection such effector compounds (Ryan and Tolbert, 1975; Walker, 1976) could undergo transient accumulation in the chloroplast and thus reduce photorespiration by inhibiting the oxygenase activity but not the carboxylase component of ribulose diphosphate carboxylase. These interactions of the two activities of ribulose diphosphate carboxylase could bring about a transient increase in net photosynthesis. Alternatively, it is possible that the oxygenase activity declines more rapidly than the carboxylase activity. Such is the case in tomato fruit, in which the ratio of ribulose diphosphate carboxylase/oxygenase changes with ripening (Bravado et al., 1977). These results suggest that in diseased tissue, which represents senescing tissue, similar changes in oxygenase–carboxylase activity could occur (see Chapter 18, this volume).

B. Loss of Energy through Respiration in Infected Green Tissues

Respiration in green tissues involves oxidation of carbohydrates and the resulting generation of NADH and ATP by the Embden-Meyerhof (EM) pathway, the tricarboxylic acid (TCA) cycle, and oxidative phosphorylation; and finally the oxidation of glucose via the oxidative pentose phosphate pathway with accompanying production of NADPH. A by-product of these activities is the production of intermediates for synthesis of cellular components. During active photophosphorylation, a main function of the EM pathway, the TCA cycle, and the oxidative pentose phosphate pathway is to contribute intermediates for biosynthesis. Since photophosphorylation does not occur in the dark, these pathways take over the essential process of energy production during periods of darkness.

Whereas an increase in respiration in diseased host tissue appears to be a universal phenomenon, there have been many speculations about its mechanism. The sections below will offer possible explanations for the cause of increased oxygen uptake and its consequences for the host.

1. Dark Respiration

Inasmuch as the EM pathway has the dual function of energy production and production of intermediates for synthesis of essential cellular constituents (Preiss and Kosuge, 1976; Turner and Turner, 1975), it follows that flow of carbon through the pathway is influenced by several control mechanisms which in turn are responsible to cellular demands for energy and biosynthetic intermediates (Turner and Turner, 1975). A major point for control of the pathway resides at phosphofructokinase which is inhibited by high concentrations of ATP, other products of glycolysis, and the TCA cycle intermediate, citrate. Increasing concentrations of fructose 6-phosphate diminish the inhibitory effects of ATP. Inhibition by ATP is the signal that the energy-yielding function of the pathway is being fulfilled and flow of carbon in the pathway should be curtailed to prevent waste of carbon or harmful accumulation of products of the pathway. Inhibition of ATP could act in concert with control of pyruvate kinase where lack of ADP, the consequence of high ATP production in the cell, would also serve to regulate flow of carbon through the EM pathway. In addition, there are controls at every point where branches lead to the diversion of metabolites into biosynthetic and ancillary pathways (Turner and Turner, 1975; Preiss and Kosuge, 1976).

In a sense, the activity of the TCA cycle is conditioned by the amount of carbon diverted into the pathway from glycolysis. The entry point of the cycle, the citrate synthase reaction, also is inhibited by ATP. The major energy yield by the TCA cycle occurs by ATP production via the electron transport system.

In actively photosynthesizing tissue, there will be abundant production of NADPH and ATP and almost immediate transfer of that reducing power and energy into the cytoplasm by various shuttle systems (Heber, 1974). The resulting high ATP/ADP ratio in the cytoplasm would exert strong control over phosphofructokinase and pyruvate kinase and would lessen activity of the EM pathway (Walker, 1976). In addition, various shuttle systems occur in the cytoplasm that convert NADPH to NADH which then can be oxidized via the mitochondrial cytochrome system to yield ATP. On the other hand, in chlorophyll-containing tissues in the dark, neither ATP nor NADPH would be produced via photophosphorylation. Thus control over the EM pathway would be achieved by ATP produced by glycolysis and oxidative phosphorylation.

In diseased tissue, respiration usually is stimulated and energy is wasted. Necrotrophic organisms almost always stimulate respiration (Daly, 1976) although certain toxins produced by pathogens apparently inhibit electron transport and mitochondrial oxidation of TCA cycle intermediates (Flavell, 1975). Generally speaking, increases in respiration are attributed to loss of control at one or more sites in metabolic pathways. Perhaps the most accepted explanation is that the increased respiratory activity is due to uncoupling of oxidative phosphorylation. In healthy tissue, respiration is controlled via "tightly" coupled mitochondria. In infected tissue, however, mitochondrial phosphorylation is uncoupled, perhaps by action of toxins or disruption of mitochondrial structure. Simultaneously, chloroplasts may rupture and the capacity for photophosphorylation may be lost. With either or both events, the consequence would be the same, namely that carbohydrate catabolism would continue without ATP production and the plant tissue would be starved for energy. Continued uncontrolled dark respiration in the absence of photosynthesis would decrease reserves of energy-yielding carbohydrate and cause accumulation of undesirable by-products. The increase in rate of respiration due to infection is estimated to be about 3 to 6 μmoles CO_2 hr^{-1} g^{-1} fresh weight of tissue (Bateman and Daly, 1967). If one assumes that this rate of respiration is maintained for 24 hr in diseased tissue, and that storage carbohydrates and other photoassimilates are being respired, there would be a 2% reduction in dry weight of that tissue (assuming 80% water) and a loss of up to 16.5 calories of stored energy per gram fresh weight per day. While it is unrealistic to expect that host enzymes would remain active for long periods of time in severely infected tissues, these hypothetical calculations nevertheless indicate the potential magnitude of the effect of disease on yield.

2. Oxidative Pentose Phosphate Pathway

Isozymes of the first two enzymes of the oxidative pentose phosphate pathway, glucose-6-phosphate dehydrogenase and 6-phosphogluconate dehydrogenase, exist in the chloroplast and the cytosol (Schnarrenberger *et al.*, 1975). In light, the chloroplast form of glucose-6-phosphate dehydrogenase would be effectively controlled being inhibited by high concentrations of ATP and NADPH. Since cytoplasmic glucose-6-phosphate dehydrogenase also is inhibited by NADPH and ATP, the transfer of energy and reducing power from the chloroplast to the cytoplasm during photosynthesis would effectively control the activity of that enzyme and the operation of the oxidative pentose phosphate pathway. It is commonly observed that enzymes of the oxidative pentose phosphate pathway increase in activity together with respiratory increases in diseased tissue, particularly at sites of leaf infection by biotrophic fungi. These tissues become sinks for photoassimilates and centers of intense metabolic activity. For this reason increased respiration in host tissue surrounding infection sites of biotrophic fungi often show both increased respiration and an increase in dry weight. This is to be expected since there will be increased demands on host tissue for energy and biosynthetic intermediates for all activities of the pathogen, whether growth or sporulation is involved. Host–parasite systems have evolved into interactions in which stimuli provided by infection bring about mobilization of nutrients to the site of infection and increased host metabolic activity favoring infection. In the above examples, reducing power is not wasted in the usual sense, as discussed below.

3. Wasteful Loss of Reducing Power

There are other possible routes by which NADPH generated by the oxidative pentose phosphate pathway can be reoxidized. The sequences listed below could occur in tissues where the metabolic compartmentation is altered and oxidation of reduced pyridine nucleotides could be accomplished by coupling with oxidase systems that are active in infected plant tissue.

Reaction Sequence 1
   ascorbate oxidase
   ascorbate $+ \frac{1}{2}O_2 \rightarrow$ dehydroascorbate $+ H_2O$ (1)
   dehydroascorbate reductase   (2)
   dehydroascorbate $+$ reduced glutathione $\rightarrow$ ascorbate $+$ oxidized glutathione
   glutathione reductase
   oxidized glutathione $+ NADPH + H^+ \rightarrow$ reduced glutathione $+$ NADP   (3)
Reaction Sequence 2
   peroxidase
   $NADPH + H^+ + \frac{1}{2}O_2 \rightarrow NADP + H_2O$ (4)

Reaction Sequence 3

   phenol oxidase

   ortho-dihydroxyphenol $+ \frac{1}{2}O_2 \rightarrow$ orthoquinone $+ H_2O$ (5)

   quinone reductase

   orthoquinone $+$ NADPH $+$ H$^+$ $\rightarrow$ ortho-dihydroxyphenol $+$ NADP (6)

NADH generated by the EM pathway also can be oxidized in the peroxidase and quinone reductase reactions.

In all cases reducing power is used wastefully and the original oxidant, NADP, is regenerated. These systems if operative would be very efficient in dissipating reducing power. Evidence that any of the sequences function *in vivo* is still circumstantial, although increases in activity of peroxidase, polyphenol oxidase, and ascorbate oxidase in infected tissue have been reported in a variety of diseases and are characteristic of diseases involving necrotrophic organisms (Maxwell and Bateman, 1967; Tomiyama and Stahmann, 1964; Batra and Kuhn, 1975).

C. Respiration in Nonchlorophyll-Containing Tissue

Increases in respiration in nonchlorophyllus tissue could be attributed to mechanisms similar to those occurring in chlorophyll-containing tissue. In those tissues, there would be no complications provided by photorespiration and photosynthesis. The EM pathway, TCA cycle, oxidative phosphorylation, and the oxidative pentose phosphate pathway are the principal routes of carbohydrate utilization and energy production. These, together with oxidative enzymes such as those described in the preceding section, could contribute to increased respiratory activity in response to infection.

D. Net Respiration of Host and Pathogen

Past studies on respiration usually have focused on the changes that occur in the host as a result of infection. In such research, great pains have been taken to work with host tissue which are not invaded by a pathogen but nevertheless show effects of the disease. By such procedures, many past researchers may have overlooked the most important part of the phenomenon, namely the interaction of the host and pathogen. Unfortunately, available experimental methods do not permit distinction between the contributions of the host and the pathogen except in the case of virus-infected tissues. With all living pathogens, measurements of the respiratory activity of infected host tissue indicate the net contributions of both host and pathogen.

Overall, such distinctions may not be important if the ultimate goal is to measure the effect of disease on yield of a crop. It should be possible to measure the gross photosynthesis of infected tissue and to deduct losses due to the combined effects of host and pathogen with appropriate corrections for photorespiration. Thus, net photosynthesis could be calculated and the effect on yield could be estimated.

VII. EFFECT ON ENERGY UTILIZATION

A. Competition for Energy by the Host and Pathogen

1. Diversion of Energy for Pathogen Metabolism

Virus multiplication in host tissue requires energy in the form of ATP for synthesis of viral protein and nucleic acid. Depending on the site of virus synthesis, the energy source would be photophosphorylation or mitochondrial phosphorylation or a combination of both. Based on the amount of virus synthesized, it should be possible to calculate the expenditure of host energy to synthesize a given quantity of virus. Daily yields of tobacco mosaic virus have been reported as high as 2.5 mg/g fresh weight (Hirai and Wildman, 1969). This requires 2.375 mg protein and 0.125 mg viral nucleic acid synthesized per gram of host tissue per day or 135 μmoles of ATP and 1.25 mg of carbon per day per gram fresh weight when glucose is the starting material. Total energy requirements for virus synthesis starting from CO_2 would be about 243 μmoles of host ATP per day per gram fresh weight of leaves. This together with requirements for host maintenance and growth would not stress a plant system capable of producing up to 1.5 mmoles of ATP per gram fresh weight per hour via photophosphorylation. Similar calculations could be made for energy needs for bacterial and fungal multiplication. Bacteria grown in culture convert 20–50% of the glucose utilized to cellular carbon and release the remainder in the form of CO_2. In the severest infections, pathogenic bacteria in plant leaves reach populations as high as 10^{10} cells (total 2 mg) per gram dry weight of leaves (Ercolani and Cross, 1966) requiring no more than 10 mg or 1 to 2% of total host carbohydrates for energy and cellular needs. If that cell population were attained in 4 days, consumption of host carbohydrates by the pathogen could reach 17 μmoles per gram fresh weight host tissue per day. Concentrations of soluble carbohydrates, largely mono and dissacharides, may be as much as 0.25 M in phloem exudates and fluids of the intracellular space in leaves, 1 M or higher in cell sap of some fruits, but only transiently detectable in xylem fluids. Thus, depending upon the site of infection, some bacterial pathogens could cause transient competition for host nutrients.

However, in most cases, populations of pathogens in infected tissue undoubtedly would be substantially less than the above number of 10^{10} cells per gram fresh weight and, from a whole plant perspective, their needs for nutrients would have little competitive impact on nutrient requirements of the host. Instead it seems likely that the most harmful effects of the disease are brought about by the action of extracellular products of pathogens such as enzymes, and primary and secondary metabolites, which have direct deleterious effects on the host.

2. Alteration in Biosynthesis in Host Tissue

The calculations in the foregoing section reveal that competition by the pathogen for biosynthetic intermediates is not a major factor limiting biosynthesis in the host. The most deleterious effects are the losses of reducing power and energy by disruption of the host's metabolic machinery. One of the principal effects of infection in certain host parasite systems is the activation of host systems for repair of injury (see Volume V). Thus, systems for the synthesis of phytoalexins, phenolic compounds, lignin, or starch are activated. In other cases, biosynthesis in host tissue is completely interrupted by the destruction or partial inactivation of the host's energy-producing systems.

B. Waste of Stored Energy in the Host

In some diseases host storage organs are attacked by plant pathogens and storage carbohydrates such as starch are broken down by action of the pathogen enzymes. For every gram of starch hydrolyzed and utilized by the pathogen, 4.1 kcal of energy is lost. For man, the loss would be even greater since a partially decayed potato, for example, would be unsuitable for consumption and discarded.

VIII. EFFECT OF DISEASE ON COMPONENTS OF YIELD

Rarely does disease affect only a single component of yield; instead, it is likely that at least two and perhaps all components are affected simultaneously. In many virus diseases, general stunting of the plant occurs. Leaves become thick and brittle and contain much free carbohydrate (Diener, 1963). If the source is affected by reduced photosynthesis, it is likely that transport is similarly affected. Nevertheless, it is difficult to determine the cause–effect relationship of the system. It has been suggested that virus infection causes impaired phloem transport which in turn causes assimilates to back up. This is presumed to account for accumulation of assimilates in the leaves. The reduced capacity of the phloem to transport assimilates prevents development of storage organs.

Another explanation is equally plausible, however. Disease causes disruption of the storage component as, for example, in the capacity of a beet root to store sugar. There would be less demand for assimilates which would accumulate and produce conditions that cause feedback inhibition of photosynthesis (Hall and Loomis, 1972; Walker, 1976). The overall effect in crops such as cereals will be reduced photoassimilation, accumulation of carbohydrates in leaves, and reduced grain yields (Jensen and Van Sambeek, 1972). For further discussion of disruption of the flow of food see Chapter 5, this volume.

Storage capacity varies with the crop and with the type of component used for yield measurements. In the case of wheat, storage capacity depends on the number of ears per unit area, number of spikelets per ear, grains per spikelet, and on individual size of grain (Evans *et al.*, 1975). There is a sequential determination of these components during crop development. Both ear and spikelet numbers are determined well before anthesis, grain number is determined at or near the time of anthesis, and grain size after anthesis. Thus the storage capacity of wheat will be affected by disease through much of the development of the plant. Unfortunately, in the field it will be difficult to distinguish between the effects of environment and the effects of disease on components of yield because various environmental factors, singly or in combination, will have varying effects on each yield component. Those same environmental factors also will have profound effects on disease development in the plant.

Components of yield must be balanced for realization of the full yield potential of a crop. For cereals, an increase in sink capacity, namely storage capacity in the grain, must accompany an increase in photoassimilation. Otherwise grain development will limit utilization of photoassimilates. It has been calculated that 90–95% of the carbohydrate in grain is derived from photoassimilation after anthesis (Evans *et al.*, 1975). However, photoassimilation during ear development significantly influences the components of storage capacity. Generally speaking, there is a close positive correlation between leaf area duration (duration and extent of photosynthetic tissue) and grain yield after ear emergence of wheat. Also there is an inverse correlation between the leaf area index (LAI) and grain weight. When the LAI is 7 and one-half of the leaves are removed at heading, no reduction in yield occurs. Earlier defoliation may cause significant reductions in yield. Thus it is evident that interference in photoassimilation by foliage infections after heading would have substantially less effect on yield than disruption of the process by early infections.

Maintenance respiration would increase during the response of the plant to repair the injury caused by infection. Virtually all injuries stimulate respiration, synthesis of essential cell components, and several secondary metabolites (Uritani, 1976). If these responses are stimulated by the need to repair the injury induced by the pathogen, the energy expended must be classed as maintenance respiration. Undoubtedly, a considerable portion of the increased respiration may be wasteful respiration, that is, respiration uncoupled from oxidative phosphorylation or "wasteful" respiration by one of the several mechanisms proposed in section VI,B,3. The increased respiration ultimately will reduce the stored energy available for growth respiration. Estimates of increased oxygen uptake attributed to the phenol oxidase system were 3 μmoles/hr/g fresh weight (Tomiyama and Stahmann, 1964). Assuming that this rate of oxygen uptake is maintained for 2 days, the ultimate reduction in assimilates and storage compounds will be equivalent to 144 μmoles of glucose per gram fresh weight of tissue.

IX. THE GENERAL IMPACT OF REDUCED ENERGY CAPTURE BY PLANTS

On a global basis in 1970 the yield of cereals in metric tons of dry matter was 104.9×10^7 (Evans, 1975). This constitutes more than two-thirds of the edible food produced in the world. In addition, over 80% of the world's food production comes from only 11 species of plants. With so much invested in so few crops, it is essential that scientists know as much as possible about these crops. Plant pathologists in particular must join in that effort and determine the effects of disease on each component of yield of those important food crops.

The total solar power available at the earth's surface is 7×10^{13} kW; that fixed by terresterial photosynthesis is 9.2×10^{10} kW. Clearly, plants harvest only a fraction of the total solar energy received by the earth. Currently only a fraction of that energy is consumed by human beings for the maintenance of life. Yet, the availability of food in some countries becomes so precariously low that a 10% reduction in production may mean increased human starvation. Fortunately, disasters in food production have not reached global proportions, and surpluses in one part of the world have substantially offset crop failures in another.

One wonders how long we can depend upon good fortune. Unless substantial advances are made in the control of diseases of crop plants, and in crop yields in developing countries, the world some day may be faced with famine of global proportions. It is sobering to learn that 80% of the world's food is contributed by only 11 species of plants. Mangelsdorf (1966) was right when he said, "11 plant species quite literally stand between mankind and starvation."

X. A LOOK INTO THE FUTURE

The process of energy capture and transfer in plants is exceedingly complex. Much more research is needed to elucidate the fundamental mechanisms involved. Nevertheless, it is clear that the primary effects of disease on energy capture, transfer, and ultimately yield loss are the result of interference and disruption of the various components of energy capture, photoassimilation, translocation, and storage; direct competition between host and pathogen for energy is not a major factor causing reduction in yield. With the substantial amount of information now available on crop physiology, it is essential that plant pathologists devote considerable effort to the study of the effect of disease on capture and transfer of energy from the "whole plant" perspective and most importantly on yield. Mathematical modeling should contribute significantly to this effort. The attention now being given to photorespiration, dark respiration, translocation, and storage mechanisms (McCree, 1976; Zelitch, 1971) bodes well for future progress in both crop physiology and crop pathology.

New approaches are being unraveled, and new information is being developed on the nature of the photochemical apparatus and photoassimilation. At last, research on crop physiology is being focused directly on problems related to food production. Increasing constraints on energy used for crop production and on chemical methods of pest control are forcing a reassessment of what constitutes the ideal plant type for each crop. In this connection, plant pathologists must join together with crop physiologists in a renewed effort to define and develop plant ideotypes to fit the future needs of agriculture.

Renewed effort and innovative studies are needed on the role of membranes in host–parasite interaction. Membrane systems contain the machinery for photophosphorylation and oxidative phosphorylation; they regulate the flow of substrates and ions among cellular organelles and from cell to cell; they provide compartments for energy-yielding and energy-utilizing cycles. Thus it is not surprising that the host membrane is the site of attack by enzymes and/or toxins produced by many plant pathogens (see Chapter 15, this volume).

It seems evident that plant pathologists have many opportunities for making important future contributions to the knowledge of bioenergetics in host–pathogen interactions. In their quest for knowledge plant pathologists must build upon the information and expertise from other disciplines to develop concepts unique to plant–pathogen interactions.

References

Akazawa, T., and Ramakrishnan, L. (1967). Change in chloroplast proteins of the rice plant infected by the blast fungus, Piricularia oryzae. *In* "The Dynamic Role of Molecular Constituents in Plant-Parasite Interaction" (C. J. Mirocha and I. Uritani, eds.), pp. 329–341. Am. Phytopathol. Soc., St. Paul, Minnesota.

Ayres, P. G., and Jones, P. (1975). Increased transpiration and the accumulation of root absorbed [86]RB in barley leaves infected by Rhynchosporium secalis (leaf blotch). *Physiol. Plant Pathol.* **7**, 49–58.

Bassham, J. A. (1971). The control of photosynthetic carbon metabolism. *Science* **172**, 526–534.

Bateman, D. F., and Daly, J. M. (1967). The respiratory pattern of Rhizoctonia-infected bean hypocotyls in relation to lesion maturation. *Phytopathology* **7**, 127–131.

Batra, G. K., and Kuhn, C. W. (1975). Polyphenoloxidase and peroxidase activities associated with acquired resistance and its inhibition by 2-thiouracil in virus-infected soybean. *Physiol. Plant Pathol.* **5**, 239–248.

Bedbrook, J. R., and Matthews, R. E. F. (1972). Changes in the proportions of early products of photosynthetic carbon fixation induced by TYMV infection. *Virology* **48**, 255–258.

Bravado, B., Palgi, A., and Lurie, S. (1977). Changing ribulose diphosphate carboxylase/oxygenase activity in ripening tomato fruit. *Plant Physiol.* **60**, 309–312.

Carroll, T. W., and Kosuge, T. (1969). Changes in structure of chloroplasts accompanying necrosis of tobacco leaves systematically infected with tobacco mosaic virus. *Phytopathology* **59**, 953–962.

Daly, J. M. (1976). The carbon balance of diseased plants: Changes in respiration, photosynthesis and translocation. *In* "Physiological Plant Pathology" (R. Heitefuss and P. H. Williams, eds.), pp. 450–479. Springer-Verlag, Berlin and New York.

Diener, T. O. (1963). Physiology of virus-infected plants. *Annu. Rev. Phytopathol.* **1**, 197–218.

Duniway, J. M. (1976). Water status and imbalance. *In* "Physiological Plant Pathology" (R. Heitefuss and P. H. Williams, eds.), pp. 430–449. Springer-Verlag, Berlin and New York.

Duniway, J. M., and Durbin, R. D. (1971). Detrimental effect of rust infection on the water relations of bean. *Plant Physiol.* **48**, 69–72.

Duniway, J. M., and Slatyer, R. O. (1971). Gas exchange studies on the transpiration and photosynthesis of tomato leaves affected by Fusarium oxysporum f. sp. lycopersici. *Phytopathology* **61**, 1377–1381.

Edwards, G. E., Huber, S. C., Ku, S. B., Rathnam, C. K. M., Gutierrez, M., and Mayne, B. C. (1976). Variations in photochemical activities of C_4 plants in relation to CO_2 fixation. In "CO_2 Metabolism and Plant Productivity" (R. H. Burris and C. C. Black, eds.), pp. 83–112. Univ. Park Press, Baltimore, Maryland.

Ercolani, G. L., and Crosse, J. E. (1966). The growth of Pseudomonas phaseolicola and related plant pathogens in vivo. J. Gen. Microbiol. 45, 429–439.

Esau, K. (1941). Phloem anatomy of tobacco affected with curly top and mosaic. Hilgardia 13, 437–490.

Evans, L. T. (1975). The physiological basis of crop yield. In "Crop Physiology— Some Case Histories" (L. T. Evans, ed.), pp. 327–355. Cambridge Univ. Press, London and New York.

Evans, L. T., Wardlaw, I. F., and Fischer, R. A. (1975). Wheat. In "Crop Physiology—Some Case Histories" (L. T. Evans, ed.), pp. 101–149. Cambridge Univ. Press, London and New York.

Flavell, R. (1975). Inhibition of electron transport in maize mitochondria by Helminthosporium maydis race T pathotoxin. Physiol. Plant Pathol. 6, 107–116.

Ghosh, H. P., and Preiss, J. (1966). Adenosine diphosphate glucose pyrophosphorylase. A regulatory enzyme in the biosynthesis of starch in spinach leaf chloroplasts. J. Biol. Chem. 241, 4491–4504.

Giaquinta, R. (1977). Sucrose hydrolysis in relation to phloem translocation in Beta vulgaris. Plant Physiol. 60, 339–343.

Goffeau, A., and Bove, J. M. (1965). Virus infection and photosynthesis. 1. Increased photophosphorylation by chloroplasts from Chinese cabbage infected with turnip yellow mosaic virus. Virology 27, 243–252.

Hall, A. E., and Loomis, R. S. (1972). An explanation for the difference in photosynthetic capabilities of healthy and beet yellows virus-infected sugar beets (Beta vulgaris L.). Plant Physiol. 50, 576–580.

Hatch, M. D. (1976). The C_4 pathway of photosynthesis: Mechanism and function. In "CO_2 Metabolism and Plant Productivity" (R. H. Burris and C. C. Black, eds.), pp. 59–81. Univ. Park Press, Baltimore, Maryland.

Heber, U. (1974). Metabolic exchange between chloroplasts and cytoplasm. Annu. Rev. Plant Physiol. 25, 393–421.

Hewitt, H. G., and Ayres, P. G. (1975). Changes in CO_2 and water vapour exchange rates in leaves of Quercus robur infected by Microsphaera alphitoides (powdery mildew). Physiol. Plant Pathol. 7, 127–137.

Hirai, A., and Wildman, S. G. (1969). Effect of TMV multiplication on RNA and protein synthesis in tobacco chloroplasts. Virology 38, 73–82.

Jensen, S. G. (1972). Metabolism and carbohydrate composition in barley yellow dwarf virus-infected wheat. Phytopathology 62, 587–592.

Jensen, S. G., and Van Sambeek, J. W. (1972). Differential effects of barley yellow dwarf virus on the physiology of tissues of hard red spring wheat. Phytopathology 62, 290–293.

Kawashima, N., and Wildman, S. G. (1970). Fraction I protein. Annu. Rev. Plant Physiol. 21, 325–358.

Kelly, G. J., and Latzko, E. (1977). Chloroplast phosphofructokinase. I. Proof of phosphofructokinase activity in chloroplasts. Plant Physiol. 60, 290–294.

Kelly, G. J., Latzko, E., and Gibbs, M. (1976). Regulatory aspects of photosynthetic carbon metabolism. Annu. Rev. Plant Physiol. 27, 181–205.

Kosuge, T., and Gilchrist, D. G. (1976). Metabolic regulation in host-parasite interactions. *In* "Physiological Plant Pathology" (R. Heitefuss and P. H. Williams, eds.), pp. 679–702. Springer-Verlag, Berlin and New York.

Livne, A., and Daly, J. M. (1966). Translocation in healthy and rust-affected beans. *Phytopathology* **56**, 170–175.

McCree, K. J. (1970). An equation for the rate of respiration of white clover plants grown under controlled conditions. *Predic. Meas. Photosynth. Prod., Proc. IBP/PP (Int. Biol. Programme/Prod. Processes) Tech. Meet., 1969* pp. 221–229.

McCree, K. J. (1976). The role of dark respiration in the carbon economy of a plant. *In* "CO$_2$ Metabolism and Plant Productivity" (R. H. Burris and C. C. Black, eds.), pp. 177–184. Univ. Park Press, Baltimore, Maryland.

MacDonald, P. W., and Strobel, G. A. (1970). Adenosine diphosphate-glucose pyrophosphorylase control of starch accumulation in rust-infected wheat leaves. *Plant Physiol.* **46**, 126–135.

Magyarosy, A. C., Buchanan, B. B., and Schurmann, P. (1973). Effect of a systemic virus infection on chloroplast function and structure. *Virology* **55**, 426–438.

Magyarosy, A. C., Schürmann, P., and Buchanan, B. B. (1976). Effect of powdery mildew infection on photosynthesis by leaves and chloroplasts of sugar beets. *Plant Physiol.* **57**, 486–489.

Mangelsdorf, P. C. (1966). Genetic potentials for increasing yields of food crops and animals. *Proc. Natl. Acad. Sci. U.S.A.* **56**, 370–375.

Maxwell, D. P., and Bateman, D. F. (1967). Changes in the activities of some oxidases in extracts of Rhizoctonia-infected bean hypocotyls in relation to lesion maturation. *Phytopathology* **57**, 132–136.

Mohamed, N. A., and Randles, J. W. (1972). Effect of tomato spotted wilt virus on ribosomes, ribonucleic acids and Fraction 1 protein in Nicotiana tabacum leaves. *Physiol. Plant Pathol.* **2**, 235–245.

Montalbini, P., and Buchanan, B. B. (1974). Effect of a rust infection on photophosphorylation by isolated chloroplasts. *Physiol. Plant Pathol.* **4**, 191–196.

Panopoulus, N., Faccioli, G., and Gold, A. H. (1972). Translocation of photosynthate in curly top virus-infected tomatoes. *Plant Physiol.* **50**, 266–270.

Peavey, D. G., Steup, M., and Gibbs, M. (1977). Characterization of starch breakdown in the intact spinach chloroplast. *Plant Physiol.* **60**, 305–308.

Preiss, J., and Kosuge, T. (1976). Regulation of enzyme activity in metabolic pathways. *In* "Plant Biochemistry" (J. Bonner and J. E. Varner, eds.), 3rd ed., pp. 277–336. Academic Press, New York.

Ryan, F. J., and Tolbert, N. E. (1975). Ribulose diphosphate carboxylase/oxygenase. IV. Regulation by phosphate esters. *J. Biol. Chem.* **250**, 4234–4238.

Saeki, T. (1963). Light relations in plant communities. *In* "Environmental Control of Plant Growth" (L. T. Evans, ed.), pp. 79–94. Academic Press, New York.

Schipper, A. L., Jr., and Mirocha, C. J. (1969). The mechanism of starch depletion in leaves of Phaseolus vulgaris infected with Uromyces phaseoli. *Phytopathology* **59**, 1722–1727.

Schnarrenberger, C., Tetour, M., and Herbert, M. (1975). Development and intracellular distribution of enzymes of the oxidative pentose phosphate cycle in radish cotyledons. *Plant. Physiol.* **56**, 836–840.

Thimann, K. V., Tetley, R. R., and Van Thanh, T. (1974). The metabolism of oat leaves during senescence. *Plant Physiol.* **54**, 859–862.

Tomiyama, K., and Stahmann, M. A. (1964). Alteration of oxidative enzymes in potato tuber tissue by infection with Phytophthora infestans. *Plant Physiol.* **39**, 483–490.

Turner, J. F., and Turner, D. H. (1975). The regulation of carbohydrate metabolism. *Annu. Rev. Plant Physiol.* **26,** 159–186.

Turner, N. C., and Graniti, A. (1976). Stomatal response of two almond cultivars to fusicoccin. *Physiol. Plant Pathol.* **9,** 175–182.

Uritani, I. (1976). Protein metabolism. *In* "Physiological Plant Pathology" (R. Heitefuss and P. H. Williams, ed.), pp. 509–525. Springer-Verlag, Berlin and New York.

Walker, D. A. (1976). Regulatory mechanisms in photosynthetic carbon metabolism. *Curr. Top. Cell. Regul.* **11,** 203–241.

Zaitlin, M., and Jagendorf, A. T. (1960). Photosynthetic phosphorylation and Hill reaction activities of chloroplasts isolated from plants infected with tobacco mosaic virus. *Virology* **12,** 477–486.

Zelitch, I. (1971). "Photosynthesis, Photorespiration, and Plant Productivity." Academic Press, New York.

Zelitch, I. (1975). Pathways of carbon fixation in green plants. *Annu. Rev. Biochem.* **44,** 123–145.

Waggoner, P. E. (1969). Predicting the effect upon net photosynthesis of changes in leaf metabolism and physics. *Crop Sci.* **9,** 315–321.

Chapter 5

Dysfunction in the Flow of Food

MARTIN H. ZIMMERMANN AND JANET McDONOUGH

I. INTRODUCTION

One of the adaptations land plants had to undergo during evolution was the development of structures and mechanisms for the movement of water and nutrients throughout the plant body. Once this was accomplished plants could grow to considerable size. This happened early, the first trees appearing almost 400 million years ago. Extensive forests existed during the Carboniferous era, some 300 million years ago.

The two major long-distance transport channels in vascular plants are the xylem and the phloem. These can also be regarded as compartments within which liquids can move more or less freely. Water and soil nutrients move from roots to all transpiring organs via the xylem and products of photosynthesis, primarily sugars, move via the phloem from leaves or from mobilizing storage organs (functionally referred to as sources), to all nonphotosynthesizing organs where growth or storage takes place (sinks). Phloem and xylem channels usually lie parallel, very close together. In woody gymnosperms and dicotyledons, radial transport between xylem and phloem takes place in the rays; ray transport and storage has recently been summarized by Höll (1975). Transfer from xylem to phloem and vice versa enables the plant to circulate nutrients and move them to where they are needed.

There are so many diseases affecting the translocation systems that it is impossible to discuss them all in detail within the space of this chapter. Instead, an attempt has been made to look at specific types of malfunctions of the translocation mechanism regardless of the pathogens causing them.

II. STRUCTURE AND FUNCTION OF THE XYLEM

A. The Apoplast Concept

Water can move easily from the conducting channels of the xylem through all cell walls except where walls are suberized as in the root endodermis or dermal cell layers. Thus all living cells are surrounded by xylem liquid penetrating their walls. This whole area is referred to as the "apoplast" or "free space," which is freely accessible to a solution into which a piece of tissue may be immersed for experimental purposes.

The long-distance transport channels are tracheids and vessels. In coniferous trees, for example, the only water-conducting cells are tracheids; these are about 3 mm long and some 50 μm in diameter. Water flows through them on its way up, but every 3 mm it must pass from one cell into the next through special structures of the walls, the bordered pits. Flowering plants (angiosperms) and ferns evolved multicellular water-conducting units. Individual elements are aligned end-to-end, with end walls partly or wholly dissolved, thus forming long capillaries, the vessels. Vessels are of finite length, their ends tapering out along other, continuing vessels. Vessel length distribution can be measured. In *Eucalyptus obliqua*, for example, most vessels are less than 50 cm long, few are longer than 1 m, and some are longer than 3 m (Skene and Balodis, 1968; see also Figs. IV-2 and 3 in Zimmermann and Brown, 1971). Water passes from vessel to vessel through bordered pit pairs. Vessels form a complex three-dimensional network. The principle of vessel arrangement, very much simplified, is shown in Fig. 1 (left). Cinematographic analysis (Zimmermann, 1976) permits a detailed elucidation and reconstruction on paper of the vessel network (see Fig. IV-6 in Zimmermann and Brown, 1971). Indeed, the only absolutely clear and unambiguous way of show-

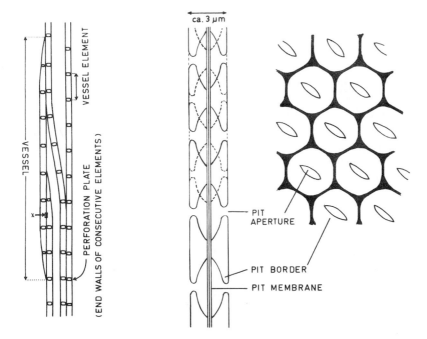

Fig. 1. *Left*: Diagrammatic view of a vessel network. The functional unit of water conduction is the vessel, not the vessel element. Water moves freely in the vessel, but must pass from one vessel to the next on its way up to leaves. Overlapping vessel ends are dozens of elements long, not merely three or four as shown here. In case of injury, the whole vessel is embolized. (From Zimmermann, 1976.) *Center*: A diagrammatic section through the wall between two adjacent vessels as, for example, in the boxed area at X in the left-hand drawing. Passage of water from vessel to vessel is through the pit membranes of the bordered pit pairs. If one of the vessels embolizes, the gas–water interface cannot enter the neighboring vessel. The dashed lines in the upper part of the drawing show the section at another level. *Right*: Vessel-to-vessel pits in surface view to show how they are often arranged in honeycomb fashion. Pit apertures are usually slitlike because cellulose microfibrils in the secondary wall have a high degree of parallel orientation. The secondary wall, which gives the vessel elements strength, touches the primary wall (i.e., the pit membrane) only in a limited area, indicated here as a black hexagonal pattern. A very large percentage of the primary walls of the two adjacent vessel elements is exposed and thus available for water movement from vessel to vessel.

ing vessel ends is with a motion-picture film, made up of serial images of transverse wood sections. Such a film is available (Zimmermann, 1971).

The construction of the bordered pit pairs between adjacent vessel elements is of crucial importance (Bailey, 1953). They are circular or elongated spots in the wall where the secondary wall opens to form the pit aperture and arches over the primary wall which forms the pit membrane (Fig. 1, center and right). The pores in the pit membranes are extremely small, but the overlapping areas of vessel ends are so large, the pits between vessels so numerous, and the membrane area so extensive, that the resistance to flow through dicotyledonous wood is only about twice, and coniferous wood about three times, the resistance through an equal number of capillaries of infinite length (see Zimmermann and Brown, 1971). Although limited length of vessels does increase the resistance to flow compared with endless capillaries, it gives the plant a considerable advantage: accidental embolism, whenever it occurs within a compartment like a tracheid or a vessel, will remain confined within the compartment in which it occurs.

B. The Cohesion Theory

Water enters the xylem of the roots osmotically, following metabolic solute uptake. It ascends via the xylem into the leaves, where it spreads from the xylem ends through all cell walls, thus reaching the stomatal cavities into which it evaporates (Strugger, 1953).

Movement in the xylem is passive along pressure gradients, comparable to laminary flow in capillaries, and can be described by the Hagen-Poiseuille Eq. (1). Volume flow (dV/dt) is proportional to the pressure

$$\frac{dV}{dt} = \frac{\pi r^4}{8\eta} \cdot \frac{dP}{dl} \tag{1}$$

gradient (dP/dl) and the fourth power of the capillary radius, and inversely proportional viscosity (η). The fourth-power relationship is very important, because it means that most transport is in the widest vessels. Vessels that are 10% wider transport 46% more water ($1.1^4 = 1.4641$), all other factors held constant, and vessels four times as wide move 256 times as much water. For example, if one sees on a transverse section two vessels side by side, one with a diameter of 200 μm, the other with a diameter of 50 μm, the wider one will carry 99.6% of the volume. The evolution of wider vessels makes the xylem very much more efficient. However, evolutionary success depends as much on safety as on efficiency. Wider vessels embolize more easily, thus, historically speaking, limiting the evolution of efficient water-conducting systems (Zimmermann, 1978).

Pressures in the xylem are usually (and in trees almost always) below atmospheric pressure, often negative. Because relative stability can be maintained only in small volumes, wide vessels are less safe water conductors than narrow ones. Water can only exist in a state of tension if the liquid is bubble free (see Hammel and Scholander, 1976). Flow always occurs along a pressure gradient, i.e., from higher to lower pressures, no matter whether these are positive or negative.

C. Conducting Units and Embolism

The safety of the water-conducting system is assured by a large number of relatively small individual conducting units (tracheids in conifers, tracheids and vessels in flowering plants and ferns). The number of conducting units in a large plant is astronomical. All these units are arranged in such a way that passage from one to the next can take place with ease, and that in the case of failure of some of them, alternate routes are readily available (see Zimmermann, 1971; Zimmermann and Brown, 1971).

Water-conducting units often embolize when they are injured, for example, by a burrowing insect. Water pressure inside a tracheid or vessel is low; air under ambient pressure ($+1$ atm) is therefore sucked into it through the injury and the water retreats into intact neighboring units. Air cannot be sucked into an intact vessel through pits because pit pores are very small. An air–water interface can penetrate a wet pore only when

$$\text{pore diameter (in } \mu\text{m)} \times \text{pressure difference (in atm)} > 3 \qquad (2)$$

(Zimmermann and Brown, 1971, p. 207). Pit pores are so small that several hundred atmospheres of tension would be needed to pull air through the pit membrane. If, however, an insect takes a bite of the cell wall the hole could be sufficiently large to admit air. A 1-μm puncture would allow ambient air ($+1$ atm) to pass into the vessel when the water pressure drops below -2 atm. However, entry into the insect-damaged unit is limited to the length of that vessel or tracheid.

D. Xylem Transport in Trees and Herbs

Whenever there are leaves on trees, xylem pressures are practically always below atmospheric or negative. Many herbs, however, guttate early in the morning. This is an indication of positive root pressure developing during the night. Xylem pressures in trees increase also during the night, but do not ordinarily reach positive values. Milburn (1973) found that the xylem of herbaceous plants may embolize rather regularly during hot summer days. Each breaking water column produces a vibration which can be made audible by "listening" with a sound transducer. Under high stress conditions the xylem may be "routinely" embolized; xylem transport is thus slowed down and transpiration regulated. Positive pressures refill the xylem during the night. Such a mechanism can work only within brief time intervals, e.g., during a single diurnal cycle. If vessels remain embolized for a longer period of time, as in trees, embolism becomes permanent because zero pressure in an empty vessel extracts dissolved gases from the wet walls. These gases would redissolve only with great difficulty even if xylem pressure later becomes positive.

III. STRUCTURE AND FUNCTION OF THE PHLOEM

A. Phloem Structure in Relation to Function

Phloem structure is described in anatomy textbooks (see Esau, 1965; or Fahn, 1967). Esau (1969) also wrote a very comprehensive book on phloem structure. Phloem function has been the topic of books by Crafts and Crisp (1971), Canny (1973), and Peel (1974). A very comprehensive multiauthored work, giving a broad view of experimental work and the major proposed mechanisms, has been edited by Zimmermann and Milburn (1975).

There are some obvious structural similarities between xylem and phloem, which is not surprising because both tissues originate from the same initial cells of the procambium and cambium. The conducting elements of the phloem, the sieve cells, resemble in size and shape the tracheids in the xylem, and the phloem's sieve tubes resemble the xylem's vessels. Mature sieve elements are not quite as specialized as vessel elements are (Parthasarathy, 1975). End walls of sieve elements are never completely dissolved as are those of vessel elements, but their walls are reduced to the sievelike structure from which their name is derived. From the point of view of functional evolution we can assume that the increase in conductivity (i.e., the reduction of the end wall) was not as important for survival as the presence of the end walls as some sort of safety valves. There are two mechanisms that can rapidly plug end walls (sieve plates) in case of injury: the displacement of P-protein against sieve pores and the deposition of a carbohydrate, callose, in the sieve pores (Eschrich, 1975). Since sieve tubes are usually under positive pressure, a sealing mechanism like the one described for the xylem would not work. Upon injury, sieve tubes leak and exude their contents until the sieve plates are completely sealed off.

B. Mechanism of Phloem Transport

When an incision is made into phloem, one can obtain sieve tube exudate unless the phloem seals too rapidly. A similar exudate can be obtained from the stylets of aphids which have been cut off at their mouthparts while feeding. Internal pressure of the sieve tubes maintains exudation (see, e.g., Dixon, 1975; Peel, 1974). There is good evidence that sieve tube exudate is the solution that is normally transported via the phloem. It contains sugars in high concentration (of the order of 10–20% w/v). The most important of these is sucrose, but in some plant families one finds, in addition, oligosaccharides of the raffinose series (raffinose, stachyose, and verbascose) and sometimes sugar alcohols and lesser amounts of other substances (Ziegler, 1975). In intact sieve tubes this liquid is normally under high positive pressure (of the order of +10 atm under transpirational conditions). Movement has been explained by Münch (1930) as an osmotic phenomenon, referred to as the "pressure flow hypothesis" (Milburn, 1975). It is as yet by far the best explanation of phloem transport and will be used in this chapter to describe transport and its regulation in the phloem. Alternative explanations have been and are still being proposed by a number of authors. The reader is referred to Zimmermann and Milburn (1975) for a description of other proposed mechanisms.

According to the pressure flow hypothesis, sugar is secreted into the sieve tubes in the area of a source. This is referred to as "loading" (Geiger, 1975). Loading causes water to enter the sieve tubes osmotically and thus raises the turgor pressure in the sieve tube. In a sink, the opposite happens; sugars are unloaded (removed), water is lost, and the turgor drops. A pressure gradient from source to sink results along which transport takes place, because sieve tubes are highly permeable in the longitudinal direction (as, for example, the phenomenon of exudation manifests).

IV. CIRCULATION THROUGHOUT THE PLANT

Sugar is the major osmotic substance in the sieve tubes; its distribution via phloem is easily explained by Münch's pressure flow mechanism. However, the plant must also distribute all other substances to places of need, and it is obviously not possible to explain all distribution patterns by pressure flow in the phloem alone. One could assume that other (quantitatively less effective) transport mechanisms exist in sieve tubes, operating in addition to pressure flow (see, e.g., Fensom, 1975). However, this is not necessary if the combination of xylem and phloem transport is considered. The example of nitrogen transport, which has been the subject of research and speculation for several decades, may illustrate this.

Carbohydrates moving downward in the phloem are necessary for the assimilation of nitrogen by the roots (e.g., Kursanov, 1963). If this downward flow of sugars is stopped, roots are starved and are unable to supply (via xylem) the crown with root nutrients such as nitrogen. Therefore, interruption of the phloem may decrease transport of nitrogen and other nutrients from roots to crown indirectly. The question of nitrogen movement is still not fully resolved. Most nitrogen originates from the roots. We now assume that all of this ascends via the xylem. This means that most of it must move into mature leaves where most of the xylem water goes for transpiration. A mature leaf does not need all this nitrogen and does not seem to accumulate it throughout the summer. Nitrogenous substances are needed in growing organs. One would then assume that nitrogenous substances, arriving in the mature leaves via xylem, are immediately reexported via phloem, and directed toward phloem sinks. This is a reasonable thought, but the difficulty is that sieve tube exudate of trees contains very low concentrations of nitrogenous substances, except during a brief period in autumn when chlorophyll and other substances are broken down in the blade and the raw material is reexported into the stem (Ziegler, 1975). But what happens to nitrogenous substances in the summer? The nutrient-cycling studies of Pate (1976) offer one interesting answer: nitrogen is transferred from xylem to phloem via transfer cells in the nodal area. Thus, amino acids, or whatever comes from the roots, are removed from the xylem by transfer cells, secreted into the phloem, and carried into the growing shoot tips, fruits, etc. This explanation is quite satisfactory for herbaceous plants. In the case of trees nitrogen may be transferred from the xylem to the cambial area all along the stem, so that xylem sap reaching the mature leaves is practically nitrogen free.

V. FAILURE OF THE TRANSPORT SYSTEMS

A. Interruption of the Transpiration Stream

Practically all failures of the xylem transport system can be traced to the same cause: embolism of the conducting units. The many different reasons why this can occur are discussed below.

1. Physical Injury

If xylem channels are severed, be it experimentally by man, or by a beetle chewing its way through wood, water is withdrawn into contiguous, intact vessels via vessel-to-vessel pits and air is sucked into the injured vessels. (Visualize this happening in the vessel to the far left in Fig. 1.) The embolism normally remains confined within the injured vessels for reasons described above in Section II,C. The water-conducting system of a healthy plant is usually overefficient and the vessel network is so complex that, if anything happens to one part, many alternate pathways via intact vessels are available.

Some of our trees, the so-called ring-porous trees, have evolved very wide vessels (up to almost 500 μm). Because of the fourth-power relationship in capillary flow [see Eq. (1)], water conduction via wide and long vessels is extremely efficient, but at the same time involves high risk. The functional life of wide vessels is short; in ring-porous trees only the most recently produced (i.e., outermost), large, early-wood vessels are conducting (Huber, 1935). These are so efficient that they can easily carry the entire water supply for the crown, but they are so vulnerable that they hardly ever survive the winter. Their superficial location in the outermost growth ring makes them particularly vulnerable to physical injury. The large vessels of ring-porous trees are among the first derivatives of the cambium, and therefore appear as a ring inside each growth ring. They have to be produced by the cambium before the leaves expand; leaves cannot grow before a sufficient water supply from the roots permits transpiration. Additional wood is produced later during the growing season, but this consists of other woody cells such as parenchyma, fibers, tracheids, and small late-wood vessels that carry a very insignificant portion of the transpiration stream. The late-wood vessels, on the other hand, often remain functional over many years and provide the tree with emergency channels that are not very effective. But they are responsible for the ascent of sufficient water for the growth of the large vessels in the spring. In many of our quick-killing diseases (chestnut blight, Dutch elm disease, oak wilt, etc.) damage is probably due to injured and thus embolized early-wood vessels: the first fungal hypha to dissolve a spot in a vessel wall causes embolism in that vessel. It has been noted by S. M. Jutte (personal communication) that chestnut, oak, and elm are ring-porous trees whose vessels are surrounded by tracheids (i.e., dead cells) (see Braun, 1963). It was speculated earlier by Huber (1956) that paratracheal parenchyma, i.e., living cells surrounding vessels, have a protective effect on vessel function. It may be more difficult for a fungal hypha to reach a vessel through a layer of living cells, or, if it does reach the vessel, living cells may seal around the hypha and prevent air from being sucked into the vessel (Table I).

Embolism is probably the major cause of failure of water conduction, but this has rarely been recognized by plant pathologists. A notable exception is Mathre (1964) who described the pathogenicity of *Ceratocystis* species to *Pinus ponderosa*. He found that water conduction failed in an area beyond the infection site. The fungus probably renders tracheids leaky, perhaps by affecting pit membranes enzymatically.

TABLE I

Vulnerability of the Water Conduction System of Trees

Large vessels surrounded by tracheids (ring-porous trees like *Castanea, Quercus* and *Ulmus*)	Most vulnerable
Large vessels surrounded by living parenchyma (ring-porous trees like *Fraxinus*)	Vulnerable
Small vessels only (diffuse-porous trees)	Less vulnerable
Tracheids only (conifers)	Least vulnerable

2. Embolism in the Absence of Injury

Vessels can embolize even if their walls remain intact, i.e., without entry of air. This happens when, for some reason, a water vapor bubble appears within the vessel lumen. The result is the displacement of water by the expanding vapor bubble. If the pressure becomes positive shortly thereafter, as it does overnight in certain herbs, this kind of embolism may be reversible and the damage is repaired. If, however, tensions persist in the xylem for a long period of time, as in trees, dissolved gases are extracted from the wet walls and embolism becomes permanent. A common cause of a similar type of embolism in trees is winter freezing of xylem water. This problem is discussed in Zimmermann and Brown (1971).

Another very common cause of embolism in intact xylem vessels is excessive water stress. If tension exceeds the tensile strength of water, the water columns break. Wilting caused by drought conditions becomes permanent when the channels of water conduction are interrupted by embolism. Water stress can be brought about under normal soil–water conditions, i.e., in the absence of drought, if a wilt toxin interferes with the regulation of the stomatal mechanism, as suggested for diseases like Fusarium wilt.

3. Plugging of Xylem Vessels

One occasionally reads in the literature that the transpiration stream is interrupted by the plugging of xylem vessels by tyloses, gums, excessive growth of bacteria or fungal hyphae within the vessel lumen, or polymers produced by fungi, etc. These can often be observed microscopically as a result of infection of the xylem by a pathogen (Dimond, 1970; Rudolph, 1976).

Tyloses have been known to botanists for a long time. The remarkable investigations of Hermine von Reichenbach (1845) established that they are paratracheal parenchyma cells that grow through the pits into the vessel lumen, often completely filling it. They are a regular feature in the nonconducting xylem vessels of some species. Many diseases result in premature tylosis formation, but there is good evidence that tylosis is always the result of embolism and not the cause of cessation of water conduction (e.g., Klein, 1923). The plant seals off old, nonfunctioning, or leaky vessels with gums or tyloses.

The widespread belief that bacteria or fungal hyphae grow within the transpiration stream and finally interrupt it by "plugging" seems to be based, first, upon complete disregard of the fact that vessel water is in a metastable condition most of the time, and second, upon the observation that fungal spores can be carried by the transpiration stream. We shall deal with the second question in a later section. There is little, if any, evidence in support of the view that bacteria or fungal hyphae can enter intact vessels containing water under tension, without breaking the water column. It is also difficult to visualize how polymers, produced by pathogens, could be released into functioning vessels in sufficient quantities to affect flow seriously. Experiments in which solutions of toxins of high molecular weight are supplied to cuttings are not very informative because vessels have been opened at the cut ends (e.g., Van Alfen and Turner, 1975). Any suspension will reduce conductivity in such a situation.

Rickettsialike bacteria (RLB) are of particular interest (Hopkins, 1977). For example, in Pierce's disease of the grapevine, they cause symptoms similar to vascular wilt diseases. Leafhoppers seem to transmit RLB into xylem tissue where eventually gums and tyloses can be seen in the vessels, as well as aggregations of RLB. Again it seems unlikely that gums, tyloses, or RLB aggregates are the cause of xylem interruption. The RLB of Pierce's disease are rod-shaped, 0.25 to 0.5 μm in diameter, and 1 to 4 μm long. This is much larger than the pores of pit membranes. The only way an RLB can enter a vessel is through a puncture large enough to admit air for embolism [see Eq. (2)]. The first one is made by the leafhopper, but movement from one vessel to the next must be effected by the RLB itself.

4. Interference with Stomatal Regulation

Guard cells form a turgor-operated valve that opens or closes the stomatal opening. The degree of opening is dependent upon the water potential of the plant as a whole. However, this hydropassive regulation is usually overridden by two active feedback mechanisms, one sensitive to CO_2 and the other sensitive to water potential. The actual turgor adjustment of the guard cells is thought to be regulated by ion pumps.

Much of this is still not fully understood. The hydroactive feedback mechanism seems to operate between mesophyll and the guard cells, with the involvement of hormonal messengers such as abscisic acid (Raschke, 1975). It is possible that certain pathogens release substances that diffuse into the transpiration stream, ascend into the leaves and there interfere with the delicately balanced mechanism of stomatal regulation. The result could be that stomata are either opened or closed. Some wilt toxins cause stomata to open (see Chapter 6 by Talboys, this volume). The resulting excessive transpiration can overtax the capacity of the water supply system, and drop the pressure in the xylem to a level where water columns break; this causes permanent wilting.

There may be toxic substances that have the opposite effect, causing the stomata to close. This seems to be the case in lethal yellowing of coconut palms, where relatively high pressures (i.e., low tensions) are measured in the xylem after infection (McDonough, 1977). Stomatal closure may prevent photosynthesis, but in addition there are other forms of damage.

Interference with the regulation of stomatal opening could also result from air pollutants. If this happens in conifers during the winter months while the ground is frozen, extremely low concentrations of substances not toxic per se could have a damaging effect by causing increased water loss from needles.

B. Disturbance of Phloem Transport

A large percentage of the captured solar energy moves in the form of sugars in the phloem of higher plants. It is therefore not surprising that innumerable organisms, ranging from microscopic forms to aphids and even to large animals like porcupines and beavers, have learned to feed on the sieve tubes or on the phloem as a whole. Plants are quite over-efficient and can tolerate a certain degree of parasitism. However, when we are planting crop plants, we are not only interested in plant survival, but in the highest possible productivity.

Complete interruption of the phloem stops the supply of photosynthetic products (sugars) to areas of growth and storage. The effect of this blockage can be quite variable. For example, leaves whose phloem is blocked may continue to live as long as they are supplied with water and root nutrients via the xylem; they merely cannot export sugars. In the healthy plant, the photosynthetic rate exceeds the rate of phloem export; excess sugars are stored during the day and exported during the night when there is no photosynthesis. However, the storage capacity of leaves is very limited and interruption of phloem export causes a dramatic decrease in photosynthesis (e.g., Geiger, 1976). Root systems of perennial plants are to some degree self-sufficient, because considerable amounts of carbohydrates are usually stored in underground axes; these can therefore survive for some time after the phloem supply is cut off. Thus, apple trees whose stem phloem is interrupted by inverting rings of bark may remain small, may flower and fruit earlier, but do not die or look diseased (Sax, 1954). Interruption of the phloem by a pathogen is much more serious because it is usually accompanied by damaging side effects.

If the phloem is interrupted at any one point, the sugar concentration in the sieve tubes increases between the source and the interruption, and decreases between the sink and the interruption (Zimmermann, 1960). Corresponding effects result from the removal of either sink or source. Sieve tube sugars are, of course, in direct equilibrium with tissue sugar; translocation therefore affects sugar levels in the plant. There is good evidence that resistance to certain diseases is dependent upon tissue sugar levels. One could thus say that, under certain circumstances, translocation affects diseases (Horsfall and Dimond, 1957).

1. Phloem Interruption in the Leaves

Viruses, mycoplasmalike organisms (MLO), and flagellates may cause phloem necrosis, sometimes preceded by tissue hyperplasia (Schneider, 1973). Damage can be primarily in leaves, causing leaves to curl and turn red as, for example, in potato leaf roll (a viral disease), X-disease of wild chokecherry, and pear decline in the Pacific Northwest (the latter are both mycoplasma diseases). Red leaf color, anthocyanin, often results from an increased sugar content due to phloem blockage. The anatomical deformations causing the leaf to curl, however, are not due to phloem blockage. We know from many examples, such as leaf-mining insects (Schneider-Orelli, 1909), that mere interruption of the phloem track does not damage the leaf. It must be remembered that harmful metabolites, leaking into the xylem, move distally from the diseased point and may cause secondary effects. Damage to the phloem interferes with the normal flow of sugars to sinks which may be storage organs, i.e., the product intended for harvest in the case of cultural plants. The potato tubers—or whatever it may be—remain small.

2. Phloem Interruption in the Stem

Interruption of the stem phloem is called girdling and has been used as a method to study the effect of phloem transport interruption. Girdled trees can survive for many years (e.g., Sax, 1954). When trees are killed quickly by girdling, as in the case of canker formation by the chestnut blight, the immediate death is caused not by phloem but by xylem failure.

The particular sensitivity of ring-porous trees to girdling should be evident from what has been said about interruption of the transpiration stream.

The phloem not only transports photosynthates, but also hormonal stimuli effective in growth correlation, the most important of which is probably auxin. Stem girdling therefore releases dormant buds below the point of girdling. This phenomenon is quite distinct in chestnut stems girdled by *Endothia parasitica* cankers. If the water supply is not cut off by girdling, roots may appear above the girdle.

Rhizoctonia solani causes cankers on emerged stems of potato plants. This restricts the downward flow of photosynthates and may cause the formation of aerial tubers in the axils of branches and petioles (Baker, 1970, see also the literature cited therein).

Organs that are pure sinks, such as flowers, shoot tips, or fruits, are often killed or abscised by interruption of the phloem supply lines. The white pine weevil (*Pissodes strobi*) lays its eggs in the bark at the top of the terminal shoot. In the spring the larvae eat their way through the phloem of the leader in a downward direction, while the terminal buds grow out. The damage gets more severe as the larvae grow bigger. When the leader is completely girdled, the young terminal shoots abort.

The neck blast disease of rice is another example of local interruption of the phloem supply of the pure sink. *Pyricularia oryzae* produces a lesion, girdling the stem just below the inflorescence. Instead of moving into the rice kernels, sugars accumulate in the vegetative plant parts below.

3. Disturbance of Transport Regulation

Loading of sieve tubes at sources and unloading at sinks control the sugar concentration in the sieve tubes and thus the pressure gradient and direction of transport. For example, storage organs can act as either sinks or sources; they import and export at different times. On the other hand, mature leaves are normally geared to load only, i.e., sugars are always secreted into, never out of, sieve tubes. Even if mature leaves are placed in the dark, they do not unload and thus do not import sugars. However, if a droplet of a solution of kinetin is applied to a mature leaf, the loading mechanism is reversed, sugars are unloaded, and thus imported into the leaf (Mothes and Engelbrecht, 1961). The area of treatment remains vigorously green and senescence is delayed.

In a series of experiments, Pozsár and Király (1964) established that the rust *Uromyces phaseoli* affects leaf metabolism and phloem regulation the same way cytokinin does: the leaf remains green and vigorous longer and imports photosynthates at the expense of the rest of the plant. The effect begins at the chlorotic stage of rust infection but is more pronounced at the sporulation stage. The authors cite other reports that indicate similar phenomena.

The ability of a parasite to modify transport regulation to its own advantage is the most sophisticated form of parasitism. It is perhaps more common than generally believed. Gall-inducing nematodes seem to inject auxinlike substances into roots which cause hyperplasia of the tissue on which the nematode feeds (Endo, 1972). Certain insects even stimulate plant tissues to form phloem strands that supply the gall, and thereby the larvae, with photosynthate (Meyer, 1969). Similar observations have been made in tumors (Lang, 1965).

VI. TRANSPORT OF HARMFUL AGENTS THROUGHOUT THE PLANT

The pathologist not only has to worry about interruption of the two long-distance transport channels and thus the failure of normal transport processes, he also has to worry about distribution of harmful agents via these channels. Solutes may diffuse or be swept into the xylem with apoplast water and then follow the transpiration stream to mature leaves. On the other hand, solutes must be taken up actively into sieve tubes in order to be transported via phloem to phloem sinks. The pattern of movement within the plant normally indicates whether we are dealing with xylem or with phloem transport (Crafts and Crisp, 1971). Movement toward transpiring leaves indicates xylem transport, while movement in the direction of phloem sinks indicates phloem transport.

A. Movement in the Xylem

Pressure conditions in the xylem at the time of infection are of fundamental importance. Certain herbs have positive xylem pressures during the night (root pressure). In the morning, when transpiration begins, pressures become negative again. A xylem wound exudes sap during the night and sucks air into the injured vessels in the morning so long as the injured vessel is not plugged by gum or tylosis. This is a potential mechanism for xylem infection.

Since bacteria are too large to enter intact vessels, they must enter through a wound. When pressures are negative, bacteria would enter by dissolving a spot of a vessel wall. The resulting perforation sucks in air and could sweep the bacterium into the vessel with the retreating water. According to Eq. (2), a 1 μm-wide break in the vessel wall would require a pressure of -2 atm or lower in the vessel to suck in air from ambient pressure; less tension is required for larger holes. It is important to realize that water retreats to both vessel ends (i.e., up and down in the plant) from the point of injury. At the moment air enters, the water pressure in the injured vessel rises to $+1$ atm; this creates a temporary pressure gradient from the water in the injured vessel ($+1$ atm) to the tensile water in all contiguous intact vessels, until the injured vessel is completely air filled. Such a surging movement may take place within a few seconds during which a bacterium may be carried to either the upper or lower end of the vessel. Two-directional spread has been recorded (see the citations in Nelson and Dickey, 1970). Once sucked into a vessel, probably to a pit membrane near the vessel end, multiplication and egress from the xylem could begin as described by Nelson and Dickey (1970). A new hole must be made by the pathogen before it can enter the next vessel. The rate of spread is therefore limited by (1) vessel length, and (2) the time it takes to dissolve a spot in the pit membrane.

A particularly interesting question is whether viruses can be distributed via the xylem. The problem is entirely one of size. Inert particles can move in the xylem beyond the compartment into which they have been released only if they are smaller than the pores in the pit membranes and the capillary spaces within adjacent parenchyma cell walls. The diameters of the latter have been estimated by Frey-Wyssling (1959) to be of the order of 10 nm. Strugger and Peveling (1961) used metal suspensions to study this problem and found that the capillary diameters of parenchyma cell walls of *Helxine soleirolii* range from about 3 to 23 nm. Schneider and Worley (1959) presented good evidence that southern bean mosaic virus (SBMV) moved across steam girdles (which block the phloem path) via the xylem. The particle size of SBMV is estimated at 30 nm. It is entirely possible that capillary spaces in cell walls of bean are large enough to permit passage of this virus. Distribution via transpiration stream would bring the virus into direct contact with the plasmalemma of leaf parenchyma cells. Schneider and Worley (1959) refer to a local lesion host (*Phaseolus vulgaris* L. var. Pinto) and a systemic host (*P. vulgaris* L. var. Black Valentine) of SBMV. It would be interesting to investigate if the host response of the two varieties depends upon their respective cell wall structure. Larger pathogens, such as that causing Pierce's disease of the grapevine, once regarded as viruses, are now recognized as rickettsialike bacteria.

Another interesting question is how fungal spores can enter intact vessels and thus be distributed via the transpiration stream. Reports about spore movement in the xylem are quite frequent in the pathology literature. However, most investigators used some form of spore injection technique, analogous to the technique of injecting a dye, to demonstrate transport through vessels (in trees: Banfield, 1968; Pomerleau, 1970; Phipps and Stipes, 1976; Beckman *et al.*, 1953; in herbs: Gottlieb, 1944). Few of these reports consider that in nature infection is not the result of a wound bathed in a highly concentrated spore solution. Banfield (1941) recognized this problem while investigating Dutch elm disease. Although he investigated spore movement in vessels by injection from a liquid suspension, he admitted that such a sequence of events was not natural. In an alternative method he wounded the tree with a chisel cut, waited 5 to 10 min, then painted a spore suspension on the cut area with a slightly moistened brush. Pathogen spread was much slower and less predictable in this case. He offered two explanations for pathogen entry in nature. Spores carried externally on bark beetles could be swept along a retreating water column as the beetle burrows into the wood and severs functional vessels. Alternatively, mycelial growth from the wound area into functional vessels could be followed by spore formation and subsequent transport of spores in the sap stream. Phipps and Stipes (1976) reported microconidia in vessels of diseased *Albizia* trees. However, it is difficult to visualize how fungal hyphae could enter functional vessels without embolizing them and produce spores inside a vessel filled with water under tension. This problem should be specifically addressed with the full awareness of the fact that vessel sap is normally under tension and that collection of tissue by the investigator causes considerable surging movements into vessels at the moment the tissue is cut from the plant.

B. Movement in the Phloem

Organisms like viruses, rickettsialike bacteria (RLB), and mycoplasmalike organisms (MLO) are distributed within the plant mostly by the phloem (I. R. Schneider, 1965; Esau, 1967; De Zoeten, 1976; H. Schneider, 1973; Maramorosch, 1976). The question of entry into the sieve tubes is of particular interest. Viruses could use normal paths of entry, as they have been seen in lateral sieve areas and plasmodesmata between phloem parenchyma and sieve elements (Esau and Hoefert, 1971, 1972) or they may be injected with insect saliva. Primary infections then occur in the phloem along the enucleate pathway (Schneider, 1973). Mycoplasmalike organisms have been observed in the sieve pores on electron micrographs, and it is assumed that they move readily in the sieve tubes. In coconut palms infected with lethal yellowing, MLO are possibly injected by phloem-feeding insects with the saliva (slack sieve tube turgor might favor such injection) and accumulate in sieve tubes of phloem sinks such as young, unopened palm inflorescences (Parthasarathy, 1974). The fact that they do not reach the very young, immature sieve tubes seems to indicate that the young sieve plates are not permeable enough.

VII. DISTRIBUTION OF APPLIED CHEMICALS THROUGHOUT THE PLANT

Methods for introducing either spices or poisons into tree stems, to render fruits either pleasantly scented or poisonous, have been described several hundred years ago. As early as 1602, injection of poisons has been described also to show "how the wormes are to be killed, if they be already grown into the tree" (for citations of these early reports, see Roach, 1939). Whenever chemicals are applied to a plant for therapeutic or other reasons, it is important that one does it in the most effective way. This is obviously only possible if one is fully aware of the mechanisms of transport.

Three major points have to be considered: (1) the question of xylem and/or phloem mobility of the applied substance, (2) the mechanism of uptake into xylem or phloem, and (3) the best point of application.

Transport in the xylem is easy to accomplish. The simplest injection consists of wounding the xylem within the solution. The created xylem wound should not be exposed to air, because air will be sucked into the wound and the path will be partly or wholly blocked. Plants vary considerably in their ability to take up liquids into cut and air-exposed xylem. It is commonly known that some flowers can be gathered in the field, put into a vase at home, and still remain fresh, i.e., take up water in spite of embolism at the cut end, while others will wilt unless cut under water.

Two methods can partly or wholly overcome the problem of air blocks. Liquids can be injected into the xylem by positive pressure. This method is frequently used for injecting trees. However, it has considerable shortcomings. Pressure equipment is expensive and all too often much of the liquid is pumped into the nonconducting xylem. This is particularly true for ring-porous trees like elm, where the conducting xylem vessels are located in the most recent growth ring, i.e., only a few millimeters inside the cambium (Fig. 2). This outermost growth ring is easily plugged with the pressure nozzle. Expensive fungicide is then pumped into the nonfunctioning xylem where most of it is lost. A little-used method is vacuum injection. A hole is drilled into the xylem and the embolized vessels are vacuum infiltrated with the liquid. This has to be done only once, momentarily; when the vacuum is released while the wound is covered by the liquid, liquid will continue to be taken up by the xylem.

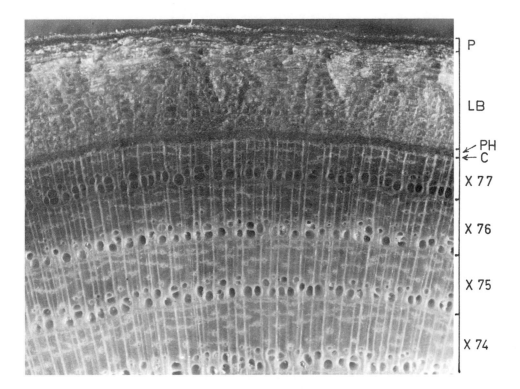

Fig. 2. Surface view of a transverse cut through the stem of a young ash tree (*Fraxinus americana* L.). The freshly cut tree was put into a solution of gentian violet at its basal end. Transpiration moved the solution through the stem, staining the pathway of water. The blue color shows as a darkly shaded area in the most recent (1977) growth ring around the early-wood vessels (X 77). Vessels in older growth rings (X 76, X 75, X 74) are nonfunctioning; in fact, some of them can be seen to be filled with tyloses. It is obvious from this illustration that the water-conducting channels of ring-porous tree species, being located so near the cambium, are very vulnerable. The photograph also shows the cambium (C), the 1977 growth ring of the phloem (PH) containing the conducting sieve tubes, the older living bark (1976 and older), serving storage function (LB), and the periderm (P).

This method is extremely useful with cut branches, but it has not been attempted on standing stems of trees so far as we know.

An interesting aspect of fungicide injection into tree xylem is that infected parts of the xylem in the tree may already be air blocked by the fungus, and fungicide can then not easily reach the infected parts of the tree. Research into this problem would be particularly useful.

Liquid always flows from higher to lower pressure. If the whole plant is under less-than-atmospheric pressure and the injection point is maintained at atmospheric (ambient) pressure, movement will be from the application point to all other points in the plant which are in direct connection via xylem channels. In any specific case such tracks should be studied first by injecting dye solutions. For a description of some patterns the reader is referred to Roach (1939), Vité and Rudinsky (1959), Kozlowski and Winget (1963), and Zimmermann and Brown (1971).

Any injection must be done in such a way that the plant does not suffer excessive damage by the injection wounds. This is particularly important in the case of palms whose stems do not have renewable vascular tissue. This brings us to the last consideration—the best point of application.

Stem injection is the most convenient but not necessarily the best procedure, because each wound damages the cambium which has to repair the stem tissue during the following spring. Ring-porous trees are most vulnerable in this respect, because each wound completely destroys all xylem tracks on the circumference part of the injury. Injection from roots or leaves (in the case of palms), twigs, branches, etc., are often less damaging to the tree, because these organs are more "disposable" than the stem. It is obvious that the experimenter has to strike a balance between desirability and feasibility of a procedure.

In order to get an exogenous substance transported via the phloem, it must be phloem mobile, i.e., it must be actively taken up or at least not be prevented from diffusing into sieve tubes. Crafts and Crisp (1971) list many phloem-mobile chemotherapeutic agents and biocides. Application of phloem-mobile solutions is best in the form of a spray to mature leaves, preferably in combination with a surfactant. Mature leaves are the major phloem sources from which transport takes place to all sinks (Crafts and Crisp, 1971). A phloem-mobile substance may also be injected into the xylem via which it enters mature leaves. It is then ideally situated for transfer into the phloem and redistribution to growing points. Tetracycline antibiotics have been successfully used in this way against MLO in lethal yellowing diseased palms (McCoy, 1976).

Treatment of plant diseases has to meet much more rigorous economic criteria than treatment of human diseases. A coconut palm may be worth $5 to a farmer in Jamaica. He obviously cannot afford a $20 cure for a single tree. High-yield agriculture is based upon monocultures, but monocultures are very vulnerable. The struggle of the plant pathologist is therefore a never-ending one. It is extremely important that agricultural scientists do not restrict their activities to purely applied research. Fundamental research is absolutely essential in the long run. The more we know about how plants function, the better position we will be in to grow them successfully.

References

Bailey, I. W. (1953). Evolution of the tracheary tissue of land plants. *Am. J. Bot.* **40,** 4–8 (reprinted in I. W. Bailey, "Contributions to Plant Anatomy," Chapter 13. Chronica Botanica, Waltham, Massachusetts, 1954).

Baker, K. F. (1970). Types of *Rhizoctonia* diseases and their occurrence. *In* "Rhizoctonia Solani, Biology and Pathology" (J. R. Parmeter, Jr., ed.), pp. 125–148. Univ. of California Press, Berkeley.

Banfield, W. M. (1941). Distribution by the sap stream of spores of three fungi that induce vascular wilt diseases of elm. *J. Agric. Res.* **62,** 637–681.

Banfield, W. M. (1968). Dutch elm disease recurrence and recovery in American elm. *Phytopathol. Z.* **62,** 21–60.

Beckman, C. H., Kuntz, J. E., Riker, A. J., and Berbee, J. G. (1953). Host responses associated with the development of oak wilt. *Phytopathology* **43,** 448–454.

Braun, H. J. (1963). "Die Organisation des Stammes von Bäumen und Sträuchern." Wiss. Verlagsges., Stuttgart.

Canny, M. J. (1973). "Phloem Translocation." Cambridge Univ. Press, London and New York.

Crafts, A. S., and Crisp, C. E. (1971). "Phloem Transport in Plants." Freeman, San Francisco, California.

De Zoeten, G. A. (1976). Cytology of virus infection and virus transport. *In* "Physiological Plant Pathology," Encyclopedia of Plant Physiology, New Series, Vol. 4 (R. Heitefuss and P. H. Williams, eds.), pp. 129–149. Springer-Verlag, Berlin, Heidelberg, New York.

Dimond, A. E. (1970). Biophysics and biochemistry of the vascular wilt syndrome. *Annu. Rev. Phytopathol.* **8,** 301–322.

Dixon, A. F. G. (1975). Aphids and translocation. *In* "Transport in plants. I. Phloem transport," Encyclopedia of Plant Physiology, New Series, Vol. 1 (M. H. Zimmermann and J. A. Milburn, eds.), pp. 154–170. Springer-Verlag, Berlin, Heidelberg, New York.

Endo, B. Y. (1972). Nematode-induced synocytia (giant cells). Host-parasite relationships of Neteroderidae. *In* "Plant Parasitic Nematodes" (B. M. Zuckerman, W. F. Mai, and R. A. Rohde, eds.), Vol. 2, pp. 91–117. Academic Press, New York.

Esau, K. (1965). "Plant Anatomy," 2nd ed. Wiley, New York.

Esau, K. (1967). Anatomy of plant virus infection. *Annu. Rev. Phytopathol.* **5,** 45–76.

Esau, K. (1969). "The Phloem." Borntraeger, Berlin.

Esau, K., and Hoefert, L. L. (1971). Cytology of beet yellows virus infection in *Tetragonia.* II. Vascular elements in infected leaf. *Protoplasma* **72,** 459–476.

Esau, K., and Hoefert, L. L. (1972). Ultrastructure of sugarbeet leaves infected with beet western yellows virus. *J. Ultrastruct. Res.* **40,** 556–571.

Eschrich, W. (1975). Sealing Systems in Phloem. *In* "Transport in plants. I. Phloem transport," Encyclopedia of Plant Physiology, New Series, Vol. 1 (M. H. Zimmermann and J. A. Milburn, eds.), pp. 39–56. Springer-Verlag, Berlin, Heidelberg, New York.

Fahn, A. (1967). "Plant Anatomy." Pergamon, Oxford.

Fensom, D. S. (1975). Other possible (phloem transport) mechanisms. *In* "Transport in plants. I. Phloem transport," Encyclopedia of Plant Physiology, New Series, Vol. 1 (M. H. Zimmermann and J. A. Milburn, eds.), pp. 354–366. Springer-Verlag, Berlin, Heidelberg, New York.

Frey-Wyssling, A. (1959). "Die pflanzliche Zellwand." Springer-Verlag, Berlin and New York.

Geiger, D. R. (1975). Phloem loading. *In* "Transport in plants. I. Phloem transport," Encyclopedia of Plant Physiology, New Series, Vol. 1 (M. H. Zimmermann and J. A. Milburn, eds.), pp. 395–431. Springer-Verlag, Berlin, Heidelberg, New York.

Geiger, D. R. (1976). Effects of translocation and assimilate demand on photosynthesis. *Can. J. Bot* **54**, 2337–2345.

Gottlieb, D. (1944). The mechanism of wilting caused by *Fusarium bulbigenum* var. *lycopersici. Phytopathology* **34**, 41–59.

Hammel, H. T., and Scholander, P. F. (1976). "Osmosis and Tensile Solvent." Springer-Verlag, Berlin and New York.

Höll, W. (1975). Radial transport in rays. *In* "Transport in plants. I. Phloem transport," Encyclopedia of Plant Physiology, New Series, Vol. 1 (M. H. Zimmermann and J. A. Milburn, eds.), pp. 432–450. Springer-Verlag, Berlin, Heidelberg, New York.

Hopkins, D. L. (1977). Diseases caused by leafhopper-borne, rickettsia-like bacteria. *Annu. Rev. Phytopathol.* **17**, 277–294.

Horsfall, J. G., and Dimond, A. E. (1957). Interactions of tissue sugar, growth substances, and disease susceptibility. *Z. Pflanzenkr. (Pflanzenpathologie) Pflanzenschutz* **64**, 415–421.

Huber, B. (1935). Die physiologische Bedeutung der Ring- und Zerstreutporigkeit. (Physiological significance of ring- and diffuse-porousness—Xerox copies of English translation available from National Translation Center, 35 West 33rd Street, Chicago, IL 60616). *Ber. Dtsch. Bot. Ges.* **53**, 711–719.

Huber, B. (1956). Die Gefässleitung. *In* "Encyclopedia of Plant Physiology," Vol. 3 (O. Stocker, ed.), pp. 541–582. Springer-Verlag, Berlin, Göttingen, Heidelberg.

Klein, G. (1923). Zur Aetiologie der Thyllen. *Z. Bot.* **15**, 418–439.

Kozlowski, T. T., and Winget, C. H. (1963). Patterns of water movement in forest trees. *Bot. Gaz. (Chicago)* **124**, 301–311.

Kursanov, A. L. (1963). Metabolism and the transport of organic substances in the phloem. *Adv. Bot. Res.* **1**, 209–278.

Lang, A., ed. (1965). Differentiation and development. Part 2. Modifications of the typical course of development and their bearing for the analysis of normal development in plants. "Encyclopedia of Plant Physiology," Vol. 15. Springer-Verlag, Berlin and New York.

McCoy, R. E. (1976). Uptake, translocation, and persistence of oxytetracycline in coconut palms. *Phytopathology* **66**, 1038–1042.

McDonough, J. (1977). An investigation of xylem pressure in coconut palm with reference to lethal yellowing. Master's Degree Thesis, Harvard Forest, Harvard University.

Maramorosch, K. (1976). Plant mycoplasma diseases. *In* "Physiological Plant Pathology," Encyclopedia of Plant Physiology, New Series, Vol. 4 (R. Heitefuss and P. H. Williams, eds.), pp. 150–171. Springer-Verlag, Berlin, Heidelberg, New York.

Mathre, D. E. (1964). Pathogenicity of *Ceratocystis ips* and *Ceratocystis minor* to *Pinus ponderosa. Contrib. Boyce Thompson Inst.* **22**(7), 363–388.

Meyer, J. (1969). Irrigation vasculaire dans les galles. *Mem. Soc. Bot. Fr.*, pp. 75–97.

Milburn, J. A. (1973). Cavitation studies on whole *Ricinus* plants by acoustic detection. *Planta* 112, 333–342.

Milburn, J. A. (1975). Pressure flow. *In* "Transport in plants. I. Phloem Transport," Encyclopedia of Plant Physiology, New Series, Vol. 1 (M. H. Zimmermann and J. A. Milburn, eds.), pp. 328–353. Springer-Verlag, Berlin, Heidelberg, New York.

Mothes, K., and Engelbrecht, L. (1961). Kinetin-induced directed transport of substances in excised leaves in the dark. *Phytochemistry* 1, 58–62.

Münch, E. (1930). "Die Stoffbewegungen in der Pflanze." Fischer, Jena.

Nelson, P. E., and Dickey, R. S. (1970). Histopathology of plants infected with vascular bacterial pathogens. *Annu. Rev. Phytopathol.* 8, 259–280.

Parthasarathy, M. V. (1974). Mycoplasma-like organisms associated with lethal-yellowing disease of palms. *Phytopathology* 64, 667–674.

Parthasarathy, M. V. (1975). Sieve-element structure. *In* "Transport in plants. I. Phloem transport," Encyclopedia of Plant Physiology, New Series, Vol. 1 (M. H. Zimmermann and J. A. Milburn, eds.), pp. 3–38. Springer-Verlag, Berlin, Heidelberg, New York.

Pate, J. S. (1976). Nutrient mobilization and cycling: Case studies for carbon and nitrogen in organs of a legume. *In* "Transport and Transfer Processes in Plants" (I. F. Wardlaw and J. B. Passioura, eds.), pp. 447–462. Academic Press, New York.

Peel, A. J. (1974). "Transport of Nutrients in Plants." Wiley, New York.

Phipps, P. M., and Stipes, R. J. (1976). Histopathology of mimosa infected with *Fusarium oxysporum* f. sp. *pernicicosum*. *Phytopathology* 66, 839–843.

Pomerleau, R. (1970). Pathological anatomy of the Dutch elm disease. Distribution and development of *Ceratocystis ulmi* in elm tissues. *Can. J. Bot.* 48, 2043–2057.

Pozsár, B. I., and Király, Z. (1964). Cytokinin-like effect of rust infections in the regulation of phloem transport and senescence. *In* "Symposium on Host-Parasite Relations in Plant Pathology" (Z. Király and G. Ubrizsy, eds.), pp. 199–210. Növényvéd Kut. Int., Budapest.

Raschke, K. (1975). Stomatal action. *Annu. Rev. Plant Physiol.* 26, 309–340.

Reichenbach, H. von ("Anonymous"). (1845). Untersuchungen über die zellenartigen Ausfüllungen der Gefässe. *Bot. Z.* 3, 225–231, 241–253.

Roach, W. A. (1939). Plant injection as a physiological method. *Ann. Bot.* (*London*) [N.S.] 3, 115–226.

Rudolph, K. (1976). Forces by which the pathogen attacks the host plant: nonspecific toxins. *In* "Physiological Plant Pathology," Encyclopedia of Plant Physiology, New Series, Vol. 4 (R. Heitefuss and P. H. Williams, eds.), pp. 270–315. Springer-Verlag, Berlin, Heidelberg, New York.

Sax, K. (1954). The control of tree growth by phloem blocks. *J. Arnold Arbor., Harv. Univ.* 35, 251–258.

Schneider, H. (1973). Cytological and histological aberrations in woody plants following infection with viruses, mycoplasmas, rickettsias, and flagellates. *Annu. Rev. Phytopathol.* 11, 119–146.

Schneider, I. R. (1965). Introduction, translocation and distribution of viruses in plants. *Adv. Virus Res.* 11, 163–221.

Schneider, I. R., and Worley, J. F. (1959). Upward and downward transport of infectious particles of southern bean mosaic virus through steamed portions of bean stems. *Virology* 8, 230–242.

Schneider-Orelli, O. (1909). Die Miniergänge von *Lyonetia clercella* L. und die

Stoffwanderung in Apfelblättern. *Zentralbl. Bakteriol., Parasitenkd. Infektionskr., Abt. 2* **24**, 158–181.

Skene, D. S., and Balodis, V. (1968). A study of vessel length in *Eucalyptus obliqua* L'Hérit. *J. Exp. Bot.* **19**, 825–830.

Strugger, S. (1953). Die Erforschung des Wasser- und Nährsalztransportes im Pflanzenkörper mit Hilfe der fluoreszenzmikroskopischen Kinematographie. (Cinematographic investigations of water and mineral transport in plants with fluorescence microscopy—Xerox copies of English translation available from National Translation Center, 35 West 33rd St., Chicago, IL 60616). *Arbeitsgem. Forsch. Landes Nordrhein-Westfalen, Nat.-, Ing.- Gesellschaftswiss. [Veroeff.]* **21a.**

Strugger, S., and Peveling, E. (1961). Die elektronenmikroskopische Analyse der extrafaszikulären Komponente des Transpirationsstromes mit Hilfe von Edelmetallsuspensoiden adäquater Dispersität. *Ber. Dtsch. Bot. Ges.* **74**, 300–304.

Van Alfen, N. K., and Turner, N. C. (1975). Influence of a *Ceratocystis ulmi* toxin on water relations of elm (*Ulmus americana*). *Plant Physiol.* **55**, 312–316.

Vité, J. P., and Rudinsky, J. A. (1959). The water-conducting system in conifers and their importance to the distribution of trunk-injected chemicals. *Contrib. Boyce Thompson Inst.* **20**, 27–38.

Ziegler, H. (1975). Nature of transported substances. *In* "Transport in plants. I. Phloem transport," Encyclopedia of Plant Physiology, New Series, Vol. 1 (M. H. Zimmermann and J. A. Milburn, eds.), pp. 59–136. Springer-Verlag, Berlin, Heidelberg, New York.

Zimmermann, M. H. (1960). Longitudinal and tangential movement within the sieve-tube system of white ash (*Fraxinus americana* L.). *Beih. Z. Schweiz. Forstver.* **30**, 289–300.

Zimmermann, M. H. (1971). Dicotyledonous wood structure made apparent by sequential sections (Film E 1735) (Film data and summary available as a reprint). Institut für den wissenschaftlichen Film, Nonnenstieg 72, 34 Göttingen, Germany.

Zimmermann, M. H. (1976). The study of vascular patterns in higher plants. *In* "Transport and Transfer Processes in Plants" (I. F. Wardlow and J. B. Passioura, eds.), pp. 221–235. Academic Press, New York.

Zimmermann, M. H. (1978). Structural requirements for optimal water conduction in tree stems. *In* "Tropical Trees as Living Systems" (P. B. Tomlinson and M. H. Zimmermann, eds.), pp. 517–532. Cambridge Univ. Press, London and New York.

Zimmermann, M. H., and Brown, C. L. (1971). "Trees: Structure and Function." Springer-Verlag, Berlin and New York (3rd printing, 1977).

Zimmermann, M. H., and Milburn, J. A., eds. (1975). "Transport in Plants. I. Phloem Transport," Encycl. Plant Physiol., New Ser., Vol. 1. Springer-Verlag, Berlin and New York.

Chapter 6

Dysfunction of the Water System

P. W. TALBOYS

I. INTRODUCTION

In the early 1950's a young and inexperienced research worker studying the role of fungal toxins in wilt diseases thought it might be instructive to examine the effects of introducing simple toxic substances such as sodium chlorate or copper sulfate into the transpiration stream of tomato shoots. He was conditioned to the belief that the osmotic properties of the leaf cells were responsible for exerting the tension required to draw water upward through the xylem. Imagine how disconcerted he was to find that, although the poisoned shoots became water soaked and collapsed, they continued for several days to take up as much water as untreated shoots. His dissatisfied conclusion, never published, was that the dead tissues must have been behaving like a "wick" of filter paper. How near he was to a correct interpretation has since become evident, if we accept as valid the concept that water movement is a function of the apoplast, not of the symplast.

141

That young worker of the 1950's now finds himself writing a chapter on how pathogens induce dysfunction in the water system. Because of the nature of this review, I shall discuss specific diseases only as examples of general principles; Duniway (1976) and Ayres (1978) have dealt in detail with a range of diseases. I am grateful to them and to earlier reviewers including Beckman (1964), Buddenhagen and Kelman (1964), Talboys (1968), and Dimond (1970).

In this chapter, I shall discuss the effects of disease on the functional rather than the structural aspects of the water system.

II. SOME GENERAL CONSIDERATIONS

Water must be extracted from the soil and lifted to the aerial parts of the plant, the amount of water transported being determined by the balance between the lifting force and the numerous resistances in the system. Pathogens can alter either one or both.

A. The Lifting Force

Many generations of students of plant physiology have been taught that osmotic forces in the leaf cells provide the lifting force. It is now clear that this is not so and that the motive force is the water potential in the leaf.

Any process or state leading to the movement of water can be related to a difference between the free energy attributable to water as it exists in the plant and that of pure water at the same temperature; the difference per unit molar volume of water is termed the water potential (ψ), and is expressed as energy per unit volume (erg cm^{-3}) or force per unit area (dyne cm^{-2}, or bar). The free energy of pure water is conventionally taken as zero, and may be modified by the presence of solutes (solute potential or osmotic potential), by imposed negative or positive hydraulic pressures (pressure potential), and by the action of molecular forces at solid–fluid interfaces (matric potential). Water moves "down" gradients of free energy and, because water potential represents a difference in free energy, it indicates the "slope" of the gradient. A detailed analysis of the thermodynamics of water movement in soils and plants is given by Slatyer (1967), and a brief account by Dainty (1969). Wardlaw and Passioura (1976) report more recent studies of water movement.

Water evaporated from the leaf by solar energy moves outward into the ambient air and loses free energy, generating a negative water potential toward which water flows. This provides the force to lift the water through the xylem from the roots and soil. The lifting force on occasion may be supplemented by "root pressure" that arises from a solute potential.

The plant has a protective feedback system: if the outflow of water vapor becomes excessive, as in drought, the stomata close and reduce the loss.

B. Resistance in the System

Water moves along two routes: (1) through the plasmalemma, cytoplasm and vacuoles of the living cells, the symplast; or (2) through the cell walls, the intercellular spaces, and the lumina of dead cells, which together constitute the apoplast. Each of these routes offers resistance to water flow in varying degrees.

It is convenient to consider the resistances in the water system as analogs of the resistances in an electrical circuit, as shown in Fig. 1. The movement through the apoplast of the root and leaf is generally faster than through the symplast by a factor of ca. $\times 50$. However, deposits of lignin or suberin in the intermicellar spaces of a cellulose wall greatly reduce its permeability. Thus, in the endodermis of the root the suberinized casparian strip increases the resistance of the apoplast so much that symplasmic movement becomes dominant. The endodermis as shown in Fig. 1 is therefore the site of greatest resistance to flow in healthy plants.

III. THE EFFECT OF DISEASE ON THE LIFTING FORCE

Pathogens may reduce the lifting force by their physical presence on or within the leaf or by their effect on the operation of the stomata. Since the water potential is determined by the difference in vapor pressure between the cell surface and the ambient air, pathogens can affect it only by influencing the vapor phase outside the cells.

Normal transpiration mainly entails the evaporation of water from mesophyll cells into the air spaces within the leaf, and diffusion of water vapor through the stomata to the external atmosphere. In many downy mildew and rust diseases the pathogen causes little obvious damage to host cells, even when haustoria are formed within them, but the air spaces within the leaf may be completely occluded by the mycelium of the pathogen. The effect of this on water relations does not seem to have been studied, but it can hardly be insignificant, and presumably must

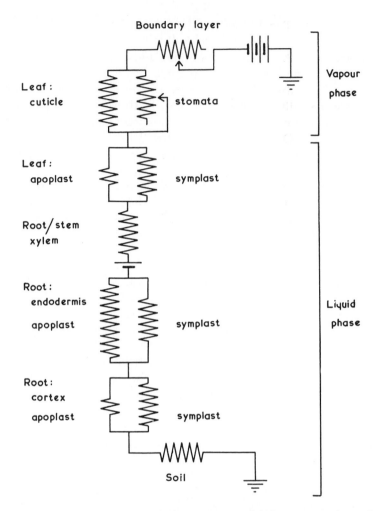

Fig. 1. The water system of a healthy plant modeled in terms of an electrical circuit of "lifting" forces and resistances in series and in parallel.

reduce transpiration via the stomata. Cuticular transpiration may not be directly affected. It is not clear whether movement of water into the mycelium takes place via the haustoria or by direct transfer through host and pathogen cell walls, or by both pathways. Further resistance to the dispersal of water vapor occurs when fungi enter the leaf by growth of germ tubes through stomata, or when structures such as sporangiophores are extended outward through them, although some water is probably lost via sporangiophores.

A. The Effect of Disease on Stomata

The apertures of the stomata determine resistance to the diffusion of water vapor from the substomatal air space into the boundary layer and then into the bulk atmosphere. The size of the aperture is determined by the relative turgidity of the guard cells and the surrounding epidermal cells. In the healthy plant the guard cells respond to the water potential in the leaf, to the CO_2 concentration in the tissues, and to the presence of increased levels of abscisic acid, which may mediate the response to water stress. A pathogen which directly or indirectly affects the water content of the tissue, or the rates of photosynthesis or respiration, or the hormonal balance of the plant is likely to have an effect on the behavior of the stomata. Most pathogens have at least one of these effects, and some change in the functioning of the water system is almost inevitable in most diseases.

The onset of water deficits in the plant, whether caused by disease or by low soil water potentials, often induces cyclic changes in stomatal aperture. A low leaf water potential causes closure of stomata. Continued uptake increases the water potential toward zero, and stomata open. Further water loss causes stomata to close again. However, reopening of stomata becomes delayed progressively after a number of such cycles, and eventually they remain permanently closed and remain so even when water becomes freely available. Accumulation of abscisic acid appears to be the proximal cause of prolonged stomatal closure, which is apparently associated with the condition known as "drought-hardening," and probably accounts for the fact that leaves detached from plants showing incipient symptoms of vascular diseases often wilt more slowly than leaves from healthy plants.

In an earlier review on water stress in vascular diseases, Talboys (1968) noted that almost no work has been done on the behavior of stomata during pathogenesis. Since that time the omission has been rectified in work on a number of diseases and these studies have been reviewed by Ayres (1978), although in many cases the stomatal behavior has been implied from measured changes in diffusional resistance, rather than by direct observation and measurement of stomatal apertures. Although the lifting force is independent of the solute potentials of the leaf cells, factors influencing their turgidity can have indirect effects on the lifting force, e.g., by modifying the K^+ ion relations between stomatal guard cells and the neighboring epidermal cells; by changing the evaporative surface area in the leaf; and by altering its orientation in relation to incident radiation and air movement. Some fungal toxins such as fusaric acid and helminthosporoside increase membrane permeability and cause leakage of water, electrolytes, and other solutes, resulting in loss of turgor and death of cells in the absence of acute water stress. These toxins do not reduce the lifting force, but they may incidentally cause a failure of the mechanism which controls it, viz., the functioning of the stomata.

Toxins implicated in vascular diseases have often been thought to cause death of cells by desiccation, but the reverse is probably more often true: the toxin kills the cells and desiccation is the consequence. One fungal toxin, fusicoccin, has an unusual effect on stomata. It reverses the closing effect of abscisic acid, causes permanent opening of the stomata, induces uncontrolled water loss, and thus induces the onset of water stress symptoms at low water potentials in the leaf by depriving the leaf of its feedback mechanism.

B. The Effect of Disease on the Cuticle

The factor most obviously affecting cuticular transpiration in leaf diseases is disruption of the cuticle by the growth of a pathogen beneath it. This commonly occurs at the completion of spore formation, and may result in the loss of this relatively impermeable layer over substantial areas of leaf surface. The consequence is an uncontrolled loss of water and localized or widespread death of leaf tissues, depending on the extent of the lesions. This form of damage is exemplified by the action of the subcuticular pathogens, *Diplocarpon rosae* and *Venturia inaequalis,* and by many rust pathogens.

C. The Effect of Disease on the Boundary Layer

Loss of water from a leaf, both through the cuticle and through stomata, is significantly affected by the thickness of the stationary boundary layer of air in contact with the surface. This is determined by the size and configuration of the leaf and by the wind speed. A fungus such as a powdery mildew growing on the leaf surface may increase the thickness of the boundary layer and so reduce transpiration, especially at low air speeds. Increased rugosity of the leaf surface as a result of hypertrophy or hyperplasia may have a similar effect, but when there is a marked reduction in the size of leaves, as in some of the vascular diseases, there may be a reduction in the boundary layer and an increased rate of water loss. However, dwarfing of plants tends to lessen the effect of air movement, because wind speed is markedly reduced near ground level.

IV. THE EFFECT OF DISEASE ON RESISTANCE IN THE PATHWAYS

The domestic water system has its built-in resistances, as valves and constrictions modify the rate of flow. Rust and dirt in the pipes are the analogs of pathogens. They constrict the lumen and reduce the flow. Similarly, water pathways in plants are subject to increases in resistance that result from reductions in the size of channels, ranging from the large vessels in the main axis to those approaching molecular dimensions in the cellulose walls. The resistances will be discussed in sequence along the route from the absorbing surfaces of the root to the evaporative surfaces of the leaves.

A. Root Epidermis and Cortex

The surfaces of roots in the soil are continually in contact with populations of microorganisms ranging from the apparently innocuous to the highly pathogenic. Roots of plants grown in sterile soil tend to be longer and less frequently branched, but probably the most significant difference is that they remain white for a much longer period than in an untreated soil. Darkening of the roots is associated with degeneration and death of the epidermis and cortex, suberinization (i.e., deposition of suberin) of the endodermis, and suberization (i.e., cork formation) in the pericycle. These are characteristics of the "aging" root, and the processes of aging are apparently advanced by the activities of the soil microbiota. All of them are likely to increase resistance to the uptake of water, by changes in cell wall permeability and by loss of contact with soil particles when root hairs collapse. These changes do not necessarily result from infection of the tissues by microorganisms. Indeed, in the absence of root pathogens there is little invasion of epidermal and cortical cells until they are moribund or dead.

The aging effect of rhizosphere microorganisms on roots is probably brought about by many different means, but production of toxic substances, including perhaps ethylene, is probably one of their modes of action. Their impact on water relations must be to shorten the duration of active uptake over a given root zone, because of advanced suberinization of the endodermis (see below) and formation of periderm. Nevertheless, this effect is a condition to which plants are fully adapted. Continual root growth provides the means both for constantly maintaining a quantity of living epidermis and cortex capable of taking up water and nutrients despite continuous microbiological activity, and for

placing those tissues in unexploited parts of the soil mass. These requirements are met by pushing lateral roots out through periderm and de-
generate cortex layers, as well as by continuous growth of the primary
roots. When the water content of a soil is less than the field capacity
there is little movement of water. Once·water has been absorbed from
the zone close to the root there can be no further uptake from that zone
until the soil has been rewetted. In such circumstances there is not necessarily any advantage in having persistent living peripheral tissues acting
as sinks for metabolites without being able to contribute to the water
and nutrient economy of the plant.

In contrast to the majority of soil organisms the soil-borne plant pathogens are adapted to life within the plant tissues, but their effects on the
tissues vary with the type of disease they cause. They often have low
competitive saprophytic ability, and are adapted in varying degrees to
avoid competition by "taking refuge" in host tissues through possessing
the ability to penetrate into, or between, living cells, although some gain
entry via wounds or moribund cells. Mycorrhizal fungi are adapted to a
dual existence within and outside the host tissues. The infection courts
for many parasites are in the zones of growth and water uptake in the
root; their effects on the water relations of the plant may arise directly
from their effects on root tissues or indirectly from other effects on the
plant.

Pathogens causing root rots are commonly characterized by the secretion of large amounts of lytic enzymes (including polygalacturonases,
pectin transeliminases, pectinesterases, cellulases, etc.) which cause loss
of coherence of the tissues, breakdown of cell walls, and death of protoplasts. Their action is rapid, the enzymes sometimes being effective in
advance of the pathogen, and death of protoplasts and water soaking of
the tissues takes place before any defensive response can occur. Disintegration of root tissues by these organisms is likely to increase resistance
to water uptake by "breaking the circuit," i.e., by removing part of the
conducting structure. Nevertheless even extensively rotted roots may
function for a limited period in very wet soil, as long as the disintegrating
tissues are sufficiently coherent to act as a "wick" and can remain continuously wet. Death of the aerial parts may result from the transport of
toxic substances from the diseased tissues, or from ionic imbalance caused
by the failure of selective absorption, rather than simply from lack of
water.

Because water moves mainly in the cell walls of the epidermis and
cortex of the root, it might be supposed that the death of the protoplasts
without disintegration of the walls would have little effect on water uptake. Indeed, it has been shown that brief immersion of a root system

in boiling water to kill the protoplasts does not immediately affect water uptake and transpiration. However, there are probably several reasons why death of the cells in these tissues adversely affects water uptake. (1) Death results in loss of turgor in the absorbing tissues, including the root hairs, thus breaking the intimate contact with soil particles and the water films associated with them. (2) Extension of roots into nonexploited parts of the soil mass cannot occur. (3) Death of the protoplasts results in the breakdown of membrane "compartments" and the mixing of enzymes and substrates, causing (*inter alia*) the "browning reaction" associated with the oxidation of phenolic compounds. Impregnation of cell walls by polymerized oxidation products may reduce the permeability of the walls in a similar manner to the deposition of lignin in them.

In other types of soil organisms their competitive ability is so low that their survival depends on gaining entry to the living tissues without rendering them susceptible to infection by wound parasites or those invading moribund cells. They are less destructive in their mode of action, and they stimulate certain physiological activities of the host cells rather than suppress them. The most obvious visible manifestation of such stimulation is in the thickening of cell walls, the deposition of callose or cell wall material around penetrating hyphae to form structures variously termed infection pegs, lignitubers, or callosities; and the formation of lignin in the original walls and in the new deposits. The cytoplasm tends to become granular and the nucleus and nucleolus become more prominent. A high proportion of incipient penetrations do not advance beyond this stage, and relatively few cells are colonized and killed. There is no evidence on whether the changes induced in the cells have any significant effect on the movement of water into the root. Thickening of cell walls could be expected to increase the free space available for water transport in the apoplast, but the deposition of lignin or related substances would decrease the permeability of the walls. However, in most circumstances these effects probably occur in such a small proportion of the root tissues that any influence on water uptake is insignificant. Nevertheless, with heavy inoculation under experimental conditions it is possible to induce cell wall modification, including the formation of lignitubers, over a large proportion of the root epidermis of suitable hosts. Such material could be used to investigate the effects of host responses on water movement.

Research on the effects of mycorrhizal fungi has largely been concentrated on their role in the uptake of ions, particularly of phosphate, but their intimate association with the roots of their hosts and the morphological changes which some types induce in the root make it probable that they have an effect on water uptake. Indeed the author's (unpublished) observations on an incidental vesicular–arbuscular mycorrhiza in a hop plant suggests that the endodermis remains in a primary state for considerably longer than in noninfected roots growing under the same conditions, and somewhat resembles that in roots growing in a sterile medium.

B. The Endodermis

The endodermis influences water uptake in three ways. (1) The presence of suberin in the casparian strip drastically limits movement along the apoplasmic pathway. (2) Movement along the symplasmic pathway depends on the permeability of the protoplast and hence on its metabolic condition. (3) After the cells have become lined with suberin during the formation of secondary endodermis, water transport through the tissue is largely eliminated. The length of root available for water uptake depends on the relative rates of root extension and suberinization. Factors which favor root growth and delay suberinization enable large amounts of root to be available for water uptake, whereas slow growth and rapid suberinization give a restricted capacity for uptake. Both environmental and genetic factors can determine the rates of the two processes, and the presence of pathogens is an important element of the environment.

Damage to the cortex, whether caused by pathogens, by the soil fauna, or by physical factors, generally causes premature suberinization of the endodermal cell walls, and this effectively excludes water and solutes from the vascular cylinder in the region of the injury. It also excludes many pathogens, and is probably one of a number of general resistance mechanisms. The cortex probably plays a significant role in delaying the pathogen and enabling the endodermal cells to respond. If a pathogen reaches the primary endodermis before suberinization has occurred there may be a lignification reaction, as in the cortex, but when rapid invasion results in death of the protoplast before any response has occurred there is presumably a breakdown of the internal membrane system, increased permeability, leakage of solutes, a free movement of water, and the nonselective passage of solutes through to the vascular cylinder. Nevertheless, the death of endodermal cells is frequently followed by a strong browning reaction associated with its high content of phenols, quinones, and oxidases. It is possible that impregnation of the cell walls with polymers of the oxidized phenolics may reduce their permeability to water.

The length of root available for water uptake can also be limited by pathogens which kill or inhibit the root tip meristem; in this case the completely suberinized endodermis will "catch up" with the root tip.

The significance of these effects of pathogens on water movement in the plant does not appear to have been assessed. Clearly, if we regard a root system as a large series of resistances in parallel, dysfunction of a single root will have little effect on the total resistance of the system. Furthermore, the whole organization of a root system is based on continuous replacement of absorptive surfaces. Nevertheless, a substantial reduction in water uptake could be a significant factor in pathogenesis when there is a high inoculum potential of a soil-borne pathogen in the soil, resulting in a number of infection sites on most roots. However, because such infections often progress either to a destructive rotting of the tissues or to a vascular disease, the effects of the superficial infections tend to be obscured. They may, perhaps, be a factor in various ill-defined "soil-sickness" conditions, or in "low grade" infections, e.g., by *Thielaviopsis basicola*.

C. The Xylem

Dysfunction of the xylem is characteristic of vascular wilt diseases in which the pathogen invades the xylem almost exclusively, either by penetration of the roots, as in Fusarium and Verticillium wilts, or by direct "inoculation" by insects, as in Dutch elm disease (*Ceratocystis ulmi*) and oak wilt (*C. fagacearum*). The bacterial wilt pathogen, *Pseudomonas solanacearum*, infects the xylem through wounds. A number of wood pathogens, e.g., *Stereum purpureum*, also infect via wounds and cause vascular dysfunction. The vascular pathogens have been the subjects of extensive and intensive studies on dysfunction of the water system because many of them cause wilting of the leaves, although this is not invariably so. They have also been the subject of much controversy on the mechanisms of pathogenesis, but improvements in techniques for the measurement of the parameters of water function in plants are helping to resolve a number of the problems (Duniway, 1976; Ayres, 1978), especially in assessing the relative importance of occlusion and toxic action in symptom development. Nevertheless a demonstration that dysfunction is a result of increased resistance or occlusion in the water system does not necessarily explain how it occurs, and it is this aspect of the problem which will be discussed here.

Resistance to mass flow of water at a constant pressure through the lumina of dead xylem elements is determined by the viscosity of the fluid and by the amount of frictional drag. The latter depends on the relation between the volume of the vessel and the area of solid–liquid interfaces within it. The radius determines this relation in a noninfected vessel, but other structures that may be present in the vessel during pathogenesis can modify the relation, having the same effect as a reduction in the radius of the vessel. Such obstructions include the mycelium of fungal pathogens and the cells of bacteria, accumulations of material resulting from partial breakdown of host tissues, or secreted by the pathogen or the host, and structures formed by renewed growth of living cells in the xylem.

Measurements have shown conclusively that, in a number of vascular diseases, the resistance to flow through infected xylem is substantially greater than that through noninfected wood. A laboratory model having the same flow characteristics as a xylem vessel showed a measurable increase in resistance when fitted with "hyphae" of appropriate dimensions. However, the reduction in flow was inadequate to account for that found in the xylem of a *Verticillium*-infected tomato plant, even though many of the vessels did not contain hyphae, and many of those that were infected contained fewer hyphae than the model. This disproportionately large reduction in flow in relation to the intensity of infection has also been noted in other vascular diseases, and it is generally assumed that the presence of hyphae per se is not sufficient to cause significant vascular dysfunction. Nevertheless, very dense masses of mycelium sometimes occur, especially in the lower parts of stems and in the main roots of plants infected with soil-borne pathogens, and their role should not be entirely disregarded. Furthermore, many fungi and bacteria yield mucilaginous polysaccharides. These may remain associated with the walls of the organism, so that the hyphae or cells with their associated mucilage form a gel-like or viscous mass, or they may become separated from the walls to form a suspension or solution in the medium. The viscous bacterial masses occurring in the vessels in bacterial wilt diseases almost certainly contribute to vascular dysfunction. It is not clear whether *Verticillium dahliae*, for example, forms *in vivo* the same gelatinous masses that it does in liquid media *in vitro*. If it does, such masses must surely occlude many vessels. The presence of such gels may account for the progressive reduction in resistance that has been reported in experimental systems when water is pumped through *Verticillium*-infected tomato stems. Possibly the polysaccharide is dissolved or physically dislodged by the water flow, or undergoes thixotropic changes. Other reports that xylem sap extracted from infected tomato plants is not significantly more viscous than that from healthy plants may indicate that the extracts from diseased plants represent mostly the content of the noninfected vessels, because of their relatively low resistance, and the fact that the flow rate is inversely proportional to the resistance.

Polysaccharides secreted by vascular pathogens have more often been regarded as "wilt toxins," dissolved or suspended in the xylem sap and exerting their action at a distance from the cells of the pathogen. If such substances are significant in pathogenesis, it is probable that they act as occlusive agents rather than as cell toxins. Their effects appear to be mainly determined by their molecular weights. Large molecules, presumably in a colloidal state, tend to cause occlusion near the point at which they originate in the vessel. Small molecules can be transported to the smaller vessels or tracheids before obstructing them, or they may even lodge ultimately in the intermicellar pores in the nonlignified areas of the terminal tracheids in the leaves. Whatever their ultimate location, their effect is essentially to increase the resistance of the water system in the infected plant. They may, nevertheless, have a synergistic action with substances that are toxic to protoplasts, giving rise to characteristic symptom patterns in the hosts (Talboys, 1968; Cronshaw and Pegg, 1976).

Many fungi and bacteria which can be grown *in vitro,* both pathogens and nonpathogens, secrete extracellular enzymes, some of which degrade the pectic and cellulosic components of cell walls. Several different roles have been proposed for these enzymes in vascular wilt pathogenesis, although studies on the cotton wilt pathogen (*V. dahliae*) have shown that mutants unable to produce various pectic enzymes have an unimpaired pathogenicity. Plugging of vessels by material composed of, or containing, pectic substances has been reported in some vascular diseases and attributed to the accumulation of the products of partial degradation of cell walls. However, there is usually little evidence of tissue breakdown at an early stage in vascular diseases, at the time when symptoms are first appearing, and there seems to be little loss of strength in woody hosts infected by vascular pathogens. Swelling and delamination of lignified cell walls occur in some herbaceous hosts infected with wilt pathogens, but mostly at a late stage of the disease.

Obstructions of vessels can also result from responses of the host to the presence of the pathogen within the vascular system. The mechanism of these responses is not fully known, but in general they correspond with the wound reaction of the host. One postulate is that both wounding and the action of a pathogen cause the formation of ethylene by damaged cells, and that the ethylene triggers the wound response. Another proposal is that wounding and infection both result in "embolism," i.e., entry of air into vessels, and that this provides the stimulus. Whatever the mechanism, the response is either the production of gums or the formation of tyloses. One type of response is usually characteristic of the host; if both occur, one predominates. Both result from the activity of living xylem parenchyma and ray cells adjacent to the vessels.

Gums vary considerably in their composition, but are mainly poly-saccharides based on glucuronic acid with associated hexose and pentose sugars. However, they usually also contain phenolic substances, the oxidation of which causes the red-brown coloring that is often a feature of occlusive gums. The release of the phenols may be associated with the death of the reacting cells, which also show the browning reaction, although in many vascular diseases they are not infected by the pathogen either before or after the response. There appears to have been no suggestion that this is a hypersensitive response, linked perhaps with phyto-alexin formation; such responses are usually associated with some specificity in host–parasite relations.

The gum is extruded through the pits connecting the parenchyma cell with the vessel, giving granular or amorphous masses of various forms, either lining the vessel and reducing its radius, or forming complete occlusive plugs.

Tyloses similarly develop from parenchyma cells via the pits, but in this case the pit membrane is distended to form a balloonlike intrusion into the vessel. Considerable wall synthesis occurs, because a tylose has a well-developed cell wall which may later show reticulate thickening and lignification. The nucleus generally migrates through into the tylose and starch grains may accumulate in it. Discoloration with oxidized phenolics generally occurs in the vessel walls, but the reacting paren-chyma cells appear to maintain continuity with the tyloses and remain alive. Neither parenchyma cells nor tyloses are infected until the whole plant becomes moribund. Tyloses are frequently so large and so numerous that they become packed together to form a tissuelike mass completely occluding the vessels. Any transport of water along the walls of such "tissues" must be insignificant compared with the mass flow through an unobstructed vessel.

The gross effects of gummosis and tylosis on the plant depend on the frequency and distribution of the obstructions, and this relates in turn to the incidence and severity of infection in the vascular system. A limited distribution may have relatively little effect, the occluded parts of the system being bypassed and compensated by a higher rate of flow in the unobstructed vessels.

The effect of the location of an obstruction was demonstrated by Dimond (1966) and also by Duniway (1973) who showed that stem infections in tomato resulted in relatively small changes in water potential, whereas in the petiole, where there were no alternative pathways, infection could give rise to leaf water potentials low enough to cause wilting and death. Massive obstruction of most of the vessels in the stem generally leads to death of the whole plant. Nevertheless, mechanisms compensating for extensive occlusion occur in some plants and will be discussed later (see Section V).

Occlusion of vessels has a second important consequence: as well as limiting the movement of water it also tends to limit the movement of the pathogen through the plant, and may therefore contribute to disease resistance as well as to pathogenesis in vascular diseases.

Obstruction of the water pathway in the xylem by gums and tyloses is not only caused by pathogens within the xylem. It has occasionally been seen to occur in young hop roots infected by *Verticillium albo-atrum* before the fungus has gained entry to the xylem, thus limiting the colonization of the system at a very early stage. It can also arise from lesions in the cortex of the root and stem, such as may be caused by mechanical damage and by various "canker" diseases, such as apple canker (*Nectria galligena*) and basal canker (*Fusarium sambucinum*) of the hop.

The extent of occlusion by a small canker lesion may be insufficient to cause symptoms of dysfunction in the water system, but large lesions may cause wilting and collapse, as well as disturbances of the downward movement of water and metabolites in the phloem.

D. The Mesophyll and Epidermis of the Leaf

The leaf is the principal location of symptom development in a wide range of diseases, including those which are not caused directly by foliar infection. In some diseases the symptoms are caused by dysfunction of the water system, either in the leaf itself or in the stem or root. In others, dysfunction of the water system is a direct consequence of the action of the pathogen in the leaf, either on the resistance to the movement of liquid water in the tissue, or on the driving force, i.e., the movement of water in the vapor phase (see Section II).

The range of symptoms expressed by a leaf is limited, and a given symptom may have more than one cause. The most frequent symptoms are epinasty, wilting, chlorosis, necrosis, and abscission. Hypertrophy, hyperplasia, or both can also occur. Diseases have characteristic patterns of distribution of these symptoms, both on the plant and on the individual leaf, and characteristic sequences through time.

Epinasty is characteristic of certain vascular diseases, e.g., verticillial and fusarial wilts of tomato. Nevertheless, it can also occur above stem lesions in the tomato, e.g., those caused by *Botrytis cinerea*. Although in some hosts it causes an appearance superficially resembling wilting, it seems to be a consequence of differential changes in cell volume in the upper and lower surfaces of the petiole and lamina and is associated with high turgidity. It is not clear whether such changes arise from modification of the solute potentials of cells, or of the elasticity of the cell walls, or both.

Epinasty can be induced by ethylene. Some vascular pathogens can produce ethylene *in vitro,* but others appear to be unable to do so, even though they induce epinasty, and it is not certain whether, in vascular diseases, this symptom occurs as a consequence of tissue damage and ethylene formation caused directly by the pathogen, or as an effect of reduced water flow, or by some other mechanism. Nevertheless, its expression entails changes in water relations.

Vascular pathogens typically cause wilting, chlorosis, necrosis, and abscission of leaves, usually in that order, and commonly in sequence starting with the oldest leaves. Wilting is often initially reversible, and if an affected leaf is removed from the plant and its petiole placed in water, the leaf will recover its·turgidity and will remain turgid until it becomes senescent. Such wilting is attributable to occlusion of the xylem at a point below that at which the leaf trace enters the petiole. When blockage occurs throughout the petiole, the wilting that occurs is not reversible. If obstruction is limited to the ultimate elements of the xylem, e.g., by relatively small polysaccharide molecules accumulating in the terminal tracheids or occluding the intermicellar pores, thus preventing movement to the bundle sheath cells, the leaf may not wilt as a whole. Because most of the water in the transpiration stream evaporates from the mesophyll cell walls, extraction of water from the xylem is greatest in those elements directly associated with the mesophyll (i.e., the terminal tracheids), so any occluding substance with a small molecular weight tends to accumulate there, rather than in the larger vascular bundles constituting the main distributive network in the leaf. Water is therefore excluded first from the mesophyll nearest to the terminal elements and farthest from the main bundles, i.e., in the interveinal areas. Retention of water along the main and subsidiary veins is probably assisted by the absence of stomata in the overlying epidermis. Thus turgidity is maintained in the main skeletal framework of the leaf and the veins remain green, while progressive desiccation caused by continued evaporation in the interveinal areas results in chlorosis and necrosis. This sequence of events could give rise to the "tiger-stripe" symptom seen, for example, in the verticillial wilts of hops and raspberries.

An alternative explanation of the tiger-stripe symptom is that toxic substances formed either by the pathogen or as a host response to infection are transported in the transpiration stream and reach high enough concentrations in the region of the terminal elements to kill the adjacent bundle sheath cells, mesophyll cells, or both. In this case any dysfunction of the water system is a consequence of cell death, not a cause. Indeed, cell death may not always lead immediately to cessation of water movement through the apoplast, especially if drying out of the epidermal cells results in shrinkage of guard cells and permanent opening of the stomata. However, the effects of disruption of the evaporative system caused by loss of turgor are probably intensified by changes in cell wall permeability arising from disorganization of the protoplasts and the consequent oxidation and polymerization of the phenols and quinones present in the bundle sheath and elsewhere. The cell walls of necrotic leaf tissues invariably show staining reactions suggestive of oxidized phenolic material rather than the normal cellulose staining response of healthy tissues.

Observations on the development of tiger-striping caused by V. albo-atrum in the hop leaf suggest that the primary dysfunction is in the bundle sheath cells surrounding the terminal tracheids, and that transfer of water to the mesophyll cells is thus prevented. The sheath cells show a strong browning reaction, but it is not clear whether this is an effect of water stress or of toxic substances, or of both mechanisms operating together or in sequence. The probability that the symptoms of vascular diseases result from a combination of toxic and occlusive mechanisms has been discussed before (Talboys, 1968). The author has since found that although introducing the ethylene-generating compound 2-chloro-ethylphosphonic acid into the hop vascular system at low concentrations (at pH 6.0) caused generalized chlorosis and defoliation, the addition of a second "model" substance, e.g., polyethylene glycol (MW 6000) was necessary to cause interveinal chlorosis and necrosis, i.e., the "tiger-stripe" pattern, followed by defoliation while the main veins were still green. It may also be significant that both in the model system and in the Verticillium-infected hop the first change in the color of the leaf was accompanied by closure of the stomata, which did not open again until the tissue had become strongly chlorotic and the chloroplasts had disintegrated. Thereafter, they remained wide open as the tissue became necrotic. In a "striped" leaf it is possible to find closed stomata in the green area near to a vein and open ones in the adjacent chlorotic and necrotic bands, all within a distance of 1–2 mm.

Evidence of a synergistic effect between ethylene and Verticillium polysaccharides in tomato verticillial wilt pathogenesis has been shown by Cronshaw and Pegg (1976). Nevertheless, there is no reason to suppose that symptom induction in vascular diseases always has the same cause. Whereas leaf symptoms in chrysanthemum wilt caused by V. dahliae are due to local occlusion of the vascular system (MacHardy et al., 1976; Robb et al., 1975a), those caused by the same pathogen in sunflower seem to be due to a predominantly toxic action (Robb *et al.*, 1975b, 1977).

In early work on the role of toxins in vascular diseases it was commonly postulated that their effect was to increase the permeability of the leaf cell membranes causing excessive water loss. A consequence of increased permeability of the plasmalemma is a leakage of solutes, and this has formed the basis of a useful means of detecting such changes in diseased tissues. However, it is frequently found that when tissues showing incipient wilting are placed in water they regain turgidity and do not show excessive leakage of solutes; in some the leakage appears to be less than from the healthy leaf. Disruption of membranes is a normal consequence of cell death, but possibly the presence of oxidized phenols causes precipitation of proteins and this, together with impregnation of the cell walls, may reduce the permeability of the tissue to solutes. Nevertheless, some pathogens, e.g., *Pseudomonas syringae* in tobacco, cause irreversible loss of turgidity and severe leakage. It seems clear, however, that in such diseases disruption of water relations in the leaf is generally the consequence of cell death, not the cause.

Leaf spot pathogens characteristically have a localized action, causing death of leaf tissues, but frequently are limited either by host responses, which may include formation of phytoalexins or a suberinized periderm, or both, or by desiccation of the lesion. In diseases of this type any disturbance of the water system is localized and is the consequence of tissue death. However, in some foliar diseases, e.g., rose black spot (*Diplocarpon rosae*) and apple scab (*Venturia inaequalis*), the pathogen initially forms subcuticular colonies and does not immediately invade the epidermis and mesophyll. The effects of these pathogens on the cuticle have been discussed (see Section III,B) but the browning reaction which they induce in the epidermal cells may influence the movement of water toward the cuticle. The production of ethylene in black spot lesions may affect the action of stomata and thus influence water movement in the chlorotic areas beyond the limits of the infected tissues. Powdery mildews colonizing leaf surfaces and forming haustoria in the epidermal cells cause changes in the cell wall structure, e.g., formation of ligninlike material in the vicinity of haustoria and, like the subcuticular pathogens, eventually cause disorganization of the epidermal cells and oxidation of phenolic compounds. These oxidation products may, perhaps, affect water movement through the epidermis and modify cuticular transpiration. The solute balance between epidermis and guard cells may also be modified, causing changes in stomatal function and hence in the driving force. Systemic infection by downy mildews and rusts may cause structural changes in cells and tissues, e.g., thickening and lignification of cell walls, hypertrophy, and hyperplasia, but their effects on water movement in the liquid phase are probably less significant than in the vapor phase.

V. MECHANISMS LIMITING DYSFUNCTION

Plant growth and development is to a large extent a process of continuous replacement of dysfunctional structures. Reference has already been made (see Section IV,A) to the continual formation of new absorptive surfaces in the root system to replace those which become nonfunctional because of microbiological action, exhaustion of available water, and other environmental factors. The root system readily produces new roots to replace damaged or diseased ones, and new adventitious roots may also arise from stem tissues as a response to disease. Replacement also takes place when radial growth of roots and stems sloughs off dysfunctional cortex and phloem and "buries" dysfunctional xylem, and when extension of shoots and growth from dormant buds provide replacement of dysfunctional leaves. Short-lived annual plants of limited growth have little capacity for replacement of senescent or diseased parts, but they often have a high reproductive capacity which provides complete replacement of whole plants. In the absence of obvious disease, tissue replacement in most climatic zones tends to be cyclic and seasonal, but the pattern of "disposal" and replacement can be modified by exceptional environmental and pathological conditions.

To a considerable extent, therefore, damaged and diseased parts of a perennial plant are eliminated by no more than minor modifications of the seasonal cycle, most obviously in those diseases which cause leaf damage either through direct attack or by inducing water deficits in the leaves. Extensive injury to leaves generally results in the formation of abscission layers and in premature leaf fall. If this happens early in the year there may be a second flush of leaves; later defoliation may leave the plant leafless and prevent growth until the following year, but the limited demand for water in a defoliated plant may well be met by water potentials generated by evaporation from the surfaces of stems and from lenticels.

The main problems of dysfunction in the water system arise from infection of the xylem, the only part of a plant which is not progressively eliminated, and which constitutes its permanent framework. The most recently formed xylem in the main axis is continuous with that in the newest leaves and roots. Hence, it is the outermost layers of xylem which transport the major part of the water currently in transit, i.e., the main functional xylem forms a sheath surrounding older tissues which are largely nonfunctional with respect to the water system, although they have the main skeletal function. It follows that if growth of a pathogen in the xylem can be limited, continued cambial activity will tend to "bury" the infected wood in nonfunctional tissues. This condition occurs commonly in wilt diseases caused by *Verticillium dahliae* in woody perennials. However, if there is a substantial amount of the pathogen in close proximity to the cambium, production of new xylem may be retarded, and if the effect is localized it may lead to uneven radial growth of the stem, forming longitudinal grooves. This symptom sometimes occurs in the verticillial wilt of quince.

The presence of "sheets" of xylem parenchyma and ray cells in the xylem is probably a contributory factor to the localization of the pathogen, and it is characteristic that *Verticillium* almost never grows radially from one annual ring to the next; infection in successive rings usually denotes reinfection of roots in successive years.

Other pathogens which infect the xylem, e.g., *Stereum purpureum*, tend to be isolated in relatively resistant hosts by the formation of "gum barriers" in the vessels and rays.

The most striking example of xylem replacement in a vascular disease is probably that which occurs in hop verticillial wilt (*V. albo-atrum*). In this disease a mild infection (caused by a nonvirulent strain in low-resistance cultivars, or virulent strains in high-resistance cultivars) generally results in intense and extensive tylosis of the primary and secondary xylem. These occlusions appear to limit the spread of the pathogen, and are generally most extensive in the lower parts of the stem; clearly they must also limit the movement of water. However, widespread tylosis is usually accompanied by renewed and continuous cambial growth, so that considerable additional xylem is formed, mostly as relatively small vessels. Some of the hyperplastic xylem often becomes tylosed also, but as long as new xylem is formed fast enough to compensate for progressive occlusion, the upper parts of the plant remain substantially symptomless and grow and produce a normal crop. The leaf traces of the lower primary leaves are occluded, and the leaves fall, but the xylem of the lateral shoots arising from the axils of these leaves is continuous with the outermost layers of hyperplastic xylem, and the shoots generally appear to be normal. In this host also the radial and tangential growth of the pathogen tends to be limited by the distribution of xylem parenchyma and ray cells. When infection fails to induce extensive tylosis the plant becomes heavily invaded and develops severe symptoms.

Compensatory growth of xylem appears to be associated with tylosis induced in certain species of oaks by the wilt pathogen, *Ceratocystis fagacearum,* and in some elms infected with *Ceratocystis ulmi,* though the response is not as pronounced as in the hop.

Tylosis and the formation of new xylem are common responses to physical wounding in many woody plants, and also contribute to "natural recovery" from lesions caused by canker-inducing organisms. Such lesions are caused by infection of cortical and phloem tissues by fungal or bacterial pathogens, and may extend inward to the xylem, causing occlusion of vessels and preventing formation of new xylem. Initial limitation of the lesion by single or successive periderm layers may eventually be followed by progressive overgrowth from the edges of the lesion, while the diseased tissues are sloughed off, so that eventually a continuous sheath of xylem and extravascular tissues is reestablished.

VI. CONCLUSIONS

Dysfunction of parts of the water system is being caused throughout the life of every plant in its normal habitat. Many such incidents are small, and because they frequently result in increased resistance both to water movement and to pathogens the affected parts are isolated and their function taken over by other parts of the system. This is possible because the major units in the system each consist of a series of subunits related to each other like electrical resistances wired in parallel. A large increase in the resistance of one unit (causing a large reduction in flow) has only a small effect on the resistance of the whole system, and results in a small increase in flow in the other units. Thus, dysfunction in a few rootlets, or in a few vessels in the stem xylem, or in a few leaves, has no significant effect on the whole plant. It is only when large numbers of incidents occur more or less simultaneously that the consequences are significant, and even then mechanisms may be available to minimize, or to compensate for, the loss of function. Nevertheless, the absorptive system (roots), the dispersive system (leaves), and the transport system (xylem in main roots and stems) are all "wired" in series, and a drastic rise in the resistance of any one, or a drastic fall in the driving force, will cause dysfunction of the whole "circuit." The system showing the most obvious symptoms may not necessarily be the one in which dysfunction has occurred. If the lamp goes out the dysfunction may be in the light bulb, but it may also be at the generator, or in the circuit breaker, in the power line, or in another "load" in the system.

Pathogens can modify resistances in the water system, break circuits, increase or decrease the load, and modulate the power input, either by direct effects on the host tissues, or by indirectly inducing or eliciting physiological and structural changes in the host cells.

References

Ayres, P. (1978). The water relations of diseased plants. *In* "Water Deficits and Plant Growth" (T. T. Kozlowski, ed.), Vol. 5, pp. 1–60. Academic Press, New York.

Beckman, C. H. (1964). Host responses to vascular infection. *Annu. Rev. Phytopathol.* **2**, 231–252.

Buddenhagen, I., and Kelman, A. (1964). Biological and physiological aspects of bacterial wilt caused by *Pseudomonas solanacearum*. *Annu. Rev. Phytopathol.* **2**, 203–229.

Cronshaw, D. K., and Pegg, G. F. (1976). Ethylene as a toxin synergist in *Verticillium* wilt of tomato. *Physiol. Plant. Pathol.* **9**, 33–44.

Dainty, J. (1969). The water relations of plants. *In* "The Physiology of Plant Growth and Development" (M. B. Wilkins, ed.), pp. 421–452. McGraw-Hill, New York.

Dimond, A. E. (1966). Pressure and flow relations in vascular bundles of the tomato plant. *Plant Physiol.* **41**, 119–131.

Dimond, A. E. (1970). Biophysics and biochemistry of the vascular wilt syndrome. *Annu. Rev. Phytopathol.* **8**, 301–316.

Duniway, J. M. (1973). Pathogen-induced changes in host water relations. *Phytopathology* **63**, 458–466.

Duniway, J. M. (1976). Water status and imbalance. *In* "Encyclopaedia of Plant Physiology," Vol. 4. "Physiological Plant Pathology" (R. Heitefuss and P. H. Williams, eds.), pp. 430–449. Springer-Verlag, Berlin.

MacHardy, W. E., Busch, L. V., and Hall, R. (1976). *Verticillium* wilt of chrysanthemum: Quantitative relationship between increased stomatal resistance and local vascular dysfunction preceding wilt. *Can. J. Bot.* **54**, 1023–1034.

Robb, J., Busch, L. V., and Lu, B. C. (1975a). Ultrastructure of wilt syndrome caused by *Verticillium dahliae*. I. In chrysanthemum leaves. *Can. J. Bot.* **53**, 901–913.

Robb, J., Busch, L. V., Brisson, J. D., and Lu, B. C. (1975b). Ultrastructure of wilt syndrome caused by *Verticillium dahliae*. II. In sunflower leaves. *Can. J. Bot.* **53**, 2725–2739.

Robb, J., Busch, L. V., Brisson, J. D., and Lu, B. C. (1977). Ultrastructure of wilt syndrome caused by *Verticillium dahliae* III. Chronological symptom development in sunflower leaves. *Can. J. Bot.* **55**, 139–152.

Slatyer, R. O. (1967). "Plant-water Relationships." Academic Press, New York.

Talboys, P. W. (1968). Water deficits in vascular disease. *In* "Water Deficits and Plant Growth" (T. T. Kozlowski, ed.), Vol. 2, pp. 255–311. Academic Press, New York.

Wardlaw, I. F., and Passioura, J. B., eds. (1976). "Transport and Transfer Processes in Plants." Academic Press, New York.

Chapter 7

Disturbed Mineral Nutrition

DON M. HUBER

I. INTRODUCTION

Although mineral nutrition is one of the basic processes that may be impaired by disease, it has been neglected in recent years by plant pathologists. We speak freely of mineral deficiency as disease but seldom consider that the deficiency may be induced by some pathogens. In this chapter on disturbed mineral nutrition, I shall deal with both. Also, I shall use the word "nutrient" to mean mineral nutrient—it will not mean other nutrients such as sugars or amino acids.

II. MINERAL NUTRITION IN GENERAL

This chapter is concerned with the disruption of mineral nutrition by pathogens; however, a general understanding of the source and function of mineral elements is necessary before diseased and normal conditions can be compared. Thirteen mineral elements are generally essential for the growth of plants (Epstein, 1972). These minerals function as cellular components and substrates, or activators, inhibitors, and regulators of metabolism. Only potassium, boron, and chlorine do not become incorporated into plant constituents. Many of the elements are integral components of enzymes and electron carriers. Such functions as maintaining electrical neutrality or internal osmotic pressure are not depen-

163

dent on a specific element, but are determined by all the inorganic and organic ions in the cell solution.

Electrical neutrality or buffering is maintained through the counterion function of anions and cations and influences the physical state and conformation of colloids, polymers, proteins, nucleic acids, etc., as well as the activity and speed of various physiological processes. Thus, each element functions as a part of an intricate system of delicately balanced interdependent reactions. The deficiency or excess of one element greatly influences the activity of others and exerts catastrophic effects as secondary and tertiary consequences reverberate throughout the entire metabolic network of the plant.

It should also be remembered that the presence of an element in soil does not necessarily imply its availability for plant growth. Its availability is dependent on the abundance of an element, its form and solubility; the presence of competing or toxic entities; microbial associations; the assimilative capacity of the plant; and environmental factors such as pH, moisture, and temperature: Balanced nutrition may be as important as the presence of one or another element. Nitrogen, phosphorus, potassium, and calcium are the more generally limiting nutrients; however, chlorine is the only element that has not been found limiting somewhere under natural conditions.

III. DISEASES DUE TO MINERAL DEFICIENCY

Physiological and pathological factors cause diseases involving a disturbed mineral nutrition. Both factors may have a common mechanism of action. Sufficiently severe deficiencies or excesses of essential nutrients may manifest themselves in more or less distinct symptoms. Visible symptoms, however, are late manifestations of metabolic derangements which occurred long before the effects became apparent. Diagnosis of diseases due to mineral deficiencies is complicated because: (1) reduced growth and quality may occur without being obvious, (2) the same element may induce different symptoms in different plants or under different environmental conditions, and (3) similar or identical symptoms may result from deficiencies in different elements. For example, both manganese and iron deficiencies result in a similar chlorosis of foliar tissues in alkaline, moist soils. It is important to determine whether iron or manganese deficiency is the problem in each specific situation since treatment of manganese-deficient plants with iron further suppresses manganese uptake, and symptoms become even more severe. Multiple deficiencies magnify the problem of diagnosis. Both nitrogen and sulfur deficiencies cause a similar chlorosis, and symptoms are relieved only after both elements are available in sufficient quantity. To confuse the diagnostician further, parasites or toxicants may also cause similar symptoms.

When an element is deficient, its content in the plant is reduced and tissue testing may be used as a general guide to the nutritional status of the plant. Nutritional disorders and environmental toxicants, as nonparasitic causes of disease, can generally be corrected by supplying the nutrient in an appropriate form or by removing the toxicant. Resistance to this type of disease is manifest as tolerance to the toxicant or greater efficiency in nutrient absorption or utilization.

Excess of a specific nutrient may be as damaging as a deficiency. Toxicity may result from increased availability in the soil of nutrients such as manganese or aluminum as soil conditions become more acid, or from foliar uptake from atmospheric sources, i.e., sulfur dioxide. Drastically altered ratios of some nutrients may also induce excess uptake of toxic quantities of another element. Many plants have the capacity to absorb "luxury" amounts of readily available nutrients whereby toxicity may result from excess ion accumulation. An example is the excess accumulation of ammonium nitrogen by tobacco plants growing in fumigated soils under environmental conditions restricting photosynthesis. Under these conditions, amination of organic acids may induce an energy deficit relative to other metabolic needs. Another possible explanation of ammonium toxicity after soil fumigation is the differential effects of some fumigants on *Nitrobacter* compared with *Nitrosomonas*. In these situations, toxic levels of NO_2-nitrogen accumulate because *Nitrosomonas* sp. converting NH_4 to NO_2-nitrogen recover more rapidly than *Nitrobacter* sp. which oxidize NO_2-nitrogen to NO_3-nitrogen. Under these conditions, nitrification inhibitors such as nitrapyrin [2-chloro-6-(trichloromethyl)-pyridine] that are specific for *Nitrosomonas* will prevent this NO_2-nitrogen toxicity (Huber *et al.*, 1977).

It is not always possible to distinguish clearly parasitic from nonparasitic diseases. Many of the primary and secondary symptoms associated with infectious diseases are similar to those expressed by mineral deficiencies. Stunting, chlorosis, wilting, mottle, rosette, witches' broom, die-back, leaf spot, abnormal growth, etc., may be induced by specific mineral deficiencies as well as by pathogens. Increased amino acids, auxin, and other materials commonly associated with pathogenesis are also manifestations of specific mineral deficiencies. Potassium deficiency causes soluble nitrogen compounds to accumulate and this results in necrotic leaf spots similar to symptoms induced by several foliar pathogens. The marginal chlorosis and leaf blotch of wheat leaves that is induced by severe infections of wheat by *Ophiobolus graminis* on sandy soils is sometimes incorrectly diagnosed as magnesium deficiency. Treatment with dolomitic lime, therefore, fails to correct the symptom. The dwarfing, rosetting, and bronzing effects of zinc deficiency are common symptoms of virus infection and may be alleviated by additions of zinc.

Rather than starvation, the actual cause of a symptom may be similar to the phenomenon of "limited growth" in fungi where excess ions or metabolites accumulate to toxic levels internally (Kliejunas and Ko, 1975), with the resultant expression of disease when metabolism is interrupted. Chloroses due to mineral deficiencies, except of nitrogen, are attributed primarily to the direct toxic action of excessive accumulations of metabolites, especially nitrogen compounds. In general, the marked change in concentrations of free amino acids and amides in diseased plant tissues is a distinct aspect of nutritional disorders.

Changes in amino compounds vary depending on the plant, mineral element deficiency, and whether ammonium or nitrate nitrogen is absorbed (Huber and Watson, 1974). Additional elements altering the level of amino compounds in plant tissues include calcium, potassium, magnesium, molybdenum, phosphorus, boron, manganese, zinc, copper, iron, and sulfur (Hewitt, 1963). Some of the most striking changes in amino acid patterns occur with zinc deficiency. The profound effect of nutrient deficiencies on amino acid composition should be of considerable interest in understanding pathogenesis because of the correlation of high levels of specific amino acids with resistance or susceptibility to various types of diseases (Van Andel, 1966). The intimate relationship of mineral nutrition to metabolic functions and growth, as well as the interrelationship between the various nutrients within the plant and soil, provides ample opportunity for pathogen modification of mineral nutrition.

IV. MINERAL DEFICIENCIES DUE TO DISEASE

This section deals more specifically with the induction of mineral deficiencies by disease. It should be remembered that although some cause–effect relationships are obvious, it is hazardous to imply a causal role just because the end results of a specific host–pathogen interaction may indicate a change in the concentrations of mineral elements. Nevertheless, whether from primary, secondary, or tertiary interrelations; disturbed mineral nutrition is one of the most common effects of disease. This may reflect an imposed mineral deficiency somewhere in the plant, impaired utilization, or toxicity from abnormal accumulation around the infection site.

The intimate effects of altered nutrition are felt at the cellular level.

Direct effects are magnified indirectly many times because of the intricate balancing of the perfect biochemical work required for growth. This balance varies from moment to moment as the needs, age, and specialization of the cell dictate. The uptake of minerals and their organization into new vital substances permits growth, multiplication, and reproduction. The pathological problem of altered mineral nutrition resides in the impossibility to satisfy a definite need adequately or in time. This reflects altered availability or impaired utilization at the cellular level where the complete or partial lack of nutrients can have the gravest consequences for their general metabolism.

A. Reduced Nutrient Availability

An imposed nutrient deficiency may be specific or general, localized or systematic as a result of immobilization in soil, reduced absorptive capacity, utilization by the pathogen, impaired translocation, accumulation in nutrient sinks around infection sites, or loss from exudation. Although each of these factors is discussed separately, a specific disease may affect nutrient availability by influencing all of them.

1. Immobilization in the Soil

a. Rhizosphere. Soil occupies a unique position for plant growth because of its enormous volume of living roots, populations of microorganisms and colloids and other complexities. Soil microorganisms are not merely incidental inhabitants of soil. They are an inseparable part of it. Along with the root systems of higher plants, microorganisms give soil the very dynamic property which makes it more than an inert geological deposit. The area surrounding plant roots, the rhizosphere, is a region of intense microbial activity where large numbers of saprophytic, soil-inhabiting microorganisms interact with the plant. Their interactions may be deleterious, helpful, or neutral. Although most pathogens are also parasites, some soil organisms cause disease even though they are not dependent on the living plant. These organisms can induce disease by immobilizing the nutrients in the rhizosphere via oxidation (Fe^{2+} to Fe_3^+), reduction (SO_4^{2-} to S), or by robbing the plant during residue decomposition (soil sink). One of the better known examples of microbially induced deficiency is the gray-speck disease of oats (Timonin, 1965). With this disease, manganese-oxidizing bacteria thrive in the rhizosphere of "susceptible" oat varieties where they render manganese unavailable to the plant. Since the rhizosphere environment of resistant oat varieties is not conducive to growth of these bacteria, gray-speck, i.e., manganese deficiency, does not develop.

Inoculation of clover seedlings with suspensions of rhizosphere micro-organisms greatly reduces the uptake and translocation of calcium and rubidium (Hale *et al.*, 1971). A similar phenomenon may account for "frenching" of tobacco associated with *Bacillus cereus* in the rhizosphere of affected plants. Symptoms include a distinct reticular chlorosis; stunting; thickened, straplike leaves; and the accumulation of high levels of isoleucine, hydroxyproline, and other free amino acids in leaf tissues. Although the accumulation of free amino acids is considered the chemical basis for morphological symptoms of frenching, deficiencies of calcium, magnesium, potassium, and phosphorus induce high levels of specific free amino acids in leaf tissue and reproduce the symptoms of frenching (Steinberg *et al.*, 1950). An increase in tissue amino acids and amides is common with many mineral deficiencies (Hewitt, 1963) and may make the host more nutritionally desirable (Huber and Keeler, 1977).

b. Robbing by Microbes. Soil microbes compete with the plant for mineral nutrients as they decompose organic residues in the soil. Soil nitrogen is preferentially utilized by microbes decomposing organic materials that are low in nitrogen. This robs the crop plants of this important nutrient. Potassium deficiency of rice frequently occurs because this nutrient is utilized and immobilized by microorganisms during residue decomposition even though soil tests may indicate that a high level of potassium is present (Aiyar, 1948). Other elements may also be immobilized, especially when soil levels are marginal for crop growth or nutrient imbalance is involved. High levels of phosphate, solubilized by rhizosphere bacteria, may immobilize iron as ferriphosphate and thereby induce iron deficiency in marginal soils. The opposite effect may be observed in water-logged soils where iron that is solubilized by microbes may react to induce a phosphate deficiency (Gerretsen, 1948). Even though nutrients immobilized in this manner may eventually become available through mineralization, higher than normal levels of fertilizer amendments may be required to offset the temporary deficiency imposed.

c. Symbiont Interactions in Soil. In contrast to immobilizing minerals in the soil, microorganisms may prevent deficiency diseases. In more highly specialized microbial root associations, the root becomes infected and an intimate balanced relationship is established. Each partner of these symbiotic associations provides a nutritional contribution for the other. The best understood of these associations are the mycorrhizae and nitrogen-fixing root nodules. These are covered in more detail by Bowen in Chapter 11, this volume. It should be remembered that these are not unilateral relationships so that conditions not conducive to symbiosis may result in parasitism. Thus, under conditions limiting the photosynthesis of legumes, root nodule bacteria may function as parasites as they utilize host-produced carbohydrates but fail to fix nitrogen in return. The trade-off in the legume-*Rhizobium* symbiosis is that the bacterium fixes a gaseous nitrogen in return for carbon fixed by photosynthesis in the plant. Chief among the external factors governing nodulation are a high calcium requirement, moderate pH, and inhibition by nitrate nitrogen. Nitrifiers, in oxidizing NH_4^+-nitrogen to NO_3^--nitrogen may have an effect on nodulation because of the inhibitory effect of nitrate nitrogen. Potassium and magnesium excesses may inhibit nitrogen fixation by competing with calcium. Root infection by fungi and nematodes causes a significant decrease in nodulation and nitrogen-fixing efficiency. Nodule number and weight are inversely correlated with increasing densities of the nematode, *Heterodera glycines*, that causes a severe chlorosis of soybean. *Heterodera goettingrana* on peas and *H. trifoli* on white clover also reduce nodules on their respective hosts (Lehman *et al.*, 1971). Hence, these diseases induce nitrogen deficiency.

The entire root system of an individual plant may be mycorrhizal or only a portion of it may be infected; however, relatively few plant species are completely nonmycorrhizal. Although mycorrhizal and root pathogens are similar in that they both are stimulated by feeder roots, and infect and ramify into meristematic and immature cortical tissues, root pathogens reduce the nutrient absorptive capacity of the plant while mycorrhizae increase nutrient absorption. Mycorrhizae are especially important to plant nutrition in deficient soils where they may greatly increase the availability of phosphorus and several other nutrients not readily accessible to the plant. Many reports show that mycorrhizae enhance the uptake of nitrogen, phosphorus, calcium, sodium, magnesium, iron, copper, boron, zinc, aluminum, and strontium and reduce the uptake of potassium and manganese. Mycorrhizae remain active on old roots for relatively long periods of time while highly effective absorbing portions of the uninfected root tip "move" through the soil as the root elongates. Because of the time required for ion movement through soil, an organ active in the same position over a longer period of time could obtain nutrients from a larger volume of soil (Gerdemann, 1968). Forest trees are especially dependent on mycorrhizae for survival because of the otherwise limited availability of soil nutrients. If the plants are provided with nutrients in fertilizers, however, their need for mycorrhizal nutrient absorption is low (Gerdemann, 1968; Mosse, 1973). Since adequate phosphorus nutrition is required for root nodule fixation of nitrogen, mycorrhizae may be a precondition for nodulation under deficient soil conditions.

Perhaps an even more important function of mycorrhizae is that they may deter the infection of feeder roots by soil-borne pathogens (Marx, 1972). This protection may be afforded by reducing levels of free carbohydrates and other nutrients in root exudates that stimulate pathogenic activity, by establishing a physical barrier to penetration, or by releasing inhibitory compounds into the rhizosphere (Marx, 1972). Thus, mycorrhizal tobacco plants are less damaged by *Thielaviopsis basicola* than are nonmycorrhizal ones.

Since mycorrhizae and root nodules require healthy cortical tissue to become established, they fail to develop on roots inoculated by such pathogens as *Rhizoctonia* and *Fusarium*. This may impose a severe deficiency of nutrients on the plant which would otherwise be available through action of the symbiont. Nematodes may feed on mycorrhizae without producing obvious symptoms of root damage and reduce nutrient uptake directly, or they may render normally resistant mycorrhizal roots susceptible to pathogens such as *Phytophthora cinnamomi* (Marx, 1972). Mycorrhizal roots actually may be more favorable feeding sites than nonmycorrhizal roots for nematodes because of their higher nutritional status (Ruehle, 1973).

2. Reduced Sorption

Plants absorb nutrients that are dispersed throughout the soil profile by an enormous root system. Pavlychenko and Herrington (1934) found that the length of this system for spring wheat at maturity is nearly 2 miles and for barley and wild oats it is nearly 4 miles. This indicates that any given cubic inch of soil to a depth of 27 in. contains 5.2 in. of roots. Destruction of the absorptive capacity of this system through necrosis, malfunction, or reduced growth can severely impair plant functions that depend on soil for nutrients. A young seedling may be able to tolerate less total loss of root tissue but a greater percentage loss than the same plant at fruiting because it has fewer needs. This effect is commonly observed with take-all of cereals where few, if any, above-ground symptoms of the reduced functional root system may be observed until heading and grain formation.

a. Root Disintegration. Root destruction is a common symptom induced by many soil-borne fungi and nematodes. These pathogens reduce absorption of nitrogen, phosphorus, potassium, calcium, magnesium, and sodium by reducing the amount of functional absorptive tissue of the plant. The absorptive capacity may be reduced to only 5 to 10% of normal under severe disease situations, depending on soil conditions, microbial activity, and nutrient availability. Root rots reduce the uptake of relatively immobile elements in soil such as phosphorus which require an extensive functional root system to provide for the nutritional needs of the plant more than some of the more mobile soil elements such as nitrogen. Increased availability of particular elements may compensate for much of the root damage done by root-attacking fungi and nematodes (Huber et al., 1968; Ruehle, 1973). Thus, plants with a deeper or more extensive root system may exhibit greater resistance to a specific pathogen because of enhanced access to nutrients at soil depths below high inoculum densities of a pathogen or where nutrients are more plentiful (Phillips and Wilhelm, 1971). Loss from root rots is most obvious when plants are inadequately fed, because well-nourished plants may produce new roots to replace those damaged by a pathogen.

Little-leaf symptoms of pine infected by *Phytophthora cinnamomi* result from the insufficient absorption of nitrogen by diseased trees even though essentially normal amounts of other nutrients are absorbed. Thus, a marked deficiency of nitrogen and calcium, but normal levels of phosphorus, potassium, magnesium, and other elements in foliage are observed (Roth et al., 1948). Insufficient absorption of nutrients from poorly drained rice fields is attributed to a fungal root rot (Kido et al., 1956) although reduced absorption of potassium may also result from ionic antagonism of calcium and magnesium or from its use by large numbers of rhizosphere microorganisms (Aiyar, 1948). In contrast, Sumner and Kiesling (1977) reported that foliar levels of potassium increased and calcium decreased in corn as the intensity of root rot caused by *Pythium aphanidermatum* or *Rhizoctonia* increased. The main direct effect of disease on barley and wheat infected with *Pythium* root rot is the destruction of rootlets (Bruehl, 1953). Total phosphorus in plants affected by *Pythium* root rot was considerably lower than that in healthy plants. There also is more soluble nitrogen but less protein nitrogen compared with healthy plants (Vanterpool, 1935).

Nematodes cause damage that ranges from surface necrosis, superficial browning, and splitting or blackening of feeder roots to syncytia and gall formation (Ruehle, 1973). They thereby decrease the water- and nutrient-absorbing area of the root or reduce the availability of nutrients to other plant tissues. These effects are enhanced by adverse environmental conditions such as mineral deficiency, drought, and nonoptimal pH. This explains why symptoms suggestive of mineral deficiencies often occur on leaves of plants severely attacked by nematodes (Pitcher, 1959). Iron deficiency symptoms of nematode-infected azalea plants are especially accentuated under optimum conditions for disease (Barker and Worf, 1966). Nitrogen deficiency of soybean foliage is a typical symptom of infection by the soybean cyst nematode (Bergeson, 1966) as well as by other pathogens. Chlorosis due to a deficiency of magnesium and/or iron is also commonly induced by nematode infections (Chitwood *et al.*, 1952).

b. Reduced Root Growth. Reduced root growth from pathogenic activity reduces the effective volume of soil from which minerals are absorbed. Nematodes, fungi, bacteria, and viruses may inhibit root growth and reduce absorption. "Stubby root" and similar terms have been used to describe these effects of disease. Although not as commonly recognized as root destruction, reduced root absorption of nutrients occurs with many foliar diseases when movement of photosynthates to root tissue is reduced. This results in a shortage of energy for growth as well as for ion uptake across the cell membranes. Obligate pathogens, such as *Erysiphe graminis, Puccinia* sp. and many viruses commonly interfere with normal root development and nutrient uptake. Martin and Hendrix (1974) found that the size and number of roots of wheat plants infected with stripe rust (*Puccinia striiformis*) were greatly reduced. Plants infected in the first or second leaf stage had less than 60% of the root mass and only 16% of the root weight of healthy control plants. Similar morphological changes occur with powdery mildew infection (Fric, 1975). The net physiological effect of these host–pathogen interactions is the reduced amount of photoassimilate in roots to levels detrimental to root growth. This in turn reduces the capacity to absorb nutrients from soil. A similar response of barley to powdery mildew is also observed (Minarcic and Paulech, 1975). Roots of tobacco plants infected with yellow dwarf virus are reduced 50% and this results in a marked reduction in the nitrogen content of leaves (Mandryk, 1954).

c. Increased Loss of Nutrients. Loss of nutrients by leaching is greatly increased by physiological, mechanical, or pathological injury (Blakeman, 1971; Hale *et al.*, 1971). Diseased plants frequently have increased leaf and root exudation. This results in the loss of minerals and other nutrients and may predispose plants to nutritional stress and increased susceptibility to other diseases. The frequently overlooked complexity of plant exudate interactions and their effects on nutrient availability must be considered in order to understand the mechanisms of mineral uptake (Hale *et al.*, 1971).

3. Reduced Translocation

The question of translocation is important because the ascending sap in the xylem carries all the mineral elements necessary for biosynthesis except carbon and photoassimilates which are translocated in the phloem. Thus, the most obvious impairment of translocation is due to a malfunctioning vascular system. Gums, gells, cellular slimes, and other vascular occlusions associated with fungal, bacterial, and viral pathogens interrupt translocation and thereby the utilization of nutrients, directly by interfering with water and nutrient movement. This results in starvation of tissues at a distance from the blockage. Movement of radiophosphorus can be reduced by 96–98% and sulfur 16–46% (Hale *et al.*, 1971) in the xylem of *Fusarium*-wilted tomato plants. *Fusarium oxysporum* f. *vasinfectum* reduces the level of nitrogen, but raises the level of phosphorus in leaves of infected cotton plants (Haag *et al.*, 1967). Potassium, calcium, and magnesium contents in cotton plants are also reduced (Sharoubeem *et al.*, 1967). The iron to manganese ratio increases in pigeon pea infected by *Fusarium udum*.

All pathogens that interfere with translocation generate a deficiency of mineral substances. Deficiencies of elements such as nitrogen, phosphorus, and potassium, which are necessary in large quantities, become especially obvious. Large increases in mineral substances in wilted, compared with normal, plants may occur when cells are killed through moisture stress or toxic actions that render the minerals unavailable for translocation. Thus, little benefit is realized if translocation of absorbed mineral elements to all parts of the plant is not accomplished. Translocation includes the redistribution of elements and elaborated metabolites as growth proceeds as well as direct movement after root uptake. Restricted root growth from stem girdling or other interference with downward movement of photoassimilates has already been mentioned relative to nutrient absorption. The accumulation of nutrients in root tissues and around infection sites are special aspects involving impaired translocation but will be discussed in relation to permeability and nutrient sinks.

Physiologically directed mobilization toward infection sites or changes in membrane permeability are other mechanisms influencing nutrient movement. All of these factors induce a nutrient deficiency at the cellular level even though the total quantity of nutrients in the plant may be unchanged. The anomalies generated by imbalances in distribution are essentially those of mineral deficiency or toxicity discussed earlier in this chapter, and impact on the physiology of the whole plant. Vascular blocking can be important in itself, or because of the imbalance that it causes in the normal activities of the plant. Yellowing associated with bacterial wilt caused by *Pseudomonas solanacearum* is attributed to chlorophyll degradation following the decreased supply of nutrients and water in infected tissues (Buddenhagen and Kelman, 1964). The immobilization or deficiency of nutrients affects the dynamic aspects of synthesis through inhibited or enhanced enzyme activity as well as through a substrate relationship. Subsequent changes in amino acids and amides in host tissues infected with virus, bacteria, fungi, and nematode pathogens reflect this induced nutrient deficiency and water stress (Singh and Smalley, 1968).

Cortical pathogens interfere with translocation by immobilizing nutrients, by disintegrating tissues, or by inducing host responses that occlude vascular elements. Fellows (1928) observed a disintegration of all the cells and vessels of the stele, except the xylem tubes, after penetration of wheat roots by *Ophiobolus graminis*. When this happens, the distal portion of the root ceases to function and the xylem vessels become thickened and clogged. Disintegration of cortical cells by *Fusarium solani* f. sp. *phaseoli* is followed by vascular occlusions near the infection site even though penetration of the pathogen is restricted by the endodermis (Huber, 1963).

B. Impaired Utilization

Nutrient utilization is impaired by pathogens through immobilization (nutrient sinks, chelation, etc.), alteration of cellular permeability (tissue and membrane damage, i.e., victorin or *Corynebacterium* toxins), or competitive inhibition (pseudomonad toxins). Impaired mineral utilization may stimulate or inhibit specific host responses to penetration which, in turn, influence resistance or susceptibility of the host. Impaired nitrogen utilization is a common effect of many diseases as evidenced by the stunting, accumulation of amino acids, and chlorosis frequently observed. Wildfire and other pseudomonad toxins interrupt nitrogen metabolism by acting as antimetabolites of amino acids. An increased level of potassium nitrate in tissues of rye infected with mildew (*Erysiphi graminis*), rust (*Puccinia graminis*), or root rot (*Fusarium* sp.) indicates that these pathogens inhibit the nitrate reductase activity which is necessary for the utilization of nitrate nitrogen (Piening, 1972; Younis *et al.*, 1965). These pathogens could also inhibit nitrate reductase by immobilizing molybdenum or using energy resources which would otherwise be available for nitrate reduction. The commonly observed accumulation of mineral elements and their derivatives around infection sites is ample evidence of impaired utilization by many pathogens. This accumulation of nutrients results from the increased permeability of cells adjacent to infected cells or from an altered hormonal balance (Thatcher, 1942).

1. Altered Permeability

The initial action of many obligate and facultative pathogens is generally to alter membrane permeability as discussed by Mount in Chapter 13, this volume. This modification of permeability is extremely important in pathological processes and undoubtedly plays a part in regulating the nutrients available to invading pathogens (Thatcher, 1942; Wheeler and Hanchey, 1968). Cell walls adjacent to injured or necrotic tissues become impermeable. Increased permeability of susceptible hosts is observed with *Botrytis cinerea* and *Sclerotinia sclerotiorum* on celery, crown gall (*Agrobacterium tumefaciens*), angular leaf spot of cucumber (*Pseudomonas lachrymans*), *Puccinia coronata* on oats, *Puccinia graminis tritici* on wheat, *Helminthosporum victoriae* on oats, and *Fusarium oxysporum* var. *lycopersici* on tomato (Thatcher, 1942; Wheeler and Hanchey, 1968). Some of these pathogens cause cells to leak and at the same time prevent ion uptake. This is a pronounced effect of low molecular weight fungal toxins.

Increased permeability is associated with crown gall and as many as six of the seven biosynthetic systems unblocked in tumor cells are activated by specific ions. Tumor cells take up essential ions very effectively from dilute salt solutions while normal cells do not. There is also a progressive increase in the permeability of the plasma membrane during the transition from normal to fully transformed autonomous tumor cells. Observed effects on permeability may also be reflected in nutrient sink relationships as shown in the section that follows.

2. Nutrient Sinks

During pathogenesis, minerals tend to move toward the infection site which then becomes a sink. Nutrient sinks result from the immobilization of minerals in necrotic tissues, noninduced meristematic activity at the infection site, or altered physiology characteristic of "balanced" host–parasite systems. Modification of cellular permeability contributes to the accumulation of nutrients around the diseased tissue and enhances the sink effect. Although a readily available source of nutrients may be of great value to a pathogen, the accumulation of nutrients around the infection court may also constitute a mechanism of host defense. Kliejunas and Ko (1975) have demonstrated that limited growth of fungi in liquid culture is due to the excess accumulation of ions from the medium rather than the presence of "staling" products. This affinity of fungal mycelia for specific ions that limit growth could explain the rapid cessation of growth and altered permeability following penetration of resistant and hypersensitive hosts. Mineral sinks may be passive or active in function.

a. Passive Sinks. Most mineral elements are reused several times during the life of a plant. Necrosis associated with localized infections, partial or total defoliation, or impaired translocation renders nutrients accumulated in some areas inaccessible to other parts of the plant. This constitutes a type of passive sink when nutrients in these tissues would normally be available for new growth or fruiting. Fungi causing leaf spots such as *Venturia* on apple and pear, *Fumigo* on olive and citrus, and hypersensitive reactions to viruses, fungi, and bacteria are examples of passive sinks. The reduced protein in the kernels of winter wheat infected with snowmold (*Typhula* sp. or *Fusarium nivale*) is attributed to the loss of nitrogen and other nutrients stored in older leaves that would normally be translocated to the grain (Huber and Anderson, 1976).

Direct utilization of nutrients by a pathogen and increased root and leaf exudation may constitute another type of passive sink as nutrients are "drained" from the plant and are no longer available for its metabolic functions. The accumulation of potassium, calcium, manganese, and magnesium in wood as decay sets in is also a sink effect. These increases in soluble and bound mineral elements could affect the utilization of other elements by affecting the swelling of cellulose micelles in cell walls, pH of tissues, and activity of enzymes (Tatter *et al.*, 1971).

Radioactive phosphorus and starch accumulate at lesions induced by many parasites. Altered nitrogen levels and composition (amino, amide, and protein) are commonly observed following infection of hosts with *Bacillus cereus, Cercospora apii, Ceratocystis fimbriata, Gibberella fujikuroi, Helminthosporium sativum, Phytophthora infestans, Pyricularia grisea, Pseudomonas phaseolicola,* and *P. tabaci* (Freeman, 1964; Huber, 1977). Tissues which are normally killed by pathogens early in pathogenesis, i.e., *Botrytis* on bean and *Helminthosporium* on wheat, do not initially accumulate nutrients since their mobilization from noninfected zones can take place only if invaded cells remain alive and metabolically active for some time.

b. Meristematic Sinks. Active mineral sinks are induced by many pathogens as they alter host metabolism, but do not kill tissues outright. This is accomplished through induced meristematic activity or a "balanced" nutrition to accommodate obligate or symbiotic organisms. Meristematic tissues constitute a major sink for mineral nutrients because they are needed by the enhanced metabolic activity and new growth that is produced. Pathogen-induced meristematic activity occurs through "cork formation" (cicatrization) around root and foliar pathogens, around self-limiting galls of rusts and nematodes, and around autonomous tumors of bacterial or viral etiology. Generally, the unhealthy condition of nematode-galled plants is the result of minerals, otherwise destined for foliar growth, being mobilized to the infection site as plants are stimulated to produce additional root tissues. Root absorption of specific minerals by this new tissue may increase, but translocation to other plant parts is greatly reduced (Barker *et al.*, 1971). In addition to galls, nematodes may induce syncytia, or specialized nurse cells, that become nutrient sinks similar to galled tissue (Dropkin, 1969). Thus, even though restricted in their cellular involvement, nematodes divert considerable nutrients from noninfected host tissues.

In the case of tumors produced by *Agrobacterium tumefaciens*, it is not the parasite that starves the plant directly, but rather the neoplastic tissue that is the real parasite of normal tissue. Tumor tissues are hyperauxinated and contain high concentrations of various ions which could activate the metabolic systems responsible for autonomous growth. The addition of zinc causes a rapid rise in auxin content and large amounts of auxin accumulate in copper- and manganese-deficient plants because of decreased oxidative enzyme activity (Skoog, 1940). Thus, internal management of zinc, copper, or manganese may provide metabolic control over the auxin system required for autonomous growth.

    c. Balanced Pathogen Sinks. Obligate parasites (and symbionts) create very powerful nutrient sinks whereby minerals and elaborated materials are mobilized to the infection site as permeability and metabolic activity are increased. Substances flow initially from the histological ring that surrounds invaded cells and then from relatively remote areas of the plant. It appears that the first act of the parasite is to "open the door of the larder." The mechanism for upsetting the distributive economy of the host appears similar to that for meristematic or other areas characterized by high rates of synthesis and is undoubtedly enhanced by the high level of cytokinins in mycelia and spores of some fungi (Fric, 1975). Pathogen metabolism may also maintain a concentration gradient and higher osmotic pressure and thereby ensure a continuous flow of materials to the infection site. The host can assimilate only a small part of the mineral elements (ash), nitrogenous compounds, and acid-hydrolyzable substances mobilized to the infection court because of a scarcity of carbohydrates there (Hare, 1966). Thus, nitrogen, phosphorus, sulfur, carbon, and other elements accumulate at the infection site of viruses, rusts, mildews, and potato late blight. This accumulation is in the host tissue rather than in the pathogen tissue (Hare, 1966). A fourfold increase in amino acids and other nitrogenous compounds is reported in rusted leaves of wheat plants (Shaw, 1963). The activation of localized and systemic peptidase activity in susceptible hosts infected with *P. recondita* and *Erysiphi graminis* increases the "free" amino acid pool available for translocation (Huber and Keeler, 1977). Rust- and mildew-infected leaves of many plants accumulate more radioactive phosphorus than normal leaves whether fed directly to infected leaves or through roots. Bean leaves infected with *Uromyces appendiculatus* accumulated 7,870 times more phosphorus than noninfected leaves and accumulated sulfur in a zonate pattern around infected areas (Yarwood and Jacobson, 1955). Many of these changes reflect changes in enzyme activity, although directed mobilization is also involved (Shaw, 1963). The period of most intense synthesis of TMV protein corresponds to the maximum deficiency of nonprotein nitrogen in host cells. Nineteen times as much sulfur accumulates in TMV local lesions as in normal tissues (Yarwood and Jacobson, 1955).

V. CONCLUSIONS

Mineral nutrition of the host is one of the basic processes impaired by disease. Impairment may be due to direct plundering by the parasite or to the indirect effects of the parasite on absorption, mobilization, or function. The intimate effects of these alterations occur at the cellular level, but may be reflected in altered growth of the whole plant. Although many effects of disease on enzymes and metabolic processes have been studied, our information on interactions of disease with the availability, immobilization, or toxicity of mineral elements is exceedingly limited. Such an oversight is indeed unfortunate, since mineral elements are involved as metabolic substrates and structural components and in the activation and regulation of physiological processes. Thus, the activation or inactivation of specific metabolic processes and mineral availability, balance, or form may be one and the same phenomenon as seen through different oculars of the same microscope.

References

Aiyar, S. P. (1948). The effects of potassium deficiency on rice. *Proc. Indian Acad. Sci., Sect. B* **28**, 202–226.

Barker, K. R., and Worf, G. L. (1966). Effect of nutrients on nematode activity on azalea. *Phytopathology* **56**, 1024–1027.

Barker, K. R., Lehman, P. S., and Huisingh, D. (1971). Influence of nitrogen and *Rhizobium japonicum* on the activity of *Heterodera glycines. Nematologica* **17**, 377–385.

Bergeson, G. B. (1966). Mobilization of minerals to the infection site of root knot nematodes. *Phytopathology* **56**, 1287–1289.

Blakeman, J. P. (1971). The chemical environment of the leaf surface in relation to growth of pathogenic fungi. *In* "Ecology of Leaf Surface Microorganisms" (T. F. Preece and C. A. Dickinson, eds.), pp. 255–268. Academic Press, New York.

Bruehl, G. W. (1953). *Pythium* root rot of barley and wheat. *U.S. Dep. Agric., Tech. Bull.* **1084**.

Buddenhagen, I., and Kelman, A. (1964). Biological and physiological aspects of bacterial wilt caused by *Pseudomonas solanacearum. Annu. Rev. Phytopathol.* **2**, 203–230.

Chitwood, B. G., Specht, A. W., and Havis, L. (1952). Root-knot nematodes. III. Effects of *Meloidogyne incognita* and *M. javanica* on some peach rootstocks. *Plant Soil* **4,** 77–95.

Dropkin, V. H. (1969). Cellular responses of plants to nematode infections. *Annu. Rev. Phytopathol.* **7,** 101–122.

Epstein, E. (1972). "Mineral Nutrition of Plants: Principles and Perspectives." Wiley, New York.

Fellows, H. (1928). Some chemical and morphological phenomena attending infection of the wheat plant by *Ophiobolus graminis. J. Agric. Res.* **37,** 647–660.

Freeman, T. E. (1964). Influence of nitrogen on severity of *Piricularia grisea* infection of St. Augustine grass. *Phytopathology* **54,** 1187–1189.

Fric, F. (1975). Translocation of ^{14}C-labeled assimilates in barley plants infected with powdery mildew (*Erysiphe graminis* f. sp. *hordei* Marchel). *Phytopathol. Z.* **84,** 88–95.

Gerdemann, J. W. (1968). Vesicular-arbuscular mycorrhiza and plant growth. *Annu. Rev. Phytopathol.* **6,** 397–418.

Gerretsen, F. C. (1948). The influence of microorganisms on the phosphate intake by the plant. *Plant Soil* **1,** 51–81.

Haag, H. P., Balmer, E., and de Carvalho, A. (1967). Influencia da "Murcha do algodoerio" No composicao mineral do algodoeiro. *An. Esc. Super. Agric., "Luiz de Queiroz," Univ. Sao Paulo* **24,** 333–342.

Hale, M. G., Foy, C. L., and Shay, F. J. (1971). Factors affecting root exudation. *Adv. Agron.* **23,** 89–109.

Hare, R. C. (1966). Physiology of resistance to fungal diseases in plants. *Bot. Rev.* **32,** 95–137.

Hewitt, E. J. (1963). The essential nutrient elements: Requirements and interactions in plants. *In* "Plant Physiology" (F. C. Steward, ed.), Vol. 3, pp. 137–360. Academic Press, New York.

Huber, D. M. (1963). Investigations on root rot of beans. Ph.D. Thesis, Michigan State University, East Lansing.

Huber, D. M. (1977). The role of nutrients in resistance of plants to disease. *In* "Handbook of Nutrition and Food" (M. Rechcigel, ed.), Vol. I, Sect. D. CRC Press, Cleveland, Ohio.

Huber, D. M., and Anderson, G. R. (1976). Effect of organic residues on snow-mold of winter wheat. *Phytopathology* **66,** 1028–1032.

Huber, D. M., and Keeler, R. R. (1977). Alteration of wheat peptidase activity after infection with powdery mildew. *Proc. Am. Phytopathol. Soc.* **4,** 163.

Huber, D. M., and Watson, R. D. (1974). Nitrogen form and plant disease. *Annu. Rev. Phytopathol.* **12,** 139–165.

Huber, D. M., Painter, C. G., McKay, H. C., and Peterson, D. L. (1968). Effects of nitrogen fertilization on take-all of winter wheat. *Phytopathology* **58,** 1470–1472.

Huber, D. M., Warren, H. L., Nelson, D. W., and Tsai, C. Y. (1977). Nitrification inhibitors—new tools for food production. *BioScience* **27,** 523–529.

Kido, M., Yamadori, S., and Sato, T. (1956). Physiological and ecological research on rice plants grown on well drained and ill drained paddy fields. II. On the degree of root-rot infestation and nutrient absorption in rice plants. *Proc. Crop Sci. Soc. Jpn.* **24,** 161–162.

Kliejunas, J. T., and Ko, W. H. (1975). Continuous versus limited growth of fungi. *Mycologia* **67,** 362–366.

Lehman, P. S., Huisingh, D., and Barker, K. R. (1971). The influence of races of *Heterodera glycines* on nodulation and nitrogen-fixing capacity of soybean. *Phytopathology* **61,** 1239–1244.

Mandryk, M. (1954). Suppression of rooting of tobacco cuttings by yellow dwarf virus. *J. Aust. Inst. Agric. Sci.* **20,** 50–51.

Martin, N. E., and Hendrix, J. W. (1974). Anatomical and physiological responses of Baart wheat roots affected by stripe rust. *Wash. Agric. Exp. Stn., Tech. Bull.* **77.**

Marx, D. H. (1972). Ectomycorrhizae as biological deterrents to pathogenic root infections. *Annu. Rev. Phytopathol.* **19,** 429–454.

Minarcic, P., and Paulech, C. (1975). Influence of powdery mildew on mitotic cell division of apical root meristems of barley. *Phytopathol. Z.* **83,** 341–347.

Mosse, B. (1973). Advances in the study of vesicular-arbuscular mycorrhiza. *Annu. Rev. Phytopathol.* **11,** 171–196.

Pavlychenko, T. K., and Herrington, J. B. (1934). Competitive efficiency of weeds and cereal crops. *Can. J. Res.* **10,** 77–94.

Phillips, D. J., and Wilhelm, S. (1971). Root distribution as a factor influencing symptom expression of *Verticillium* wilt of cotton. *Phytopathology* **61,** 1312–1313.

Piening, L. J. (1972). Effects of leaf rust on nitrate in rye. *Can. J. Plant Sci.* **52,** 842–843.

Pitcher, R. S. (1959). *Pratylenchus* spp. and other migratory soil nematodes. *G.B., Minist. Agric., Fish. Food, Tech. Bull.* **7,** 77–87.

Roth, E. R., Toole, E. R., and Hepting, G. H. (1948). Nutritional aspects of the little-leaf disease of pine. *J. For.* **46,** 578–587.

Ruehle, J. L. (1973). Nematodes and forest trees—types of damage to tree roots. *Annu. Rev. Phytopathol.* **11,** 99–118.

Sharoubeem, H. H., Naim, M. S., and Habib, A. A. (1967). Combined effect of nitrogen supply and *Fusarium* infection on the chemical composition of cotton plants. *Acta Phytopathol. Acad. Sci. Hung.* **2,** 40–48.

Shaw, M. (1963). The physiology and host-parasite relations of the rusts. *Annu. Rev. Phytopathol.* **1,** 259–294.

Singh, D., and Smalley, E. B. (1968). Nitrogenous compounds in the xylem sap of American elms with Dutch elm disease. *Can. J. Bot.* **47,** 1061–1065.

Skoog, F. (1940). Relationships between zinc and auxin in the growth of higher plants. *Am. J. Bot.* **27,** 939–951.

Steinberg, R. A., Bowling, J. D., and McMurtrey, J. E., Jr. (1950). Accumulation of free amino acids as a chemical basis for morphological symptoms in tobacco manifesting frenching and mineral deficiency symptoms. *Plant Physiol.* **25,** 279–288.

Sumner, D. R., and Kiesling, T. C. (1977). Root diseases, ethoprop, and mineral uptake in corn. *Proc. Am. Phytopathol. Soc.* **4,** 147.

Tatter, T. A., Shortle, W. C., and Rich, A. E. (1971). Sequence of microorganisms and changes in constituents associated with discoloration and decay of sugar maples infected with *Fomes connatus*. *Phytopathology* **61,** 556–558.

Thatcher, F. S. (1942). Further studies of osmotic and permeability relations in parasitism. *Can. J. Res., Sect. C* **20,** 283–321.

Timonin, M. E. (1965). Interaction of higher plants and soil microorganisms. *In* "Microbiology and Soil Fertility" (C. M. Gilmore and O. N. Allen, eds.), pp. 135–138. Oregon State Univ. Press, Corvallis.

Van Andel, O. M. (1966). Amino acids and plant diseases. *Annu. Rev. Phytopathol.* **4,** 349–368.

Vanterpool, T. C. (1935). Studies on browning root rot of cereals. III. Phosphorus-nitrogen relations of infested fields. IV. Effects of fertilizer amendments. V. Preliminary plant analysis. *Can. J. Res., Sect. C* **13,** 220–250.

Wheeler, H., and Hanchey, P. (1968). Permeability phenomena in plant disease. *Annu. Rev. Phytopathol.* **6,** 331–350.

Yarwood, C. E., and Jacobson, L. (1955). Accumulation of chemicals in diseased areas of leaves. *Phytopathology* **45,** 43–48.

Younis, M. A., Pauli, A. W., Mitchell, H. L., and Stickler, F. C. (1965). Temperature and its interaction with light and moisture in nitrogen metabolism of corn (*Zea mays L.*). *Crop Sci.* **5,** 321–326.

Chapter 8

Alteration of Growth by Disease

JOHN P. HELGESON

I. INTRODUCTION

Green plants are remarkably autonomous. They thrive when supplied with adequate light, water, mineral elements, air, and a favorable temperature. When one or more of these factors is unfavorable, malfunctions in growth often occur. Given an adequate environment, however, plant cultivars can grow with a remarkable uniformity. Corn plants grow uniformly row by row and all set tassels at a uniform height. Thousands of acres of wheat growing on the high plains form a uniform sea of gold at harvest time.

The simplicity of a plant's external requirements contrasts markedly with the complexity of its internal controls and processes. Within a plant, an array of metabolic controls assures adequate supplies of energy and precursors at sites of synthesis. Communication and coordination of growth between the various parts of the plant is achieved by plant growth substances. When environmental conditions permit, and when pathogens do not interfere, the internal controls allow full expression of a plant's genetic potential.

A. Unfavorable Physical and Biological Factors in the Environment

Unfortunately, the physical environment is not always ideal for plant growth. Lack (or excess) of water can severely decrease rates of growth. Cell extension appears to be particularly sensitive to water stress (Hsiao, 1973). Lack of certain essential minerals often slows or even terminates growth. Severe boron and calcium deficiencies, for example, are indicated by malformation or death of the plant apex. Nitrogen and iron deficiencies severely limit the amounts of chlorophyll available for trapping light energy.

The biological environment also can affect plant growth dramatically. Quantitative changes in plant growth, deformation of plant parts, and loss of internal coordination often accompany the battle between host and pathogen. This chapter will address changes brought about by biological influences. Emphasis will first be on the general types of responses to pathogens and then on the possible reasons for particular responses. Much of what will follow will be highly speculative. Present knowledge is inadequate to understand the control of growth in healthy plants. When another unknown, the pathogen, is superimposed on the unknown system the problem of rational explanation is compounded.

B. Types of Growth Responses

In the original edition of this treatise, Braun (1959) cataloged numerous growth malfunctions that accompany plant disease. Wood (1967) also listed many growth responses of plants to pathogens. Plants become dwarfs or giants; plant parts assume new and sometimes exotic shapes; buds that ordinarily are held in check are suddenly released; individual cells escape from organismal control and grow as cancerous masses. The causes of many of these disorders remain as unexplained today as they were in 1959 or in 1967 and reiterating or expanding the lists of known disorders would serve no useful purpose. Instead, three general categories of malfunction will be discussed: (1) quantitative changes in growth, e.g., dwarfism and gigantism; (2) qualitative changes in the form of plant parts, e.g., distortion of leaves and reproductive organs; and (3) loss of correlative control associated with plant diseases, e.g., release from apical dominance. Later in this volume, Merlo (Chapter 9) and Wilson (Chapter 10) will discuss crown gall and teratomas, respectively. Some of the effects of parasitic seed plants and, also, the induction of galls by insects will be covered in the next volume. The present discussion will include some of what is known from studies on normal plants as well as something on deviations which occur in diseased plants. Because of the apparent involvement of plant growth substances in all phases of plant growth (and also because of the author's bias) the emphasis will often be on the phytohormone balances within malfunctioning plants.

C. Plant Growth Substances and Normal Growth

Because of their central importance in the regulation of plant growth, the plant growth substances (phytohormones) have been reviewed extensively and are covered in detail in most elementary botany texts. Some more recent useful treatments include Varner and Ho (1976), Leopold and Kriedemann (1975), Leonard (1974), Milborrow (1974), Ray (1974), Yang (1974), Hall (1973), Thimann (1972), Skoog and Schmitz (1972), Paleg and West (1972), and Galston and Davies (1970).

Much is known about the occurrence of auxins, cytokinins, gibberellins, ethylene, and abscisic acid in plants. Also, extensive lists of biological effects, usually obtained from applications of exogenous materials, have been constructed. Often the effects of a given substance are varied and seemingly unrelated. For example, cytokinins promote cell division in tissue cultures (Skoog et al., 1967) and also delay senescence in detached leaves (Richmond and Lang, 1957). Furthermore, the various growth substances appear to act in concert. For example, auxin and cytokinins act together in control of apical dominance (Sachs and Thimann, 1967). The presence of one substance may be a prerequisite for the activity of another as, for example, the promotion of cell division by cytokinins and its requirement for auxin (Das et al., 1956). These interactions of several phytohormones appear to be critical factors which, at least in early studies on both normal and abnormal growth, were not adequately considered in the construction of hypotheses on modes of action. Simply put, the effect of a change in the concentration of one substance must be considered in relation to the concentrations of all other substances.

In spite of all that is known about plant growth substances, certain critical unknowns remain. The pathways of biosynthesis of many of the substances are fairly well established, but controls of these pathways are not well understood at all. Plant growth substances have dramatic effects on plants, but the mechanisms by which these effects are achieved are still unknown.

D. Growth Substances and Abnormal Growth

Applications of exogenous substances can cause abnormalities similar to those associated with some plant diseases. Also, changes in the concentrations of growth-regulating substances often occur in diseased plants. Thus, the plant growth substances have been considered important in bringing about the growth abnormalities observed in such plants. The subject has been reviewed by Sequeira (1963, 1973), Goodman et al. (1967), Van Andel and Fuchs (1972), Dekhuijzen (1976), and Pegg (1976a–d), among others. In general, the experimental approach has been to examine one growth substance at a time and to find particular instances where the concentration of the growth substance is increased or decreased over the normal concentration in the plant or where the growth substance is synthesized (or destroyed) by the pathogen. Reiteration and updating of the materials presented in the above reviews are beyond the scope of this chapter and, to some extent, not in accord with its purpose. Hopefully, for a given growth effect, the thrust here will be a synthesis and a consideration of total hormone balance rather than a fragmentation into categories of separate compounds. However, as will be clear to the reader, the approach is fraught with problems stemming from the fragmentary nature of available experimental data.

II. QUANTITATIVE DEVIATION FROM NORMAL GROWTH

In response to unfavorable physical and biological environments, plants often grow more slowly than normal and achieve lower total biomass at maturity. If achieved, reproduction may be earlier and less extensive than normal. More unusual, perhaps, is an increase, particularly in stem length, over that achieved in the favorable environment. Although the size of the plants may deviate from the norm, the form of the various plant organs often is nearly normal.

A. Dwarfism

At first glance, a decrease in yield caused by deficits of nutrients or water may seem unrelated to the question of disrupted growth and effects of disease. Under identical environmental conditions, the size of the diseased plant may deviate from that of the normal plant growing beside it. Even from great distances, the difference between two plants may be apparent. Yet, deficits in mineral elements or water could be imagined as the cause of the diminished growth of the infected plant. A severely impaired root system, resulting from the attack of a root pathogen, could be unable to supply the shoot adequately. Furthermore, Wright (1977) found that both ethylene and abscisic acid increased in excised wheat leaves under a stress of only 1–2 more bars of leaf water potential than would normally be experienced in normal diurnal fluctuation. Both compounds might be expected to slow growth and a slight decrease in root efficiency could be reflected in a slightly more negative leaf water potential. In view of the dearth of data on quantitative water relations in diseased plants (see Duniway, 1976), this area must remain highly speculative.

More significant, perhaps, could be the competition for energy that is imposed upon a host by an invading pathogen. Virus multiplication, for example, apparently consumes a substantial amount of the energy which ordinarily would be devoted to growth of the plant. Furthermore, redirection of metabolites into an infected area (Dekhuijzen and Staples, 1968; Dekhuijzen, 1976) could further deplete the available nutrients in other parts of the plants. In this regard it is interesting that cytokinins can mimic some aspects of "green island" effect of rust (Bushnell and Allen, 1962) and that applications of cytokinins can cause mobilization of nutrients into the treated area (see Mothes, 1964).

An inadequate supply of a given growth substance could also cause dwarfing. Studies with dwarf corn (Phinney, 1961) clearly indicated that an inheritable dwarfing habit was associated with insufficient gibberellin synthesis. Applications of gibberellin will alleviate the dwarfing habit and permit nearly normal growth and reproduction of the plants. It is also possible to dwarf normal plants by inhibition of the biosynthetic pathway of gibberellins.

An inadequate auxin supply within the plant may also affect size. Because of the dwarfed appearance of virus-infected plants and because of the effect of indoleacetic acid (IAA) on etiolated peas or *Avena* tissue, it might be expected that auxin would be in short supply in virus-infected plants. Such, in fact, were the findings in some cases (see Pegg, 1976a). Caution may be advised here, however. Both auxins and gibberellins can induce parthenocarpy in fruit (see Leopold and Kriedemann, 1975, pp. 314–316). Also, treatments of excised pea stem sections with IAA and gibberellin apparently cause a synergistic enhancement of elongation (Ockerse and Galston, 1967). Thus, whether auxin is specifically involved in the dwarfing in virus diseases or whether dwarfing is a reflection of a decrease in both gibberellin and auxin remains unknown at the present time.

Studies of dwarfing support two major generalizations: (1) in many disease situations a decrease in the normal growth may be expected and (2) growth substances may be involved. The assumption that minor changes in growth substances are a major cause of dwarfing may not be justified, however. Marasas *et al.* (1971) noted the T2 toxin produced by *Fusarium tricinctum* was a potent inhibitor of seedling growth. Subsequently, it was found that this protein synthesis inhibitor slowed cytokinin-controlled logarithmic growth rates of tobacco callus tissue (Helgeson *et al.*, 1973) and auxin-controlled rates of elongation of soybean hypocotyls (Stahl *et al.*, 1973). In both cases, perhaps, protein synthesis was being inhibited and hence the primary effect of the inhibitor was not related to the activity or concentration of either cytokinin or auxin.

B. Gigantism

The outstanding example of gigantism in plants is the Bakanae disease of rice. The causal agent, *Gibberella fujikuroi,* has been shown to produce gibberellin in culture. If applied to the plant, gibberellin can cause effects similar to those seen after fungal infection. This early observation led to the characterization of gibberellic acid. Since then, more than 40 gibberellins have been isolated from fungi and from higher plants (Takahashi, 1974; Varner and Ho, 1976). It is not surprising, therefore, that other cases of extreme elongation are associated with plant disease (see Braun, 1959).

In general, greater growth in infected than in healthy plants growing under favorable environmental conditions would seem to suggest a growth substance imbalance. Whether gibberellic acid alone is responsible for all of these effects is unclear, however. Bailiss and Wilson (1967) reported that both auxin and gibberellin concentrations increased in thistles infected with *Puccinia punctiformis.* Thus these two substances may act synergistically.

III. DEVIATIONS FROM NORMAL FORM

Higher plants show an amazing similarity from generation to generation. The number, size, and arrangement of floral parts, the positioning of leaves on the stem, and the growth habit of the root and the stem are so regular as to make a given plant instantly recognizable to the trained taxonomist. In many cases, however, organs of diseased and healthy plants differ markedly in form. Roots may be short and deformed, leaves may be enlarged and thickened, petals may resemble leaves, etc.

A. Changes in Vegetative Parts

A certain amount of plasticity of form does exist in healthy plants. Leaves in the shade often expand to a much greater size than those in the sun. In fact, in shade-tolerant plants, the increase in leaf area may almost compensate for the decreased net assimilation in the shade (see Leopold and Kriedemann, 1975). Also day length and night temperature can alter the growth habit of a plant considerably. Mint plants grown in a controlled environment with long days and warm nights are erect and bushy, whereas with short days and cool nights they are relatively short and prostrate (Steward, 1968). Even so, the changes due to the environment are often minor in contrast to those that occur in some diseased plants.

1. Roots

Root-knot nematodes induce the formation of giant cells on which the nematodes feed after penetrating the roots of their hosts. Viglierchio (1971) and others have shown that these nematodes produce auxin. Thus it is tempting to speculate that localized infusion of auxin could cause the localized cell expansion. However, Dropkin et al. (1969) have shown that cytokinin treatments of roots of resistant plants can eliminate resistance and that the loss of resistance is associated with the production of giant cells not normally produced in such high frequencies in resistant plants. Thus, the relative concentrations of both growth substances may be critical for the formation of the abnormally large cells.

Some particularly interesting alterations take place when roots of crucifers are infected with *Plasmodiophora brassicae*. Root tissues grow rapidly after the infection and form enlarged areas called clubs. Both auxin and cytokinin concentrations appear to be greater in the tissues of infected than in those of healthy plants (see Dekhuijzen, 1976). In fact, excised, infected tissue can be grown in culture without the addition of cytokinins to the medium as long as the pathogen is present. In healthy tissue cultures, both auxin and cytokinin are required for growth (Williams et al., 1969).

Similarly, concentrations of IAA and cytokinins are greater in nodulating tissues of legumes infected with *Rhizobium* sp. than in uninfected root tissue (Syōno et al., 1976). The concentration of cytokinins is greatest during the growth of the nodules and falls after growth slows. Thus, transformation of relatively quiescent root tissue into actively growing nodular tissue is accompanied by increases in concentrations of the same phytohormones that are required for the culture of tissue explants. It is tempting to say that localized infusion of cytokinin (and perhaps auxin) can cause such division and expansion. Again, however, caution should be observed because of the fact that the possibility that other growth substances might be involved has not been examined.

2. Stems

In darkness or in deep shade, plants often have elongated, thin stems. Rice plants infected with *G. fujikuroi* also have extremely long, thin stems. In the former case, the low activity of IAA oxidase may be a factor. In the latter case, excess gibberellic acid may be involved. In contrast, abnormally thick stems often are associated with witches' brooms and fasciations. Here an excess of cytokinins is postulated. Perhaps the most dramatic alteration of stem form is seen in crown gall (see Merlo, Chapter 9, this volume). In this disease it appears that the ability to be self-sufficient in hormone production is imparted to the transformed tissue. Normal stem tissues do not completely lack the capacity to generate gall-like tissues, however. Skok (1967) continually debudded tobacco plant stems until all of the preformed buds had been removed. Almost as if in frustration, the plants eventually produced gall-like structures on the stem. A complete examination of the growth substance balances of these tissues could be instructive.

3. Leaves

Leaves are peculiar structures which, unlike most plant parts, are determinate in growth. Many of the cells that will constitute the mature leaf are already found in the bud and rapid expansion of these cells produces normally shaped lamina. Cytokinins and gibberellins appear to promote lateral growth of the lamina, whereas IAA appears to affect longitudinal growth. The first leaves initiated from tobacco callus cultures with high cytokinin concentrations are almost round. Leaves initiated after transplanting shoots to media with lower cytokinin concentrations assume the typical lanceolate shape. Further, the loss of chlorophyll protein and nucleic acid from senescing leaves can be slowed by the applied cytokinins (Osborne, 1962). Thus, it is not surprising that pathogens that interfere with the balance of growth substances can cause substantial modification of leaves. For example, the curved deformed leaves of peach infected with *Taphina deformans* have slightly elevated auxin and cytokinin concentrations (Sziraki *et al.*, 1975).

The bean rust case mentioned previously is also of interest. Not only can the chlorophyll be retained in and the nutrients diverted to the infected area, but cell division and considerable thickening of the leaves also occur (Dekhuijzen, 1976). Bean leaves treated with cytokinins to prevent senescence may also have a thickened, leathery appearance (Fletcher, 1969). In contrast, leaves on plants suffering from root disorders may be long and narrow. Because there is some evidence that cytokinin export from roots may be important for normal leaf development, it is possible that the abnormally long narrow leaves are the result of cytokinin deficiency.

Stunting and the production of elongated, almost grasslike leaves are characteristic of the "frenching" syndrome in tobacco and other plants treated with isoleucine and alloisoleucine (Woltz and Littrell, 1968). Kefford (1959) reported that IAA concentrations in affected plants were lower than in normal plants. A reexamination of the balance of growth substances in plants showing "frenching" symptoms may be useful since Shibaoka and Thimann (1970) found that serine antagonizes the prevention of senescence by kinetin. Kefford's study was done before the demonstration that cytokinins occur naturally in plants.

B. Changes in Reproductive Structures

Reproductive structures are also modified by certain pathogens. Inflorescences may be extremely elongated; stamens may be changed into petals or into leaflike structures; anthers may contain fungal spores rather than pollen; ears of corn are greatly modified by corn smut. Examples of modifications to such extremes are very limited in noninfected plants.

Environmental conditions can modify the ultimate structure produced at a given node. Many years ago Nitsch et al. (1952) demonstrated that day length and night temperature modified sex expression in cucurbits. Long days and warm nights delayed the appearance of pistillate flowers, whereas short days and cool nights hastened their appearance. In other studies, Nitsch (1950) showed that auxin coming from the seeds, or supplied exogenously, modified development of strawberry receptacles. These examples show that reproductive parts can be modified by the environment and phytohormones as well as by disease interactions.

Although the causes of modifications in plant parts remain unclear, a few generalizations may be drawn from studies of these symptoms. In many cases the symptoms are restricted to the area of actual contact between pathogen and host. For example, deformed ears of corn are associated with spore formation by Ustilago. Thus it is tempting to speculate that a direct contribution of the pathogen to growth substance imbalance in host cells is very important in the redirection of development.

IV. LOSS OF CORRELATED CONTROL

Correlative control mechanisms appear to be disrupted in many diseased plants. Roots appear adventitiously; dormant buds in leaf axils are suddenly released; plants bolt prematurely without the cold treatments normally required; seeds germinate in the fruit rather than after fruits are shed. In deviations of these types it is particularly attractive (and sometimes probably valid) to hypothesize that the loss of correlative control results from an imbalance in phytohormones.

A. Mechanisms of Correlated Control in Normal Plants

Apical dominance has long been used as a descriptive term for the fact that the apex exerts strong control of growth of the tissue systems below it. The release of subapical buds by loss or impairment of the apex is well known. A major advance toward an explanation of this phenomenon occurred when Wickson and Thimann (1958) demonstrated that kinetin applications would stimulate growth of lateral buds. These studies were extended by Sachs and Thimann (1967) and led to the idea that relative concentrations of cytokinins and of auxin control growth of the axillary buds.

B. Release from Apical Dominance in Diseased Plants

The so-called "witches' brooms" are spectacular examples of loss of correlative control. Numerous buds on shoots are released. These form new shoots with new buds and the new buds are released. In the case of cacao infected with *Marasmius perniciosus,* Krupasagar and Sequeira (1969) showed that extracts of the fungus were particularly effective in destroying auxin. In effect, the fungus created an excess of cytokinin by lowering auxin concentrations. *Corynebacterium fascians* achieves the same ends but by very different means. Thimann and Sachs (1966) suggested that the release from apical dominance associated with fasciation disease was caused by cytokinin produced by the bacterium. Subsequently, Klämbt et al. (1966) and Helgeson and Leonard (1966) showed that N^6-(3-methyl-2-butenylamino)purine, a very potent cytokinin, was produced in *C. fascians* culture filtrates.

Excess cytokinin may also occur in the case of false broom rape of tobacco. Hamilton *et al.* (1972) showed that kinetin application to roots or detopping the stems promoted bud formation on the roots. Thus fungal or bacterial infection, or certain environmental conditions, can lead to an excess of cytokinin relative to auxin. This imbalance will tend to cause the loss of normal bud suppression of the plant. An important point to reemphasize is that an excess of cytokinin can result either from an increase in cytokinin or by a decrease in the amount of auxin. Furthermore, as indicated by Sachs and Thimann (1967), gibberellins may be involved in the growth of buds after they have been released.

C. Formation of Adventitious Roots

Auxin applications to stem cuttings have long been used to facilitate rooting. It is not surprising, therefore, that appearance of adventitious roots on stems could be ascribed to hyperauxiny in plants. This imbalance apparently occurs in tobacco plants infected with *Pseudomonas solanacearum*. Much of the increase in auxin appears to be due to the host plant (Phelps and Sequeira, 1968). An increase in tryptophan after infection (Pegg and Sequeira, 1968) seems to provide an increased pool of precursor for auxin synthesis. In the *P. solanacearum* system there is good evidence that other growth substances are involved as well. Ethylene (Bonn *et al.*, 1975) and abscisic acid (Steadman and Sequeira, 1970) are produced in excess amounts. In the former case, the bacterium has a considerable capacity for synthesis of ethylene. In the latter case, the cause of the ABA increase is not known. It is possible that the water stress associated with the wilting syndrome may trigger the synthesis by the host. The data of Wright (1977) suggest that there is not a causal relationship between increases in ethylene and abscisic acid.

An interesting observation by Lau *et al.* (1977) may bear on the involvement of growth substances in rooting. They found that cytokinin applications greatly enhanced ethylene production by mung bean cuttings provided IAA or calcium ions were also present. Cytokinins alone were ineffective in stimulating ethylene production. Hopkins and Durbin (1971) used the mung bean system as a bioassay for root promoting substances elaborated by *Agrobacterium rhizogenes*. They found that the amount of indoleacetic acid produced by *A. rhizogenes* could not completely duplicate the rooting effect of the bacterial culture filtrates. Kinetin (at 10 μm) suppressed rooting. Ethylene production was not examined by Hopkins and Durbin nor was rooting measured by Lau *et al.* A detailed examinaion of variation in both rooting and growth substances could be most productive with this host–pathogen system.

V. SYNTHESIS AND CONCLUSION

A. Characteristics of Growth Abnormalities

Abnormal growth is observed in a diseased plant because the host has time to react before it is destroyed. In many plant diseases, the host is killed or the plant tissues are disintegrated before a growth stimulus can cause a response by the plant. Rather than merely feeding on the plant, the pathogens which cause growth abnormalities direct the growth and/or metabolism of the plant to their own benefit. Viruses redirect protein and nucleic acid synthesis toward synthesis of new virus. A fungus may even use reproductive structures of higher plants to protect its own spores. Information on the mechanisms by which pathogens redirect the plant can be valuable for understanding the diseases as well as for understanding the normal development of plants.

At one extreme, effects brought about by pathogens may be localized at the point of pathogen impact; at the other extreme, these effects may be seen throughout the whole plant. In the Bakanae disease, the entire plant elongates dramatically and its leaves become chlorotic. Where dwarfing occurs, much less biomass is accumulated by the plant. By comparison, the bean rust symptoms are localized in leaves and abnormalities are less obvious in other parts of the plant. In plants inoculated with *Agrobacterium tumefaciens,* galls develop at the point of inoculation. Although secondary galls may develop elsewhere in some cases, the growth aberrations are highly localized.

The apparent localization of a disease symptom can be misleading. Abscisic acid builds up in leaves as a consequence of decreased uptake of water by roots. Nutrients in the plant are redirected to galls or rust lesions. Thus, a disease problem in one part of a plant could result in systemic effects on other parts of the plant.

B. Relationship between Disease Physiology and Plant Physiology

The dramatic growth abnormalities associated with some plant diseases provided information for research on phytohormones in healthy plants. Gibberellins might be unknown today if it had not been for Kurosawa's observation that extracts of *G. fujikuroi* caused stem elongation. Certainly the fungus has provided an excellent system in which to examine biosynthetic pathways of gibberellins. Similarly, the demonstration of the natural occurrence of the highly active cytokinin N^6-(3-methyl-2-butenylamino)purine was greatly aided by the observation that kinetin and extracts of *C. fascians* had similar effects. The demonstration that this bacterium elaborated the cytokinin prompted the search for other cytokinins in bacteria and provided impetus for studies on cytokinin metabolism.

The discoveries of gibberellins and cytokinins were made by investigators who were not plant pathologists. These chemists and plant physiologists were alerted by plant pathologists who observed the disease symptoms, but worked independently of the plant pathologists. However, much of the speculation as to how microorganisms may redirect growth of diseased plants involves extrapolations from studies of plant growth substance effects on normal plants. Carrots infected by big bud birus may bolt and flower prematurely. The classic case with gibberellic acid effects on *Hyoscyamus* certainly might suggest, as a working hypothesis, that an excess of gibberellin in the carrot plant is responsible for premature bolting. Thus, cross fertilization of research by chemists, biochemists, plant physiologists, and plant pathologists has been considerable. It is likely that a concentrated effort toward interdisciplinary cooperation rather than independent research would result in more progress, both in disease physiology and in plant physiology.

C. An Integrated Approach for Studies of Abnormal Growth

To conclude this chapter, I would like to present some personal views on a possible integrated approach toward studies on abnormal growth. In retrospect, the approach may appear to be impossibly broad in scope. It certainly is very basic in nature. Considerable time will be required before concrete results can be documented.

The influence of environmental factors should be an important component of research on abnormal growth. Increases in abscisic acid are seen with water stress; zinc deficiency interferes with auxin metabolism; temperature affects growth rates as well as disease susceptibility. Thus, the physiological status of the plant before and during challenge by, and its response to, the pathogen should be considered.

When examining the status of the plant, the whole array of growth substances should be considered. Caution should be exercised in attributing a growth deformity to a single internal factor. In tobacco plants infected with *Pseudomonas solanacearum*, auxin metabolism was first examined, then ABA was found, and finally, ethylene evolution was noted. Detailed investigations of possible cytokinin and gibberellin involvement in this plant disease are yet to come. However, it is rather likely that this area will be investigated, especially in light of the report that cytokinins can suppress the hypersensitive reaction in leaves (Novacky,

1972) and the demonstration (see Sequeira, 1976) that this reaction is associated with resistance. Unfortunately, even when such investigations have been done, the conclusions may vary because of the different conditions used in the experiments. Only when the whole spectrum of growth substance involvement in this disease is examined at the same time in the same laboratory with standardized techniques will the relative contributions of the various phytohormones become clear.

It still cannot be assumed that the full array of plant growth substances is known. For years plant physiologists have argued over the existence of florigen. In 1959, when Braun described abnormal growth in the first edition of this treatise, auxins and gibberellins were known to exist in plants, although the occurrence of the latter group was very new. At that time, cytokinins had not been isolated from higher plants. Furthermore, the importance of ethylene and the existence of abscisic acid were not recognized. Additional phytohormones, particularly inhibitors, may yet be discovered. Both studies on growth effects and quantitative chemical analyses for the known growth substances will aid in the search for new phytohormones. Discrepancies between in vitro applications of phytohormone mixtures and the observed effects of the same balance in vivo may provide leads on the kinds of compounds to be sought.

An integrated study will require careful choice of bioassays. In addition, there should be considerable effort to demonstrate that the results obtained are the effects of the active materials being assayed for and not of modified products or of additional modifiers in the system. For example, although cytokinins delay senescence in leaves, assays based on this phenomenon are not specific for cytokinins. The tobacco callus or soybean callus assays are more specific. In the tobacco system, a response in terms of an increased logarithmic growth rate is obtained with increasing concentrations of cytokinin (Helgeson et al., 1969). Total growth of the callus tissue, however, is based on the amount of sugar in the medium (Upper et al., 1970). Thus, if tissues are not weighed soon enough, they may be well out of log phase by harvest time and hence erroneously low estimates of cytokinin concentrations may be obtained. Furthermore, in the tobacco assay, gibberellic acid can modify the growth rates normally ascribed to cytokinins (Helgeson and Upper, 1970). Thus, abnormally high concentrations of cytokinins may be estimated because of the higher yields obtained with gibberellins.

Sensitive quantitative techniques should be perfected. Although amino acids and other metabolites may be present in cells in millimolar quantities, growth substances may be 1,000 times more dilute. In the past, this problem of quantitative analysis of minute quantities has been the bane of a physiologist's existence. Today considerable advances have been made. Gas–liquid chromatography–mass spectroscopy (GLC–MS) techniques for gibberellic acids, abscisic acid, IAA, and cytokinins have been developed. Ethylene GLC is now commonplace. Thus, the approach of determining the actual amounts and structures of growth substances is available today and should be utilized in quantitative integrated studies.

The outlook for future research on growth abnormalities and plant diseases is exciting. Techniques for simultaneous determinations of changes in plant growth substances in diseased plants have become available. More and more is now known about the biosynthetic and degradative pathways of specific compounds. Because of the dramatic changes in growth substance concentrations during development of diseases, there is great value in using diseased plants for studies in both plant pathology and plant physiology.

References

Bailiss, K. W., and Wilson, I. M. (1967). Growth hormones and creeping thistle rust. *Ann. Bot.* (*London*) [N.S.] **31**, 195–211.

Bonn, W. G., Sequeira, L., and Upper, C. D. (1975). Technique for the determination of the rate of ethylene production by *Pseudomonas solanacearum*. *Plant Physiol.* **56**, 688–691.

Braun, A. C. (1959). Growth is affected. *In* "Plant Pathology: An Advanced Treatise" (J. G. Horsfall and A. E. Dimond, eds.), Vol. 1, pp. 189–248. Academic Press, New York.

Bushnell, W. R., and Allen, P. J. (1962). Induction of disease symptoms in barley by powdery mildew. *Plant Physiol.* **37**, 50–59.

Das, N. K., Patau, K., and Skoog, F. (1956). Initiation of mitosis and cell division by kinetin and indoleacetic acid in excised tobacco pith tissue. *Physiol. Plant.* **9**, 640–651.

Dekhuijzen, H. M. (1976). Endogenous cytokinins in healthy and diseased plants. *In* "Physiological Plant Pathology" (R. Heitefuss and P. H. Williams, eds.), pp. 527–559. Springer-Verlag, Berlin.

Dekhuijzen, H. M., and Staples, R. C. (1968). Mobilization factors in uredospores and bean leaves infected with bean rust fungus. *Contrib. Boyce Thompson Inst.* **24**, 39–52.

Dropkin, V. H., Helgeson, J. P., and Upper, C. D. (1969). The hypersensitivity reaction of tomatoes resistant to Meloidogyne incognita: Reversal by cytokinins. *J. Nematol.* **1**, 55–61.

Duniway, J. M. (1976). Water status and imbalance. *In* "Physiological Plant Pathology" (R. Heitefuss and P. H. Williams, eds.), pp. 430–449. Springer-Verlag, Berlin.

Fletcher, R. A. (1969). Retardation of leaf senescence by benzyladenine in intact bean plants. *Planta* **89**, 1–7.

Galston, A. W., and Davies, P. J. (1970). "Control Mechanisms in Plant Development." Prentice-Hall, Englewood Cliffs, New Jersey.

Goodman, R. N., Kiraly, Z., and Zaitlin, M. (1967). "The Biochemistry and Physiology of Infectious Plant Disease," pp. 232–280. Van Nostrand-Reinhold, Princeton, New Jersey.

Hall, R. H. (1973). Cytokinins as a probe of developmental processes. *Annu. Rev. Plant Physiol.* **24**, 415–444.

Hamilton, J. L., Lowe, R. H., and Skoog, F. (1972). False broomrape: A physiological disorder caused by growth-regulator imbalance. *Plant Physiol.* **50**, 303–304.

Helgeson, J. P., and Leonard, N. J. (1966). Cytokinins: Identification of compounds isolated from *Corynebacterium fascians*. *Proc. Natl. Acad. Sci. U.S.A.* **56**, 60–63.

Helgeson, J. P., and Upper, C. D. (1970). Modification of logarithmic growth rates of tobacco callus tissue by gibberellic acid. *Plant Physiol.* **46**, 113–117.

Helgeson, J. P., Krueger, S. M., and Upper, C. D. (1969). Control of logarithmic growth rates of tobacco callus tissue by cytokinins. *Plant Physiol.* **44**, 193–198.

Helgeson, J. P., Haberlach, G. T., and Vanderhoef, L. N. (1973). T-2 toxin decreases logarithmic growth rates of tobacco callus tissues. *Plant Physiol.* **52**, 660–662.

Hopkins, D. L., and Durbin, R. D. (1971). Induction of adventitious roots by culture filtrates of the hairy root bacterium, *Agrobacterium rhizogenes*. *Can. J. Microbiol.* **17**, 1409–1412.

Hsiao, T. C. (1973). Plant responses to water stress. *Ann. Rev. Plant Physiol.* **24**, 519–570.

Kefford, N. P. (1959). Some growth regulators of tobacco in relation to the symptoms of the physiological disease "Frenching." *J. Exp. Bot.* **10**, 462–467.

Klämbt, D., Thies, G., and Skoog, F. (1966). Isolation of cytokinins from *Corynebacterium fascians*. *Proc. Natl. Acad. Sci. U.S.A.* **56**, 52–59.

Krupasagar, V., and Sequeira, L. (1969). Auxin destruction by *Marasmius perniciosus*. *Am. J. Bot.* **56**, 390–397.

Lau, O., John, W. W., and Yang, S. F. (1977). Effect of different cytokinins on ethylene production by mung bean hypocotyls in the presence of indole-3-acetic acid and calcium ions. *Physiol. Plant.* **39**, 1–3.

Leonard, N. J. (1974). Chemistry of the cytokinins. *Recent Adv. Phytochem.* **7**, 21–56.

Leopold, A. C., and Kriedemann, P. E. (1975). "Plant Growth and Development." McGraw-Hill, New York.

Marasas, W. F., Smalley, E. B., Bamburg, J. R., and Strong, F. M. (1971). Phytotoxicity of T-2 toxin produced by *Fusarium tricinctum*. *Phytopathology* **61**, 1488–1491.

Milborrow, B. V. (1974). Chemistry and biochemistry of abscisic acid. *Recent Adv. Phytochem.* **7**, 57–91.

Mothes, K. (1964). The role of kinetin in plant regulation. *In* "Régulateurs naturels de la croissance végétate" (J. P. Nitsch, ed.), pp. 131–140. C.N.R.S., Paris.

Nitsch, J. P. (1950). Growth and morphogenesis of the strawberry as related to auxin. *Am. J. Bot.* **37**, 211–215.

Nitsch, J. P., Kurtz, E. B., Liverman, J. L., and Wert, F. W. (1952). The development of sex expression in cucurbit flowers. *Am. J. Bot.* **39**, 32–43.

Novacky, A. (1972). Suppression of the bacterially induced hypersensitive reaction by cytokinins. *Physiol. Plant Pathol.* **2**, 101–104.

Ockerse, R., and Galston, A. W. (1967). Gibberellin auxin interaction in pea stem elongation. *Plant Physiol.* **42**, 47–54.

Osborne, D. J. (1962). Effect of kinetin on protein and nucleic acid metabolism in Xanthium leaves during senescence. *Plant Physiol.* **37**, 595–602.

Paleg, L. G., and West, C. (1972). The gibberellins. *In* "Plant Physiology: A

Treatise" (F. C. Steward, ed.), Vol. 6B, pp. 146–180. Academic Press, New York.

Pegg, G. F. (1976a). Endogenous auxin in healthy and diseased plants. In "Physiological Plant Pathology" (R. Heitefuss and P. H. Williams, eds.), pp. 560–581. Springer-Verlag, Berlin.

Pegg, G. F. (1976b). The involvement of ethylene in plant pathogenesis. In "Physiological Plant Pathology" (R. Heitefuss and P. H. Williams, eds.), pp. 582–591. Springer-Verlag, Berlin.

Pegg, G. F. (1976c). Endogenous gibberellins in healthy and diseased plants. In "Physiological Plant Pathology" (R. Heitefuss and P. H. Williams, eds.), pp. 592–606. Springer-Verlag, Berlin.

Pegg, G. F. (1976d). Endogenous inhibitors in healthy and diseased plants. In "Physiological Plant Pathology" (R. Heitefuss and P. H. Williams, eds.), pp. 607–616. Springer-Verlag, Berlin.

Pegg, G. F., and Sequeira, L. (1968). Stimulation of aromatic biosynthesis in tobacco plants infected with Pseudomonas solanacearum. Phytopathology 58, 476–483.

Phelps, R. H., and Sequeira, L. (1968). Auxin biosynthesis in a host–parasite complex. In "Biochemistry and Physiology of Plant Growth Substances" (F. Wightman and G. Setterfield, eds.), pp. 197–212. Runge Press, Ottawa.

Phinney, B. O. (1961). Dwarfing genes in Zea mays and their relation to the gibberellins. In "Plant Growth Regulation" (R. Klein, ed.), pp. 489–501. Iowa State Univ. Press, Ames.

Ray, P. M. (1974). The biochemistry of the action of indoleacetic acid on plant growth. Recent Adv. Phytochem. 7, 93–122.

Richmond, A., and Lang, A. (1957). Effect of kinetin on protein content and survival of detached Xanthium leaves. Science 125, 650–651.

Sachs, T., and Thimann, K. V. (1967). The role of auxins and cytokinins in the release of buds from dominance. Am. J. Bot. 54, 136–144.

Sequeira, L. (1963). Growth regulators in plant disease. Annu. Rev. Phytopathol. 1, 5–30.

Sequeira, L. (1973). Hormone metabolism in diseased plants. Annu. Rev. Plant Physiol. 24, 353–380.

Sequeria, L. (1976). Induction and suppression of the hypersensitive reaction induced by phytopathogenic bacteria: Specific and nonspecific components. In "Specificity in Plant Diseases" (R. K. S. Wood and A. Granati, eds.), pp. 289–306. Plenum, New York.

Shibaoka, H., and Thimann, K. V. (1970). Antagonism between kinetin and amino acids. Plant Physiol. 46, 212–220.

Skok, J. (1967). Tumor and teratoma induction in tobacco plants by debudding. Plant Physiol. 42, 767–773.

Skoog, F., and Schmitz, R. Y. (1972). Cytokinins. In "Plant Physiology: A Treatise" (F. C. Steward, ed.), Vol. 6B, pp. 181–212. Academic Press, New York.

Skoog, F., Hamzi, H. Q., Szweykowska, A. M., Leonard, N. J., Carraway, K. L., Fujii, T., Helgeson, J. P., and Leoppley, R. N. (1967). Cytokinins: Structure/activity relationships. Phytochemistry 6, 1169–1192.

Stahl, C., Vanderhoef, L. N., Siegel, N., and Helgeson, J. P. (1973). Fusarium tricinctum T-2 toxin inhibits auxin-promoted elongation in soybean hypocotyl. Plant Physiol. 52, 663–666.

Steadman, J. R., and Sequeira, L. (1970). Abscisic acid in tobacco plants: Tentative identification and its relation to stunting induced by Pseudomonas solanacearum. *Plant Physiol.* **45**, 691–697.

Steward, F. C. (1968). "Growth and Organization in Plants," pp. 23–25. Addison-Wesley, Reading, Massachusetts.

Syōno, K., Newcomb, W., and Torrey, J. G. (1976). Cytokinin production in relation to the development of pea root nodules. *Can. J. Bot.* **59**, 2155–2162.

Sziraki, I., Balázs, E., and Király, Z. (1975). Increased levels of cytokinin and indoleacetic acid in peach leaves infected with *Taphrina deformans. Physiol. Plant Pathol.* **5**, 45–50.

Takahashi, N. (1974). Recent progress in the chemistry of gibberellins. *In* "Plant Growth Substances 1973" (S. Tamura, ed.), pp. 228–240. Hirokawa Publ. Co., Tokyo.

Thimann, K. V. (1972). The natural plant hormones. *In* "Plant Physiology: A Treatise" (F. C. Steward, ed.), Vol. 6B, pp. 3–332. Academic Press, New York.

Thimann, K. V., and Sachs, T. (1966). The role of cytokinins in the "fasciation" disease caused by *Corynebacterium fascians. Am. J. Bot.* **53**, 731–739.

Upper, C. D., Haberlach, G. T., and Helgeson, J. P. (1970). Limitation of tobacco callus growth by carbohydrate availability. *Plant Physiol.* **46**, 118–122.

Van Andel, O. M., and Fuchs, A. (1972). Interference with plant growth regulation by microbial metabolites. *In* "Phytotoxins in Plant Disease" (R. K. S. Wood, A. Ballio, and A. Granati, eds.), pp. 227–249. Academic Press, New York.

Varner, J. E., and Ho, D. T. (1976). Hormones. *In* "Plant Biochemistry" (J. Bonner and J. E. Varner, eds.), 3rd ed., pp. 713–770. Academic Press, New York.

Viglierchio, D. R. (1971). Nematodes and other pathogens in auxin-related plant growth disorders. *Bot. Rev.* **37**, 1–31.

Wickson, M., and Thimann, K. V. (1958). The antagonism of auxin and kinetin in apical dominance. *Physiol. Plant* **11**, 62–74.

Williams, P. H., Reddy, M. N., and Strandburg, J. O. (1969). Growth of non-infected and Plasmodiophora brassicae infected cabbage callus in culture. *Can. J. Bot.* **47**, 1217–1221.

Woltz, S. S., and Littrell, R. H. (1968). Production of yellow strapleaf and similar diseases with an antimetabolite produced by Aspergillus wentii. *Phytopathology* **58**, 1476–1480.

Wood, R. K. S. (1967). "Physiological Plant Pathology," pp. 228–286. Blackwell, Oxford.

Wright, S. T. C. (1977). The relation between leaf water potential and the levels of abscisic acid and ethylene in excised wheat leaves. *Planta* **134**, 183–189.

Yang, S. F. (1974). The biochemistry of ethylene: Biogenesis and metabolism. *Recent Adv. Phytochem.* **7**, 131–164.

Chapter 9

Crown Gall—A Unique Disease

DONALD J. MERLO

I. INTRODUCTION

In 1907, Erwin Smith and Charles Townsend reported a unique plant disease which has probably piqued the imaginations and spurred the research efforts of more plant scientists than any other plant disease. This disease, crown gall, is one of several types of tumors which occur on plants. It is unique in that (1) the causative organism is a gram-negative bacterium, *Agrobacterium tumefaciens* (Smith and Townsend, Conn); and (2) the disease occurs on an extremely broad range of host plants. As it is not my intention to review the very extensive literature of crown gall, I will refer the reader at this point to recent reviews (Braun, 1972; Lippincott and Lippincott, 1975; Kado, 1976) for a background in the basic biology of this system and for discussions of the various hypotheses advanced over the years to explain crown gall tumor induction. It is desirable, however, to point out a few milestones in the 70-year journey to our present understanding of crown gall tumorigenesis. The first of these is the proof (Smith and Townsend, 1907) that the etiologic agent is a bacterium. The second notable observations were the discoveries of Jensen (1910, 1918) and White and Braun (1941) that, once formed, tumor tissue persists in the absence of bacteria and can be maintained indefinitely in axenic culture without loss of tumorous characteristics. This observation dramatically highlighted the parallelism between crown gall tumors and many animal neoplasms.

More than one-quarter of a century elapsed before the third milestone was reached; this was the discovery by Goldmann *et al.* (1968, 1969) and Petit *et al.* (1970) that unusual amino acids that are not detected in normal plants are found in most crown gall tumors. Briefly, their observations are these. Crown gall tumors are generally found to contain either octopine [N^2-(D-1-carboxyethyl)-L-arginine] or nopaline [N^2-(1,3-dicarboxypropyl)-L-arginine]. The identity of the compound produced in the tumor is dependent upon the nature of the inciting bacterial strain, and is independent of the host plant. Bacterial strains which catabolize octopine incite octopine-synthesizing tumors; tumors incited by nopaline-utilizing strains synthesize nopaline. A third class of bacterial strains utilizes neither compound, and incites tumors which contain neither compound. The original observations of those researchers have been recently expanded by others (for example, Kemp, 1976; Montoya *et al.*, 1977; Nester *et al.*, 1977; Hack and Kemp, 1977). These reports served as compelling arguments to propose that part of the mechanism of tumor induction includes the transfer of genetic information from bacterium to plant cell.

Another recent milestone in the elucidation of the mechanism of crown gall induction was the report of Zaenen *et al.* (1974) of the presence of a large plasmid (Ti plasmid) in all tumorigenic strains of *Agrobacterium* examined. It was subsequently shown that genes controlling virulence of the bacterium (Van Larebeke *et al.*, 1975; Watson *et al.*, 1975) and the catabolism of octopine (Chilton *et al.*, 1976) or nopaline (Watson *et al.*, 1975), as well as induction of synthesis of these compounds in tumors (Montoya *et al.*, 1977), are all plasmid-coded traits. Taken together, these observations support the notion that bacterial plasmid DNA is transferred to and subsequently expressed in the host cell.

Finally, we must consider the physical confirmation of the presence (Chilton *et al.*, 1977) and transcription (Drummond *et al.*, 1977) of plasmid DNA sequences in the nucleic acids extracted from cloned tobacco tumor tissues that had been maintained axenically for several years. Indeed, these reports are especially important for several reasons: (1) they conclusively demonstrate what had previously been conjecture based on circumstantial evidence, namely, that *Agrobacterium* DNA is transferred to and expressed in the transformed plant cell; (2) they demonstrate that no natural barrier exists that prevents the exchange of genetic information between prokaryotes and eukaryotes; and (3) they suggest that there may be an underlying mechanism common to biologically induced tumors, since in the case of both crown gall and animal tumor viruses analogous events (insertion of foreign DNA into the host cell) are associated with malignant transformations.

Since the induction process and resultant neoplasms of crown gall have been the most thoroughly studied of the plant tumors, it behooves us to examine the various stages in crown gall tumor induction, and to speculate on mechanisms which may result in the loss of control over morphology and division of a healthy plant cell. I will divide the process of tumor induction into several steps, and treat each step in lesser or greater detail as the current data and available models allow. The following steps of tumor induction will be considered: (1) wounding of host plant tissue followed by specific binding of bacterial cells to plant cells, (2) transfer of bacterial DNA to the host cell, and its subsequent processing, and (3) uncontrolled growth of the transformed plant cell.

II. THE ROLES OF WOUNDING AND SPECIFIC BINDING IN CROWN GALL INDUCTION

A feature common to crown gall tumors, wound tumor virus malignancies, and some genetic tumors in tobacco is that wounding is required for the tumorous state to be initiated (Wood, 1972). In the case of crown gall, Braun has demonstrated (1952) that the period of maximum susceptibility of *Kalanchoë daigremontiana* cells to transformation to the tumor state occurs not immediately following wounding, but rather at 2 to 3 days after wounding, if the plants are held at 25°C. Lipetz (1966) found that this period of maximum susceptibility occurred about 8 hr before the first cell divisions of wound healing. It is apparent, therefore, that the wounding process does not provide merely a portal of entry for passage of the bacteria into the plant tissues.

It has been suggested (Lippincott and Lippincott, 1975) that wounding may facilitate tumorigenesis by exposing specific sites located on the plant cell wall or the cell membranes to complementary binding sites on the bacterium. Recent evidence (Lippincott and Lippincott, 1977) suggests that specific sites found in the lipopolysaccharide (LPS) fraction of the *Agrobacterium* cell envelope are important as recognition factors in specific binding to plant cells, a critical step in the infection process (Lippincott and Lippincott, 1969). Additional work by these researchers has revealed that the specific binding sites of the plant cell are found in the pectin of the cell wall middle lamella. The results further suggest that the pectin of cells susceptible to *Agrobacterium* infection differs from that of nonsusceptible cells (e.g., corn or crown gall tumor cells) by the degree of methyl esterification of the pectin, and that this modification of the pectin prevents binding of the bacterial cells to these plant cells. Plants regenerated from teratoma shoots are resistant to superinfection by *Agrobacterium* (Turgeon *et al.*, 1976), although it has yet to be determined if the mechanism of resistance involves the exclusion of bacterial binding. In any event, these results suggest that, in addition to enzymes catalyzing altered arginine metabolism (i.e., production of octopine or nopaline) crown gall cells may also contain new enzymes regulating the methyl esterification of cell wall pectins. It is not apparent what functions these altered cell walls serve in the tumorigenic process but, if one assumes that tumor induction is beneficial to *Agrobacterium* from an evolutionary standpoint, it might be advantageous to limit the number of bacteria infecting a single plant cell by such a mechanism.

Two other possible functions of wounding of plant tissues in tumor induction are: (1) provision of wound sap containing compounds that affect adjacent host cells, or nutrients utilizable by the bacteria; and (2) stimulating division of cells surrounding the wound that would normally participate in healing. Klein (1965) found that washing wounds to remove the wound sap prior to bacterial inoculation or in the first 6 hr after inoculation inhibited tumor formation, and therefore suggested that wound sap serves primarily as a stimulus for host cell conditioning and division. On the other hand, Lippincott and Lippincott (1969) found that auxotrophic mutants of *Agrobacterium* were more able to form tumors in large wound sites than in small ones. This suggested that the greater amount of wound sap in large wounds provided more of the required nutrients for the bacteria, thus allowing any metabolism necessary for tumor induction to occur. The exact nature of the requirement for wound sap has not been determined, but a possibility which has not yet been examined is the stimulation by wound sap of the formation of sex pili by the bacteria. These pili are hairlike structures responsible for the transfer of DNA between gram-negative bacteria, and may also play an important role in the transfer of plasmid DNA from *Agrobacterium* to plant cell.

The morphology of unorganized tumors suggests that the overgrowth represents a type of wound healing which started out normally, but by means of an unknown mechanism the genes controlling healing were "locked on" so that callus proliferation escaped control by the normal plant systems. Wounding would thus provide the stimulus to a large number of cells to begin dividing in a normal healing response. The insertion of bacterial DNA could then maintain the cells, either directly or indirectly (see Section IV) in the rapidly dividing state.

To summarize, there is a need for more detailed information concerning the very early stages in crown gall induction. We can easily define what is necessary for tumor induction, but we have not yet any clear ideas of the biological importance of the various features required for a successful induction. Perhaps only when we know the molecular biology of how the healthy plant responds to a wound will we be able to ascertain which mechanisms have gone awry in the formation of a tumor. Finally, if one were to search for an outstanding difference between binding of *Agrobacterium* to plant cells, and the analogous processes in other plant pathogen or plant symbiont interactions, the salient feature is the apparent noninvolvement of lectins in the binding process. Perhaps the involvement of pectin of the plant cell wall in the binding reaction is a feature which can help to explain the very broad host range of crown gall disease. This is in contrast to the specific, lectin-mediated interactions implicated in the binding of *Rhizobia* species to their leguminous symbionts (Bhuvaneswari *et al.*, 1977).

III. TRANSFER OF BACTERIAL DNA TO THE HOST CELL AND ITS SUBSEQUENT PROCESSING

Whatever the details of the early stages of infection, the works of Chilton *et al.* (1977) have shown that bacterial plasmid DNA persists through many generations of tumor cells maintained in axenic culture. It is not yet known what the intracellular location of the foreign DNA is, and one can only speculate on the steps that occur between the initial adherence of the bacterium to the plant cell wall and the final state of the foreign DNA in the cell. For simplicity, we will divide this process into three steps: (1) entrance of plasmid DNA into the plant cell, (2) replication of the plasmid DNA in the tumor cell, and (3) expression of the plasmid DNA in the tumor cell. The order of steps (2) and (3) may or may not reflect their chronological order during infection.

A. Entrance of Plasmid DNA into the Plant Cell

Following adherence of the bacterial cell to the plant cell wall, a number of biochemical events must occur. First is the provision of any physical structure(s) necessary to transfer the DNA from the bacterium to the host cell. In the simplest case, in which the plasmid DNA is released into the space immediately adjacent to the host cell, these physical structures may be minimal or nonexistent. It is possible that, as a result of adherence to the host cell, certain processes are initiated that result in lysis of the bacterial cell [e.g., induction of a lysogenic phage (Parsons and Beardsley, 1968; Tourneur and Morel, 1970)]. Plasmid DNA would therefore be released and could be engulfed by the host cell. This mechanism demands that the plasmid DNA escape the action of whatever nucleases may be present in the milieu, a protection which could be afforded by enclosing the nucleic acid in a protein coat (perhaps a crown gall virus—see review by Beardsley, 1972). This would then constitute the minimal physical structure necessary for infection.

At the other extreme, one could envision the involvement of elaborate sex-piluslike structures which would serve as vehicles for the transportation of the DNA through the membranes and cell walls of both bacterium and host cell. Synthesis of these structures would presumably be triggered by some signal from the plant and should be detectable in the early stages of infection by means of the electron microscope. Failure to detect these structures in studies to date may be due to a very low frequency of initiation of the structure during infection.

Two recent lines of evidence support the conjugation hypothesis. Kerr et al. (1977) and Genetello et al. (1977) have found that Ti plasmids carry genes which promote conjugation and transfer of Ti plasmids between bacteria. In addition, Tempé et al. (1977) have found that the transfer of the Ti plasmid during conjugation by *Agrobacterium* is inhibited at 30°C, the temperature above which tumor initiation cannot occur. This finding supports the notion that the two processes, tumor initiation and conjugal transfer of Ti plasmids, share common mechanisms. Conjugal transfer of Ti plasmids is also enhanced by the presence of octopine, and probably nopaline, in the mating medium. This result suggests an evolutionary role for the presence of these compounds in crown gall tumors. In light of all this circumstantial evidence, the author feels that some type of specific structure will be found to be essential for the efficient transfer of the plasmid DNA from bacterium to plant cell, and that the mechanism may be similar to that described for the transfer of DNA between gram-negative bacteria.

Another question to be resolved concerns the amount of plasmid DNA initially transferred from bacterium to the host cell. Chilton et al. (1977) reported that less than 4% of the plasmid DNA persists after many generations of growth of a tobacco tumor. None of the tobacco tumors examined thus far in our laboratory contain more than 8% (10×10^6 daltons) of the plasmid DNA (D. J. Merlo and R. C. Nutter, unpublished observations). The entire plasmid is evidently not required for maintenance of the tumor state, but paradoxically a large amount of genetic information seems to be necessary to confer virulence on *Agrobacterium* strains. In examinations of sizes of plasmids from many tumorigenic strains isolated around the world, no Ti plasmid smaller than about 100×10^6 daltons has been reported (Zaenen et al., 1974; Sciaky *et al.*, 1978). Since less than 5% of this DNA is required for the maintenance of the tumorous state, it is obvious that a great deal of the coding capacity of the plasmid must concern functions such as maintenance of the plasmid in the bacterial cell, transfer of the plasmid during bacterial conjugation, and the other genetic traits known to be plasmid controlled. It is possible that the plasmid genes also include functions which are expressed only in the early stages of transformation. These functions could be of two types, depending upon whether the entire plasmid or only a portion of it enters the plant cell.

In the first case, assuming that only plasmid DNA and no bacterial proteins enter the plant cell, initial manipulations of the plasmid DNA must occur by means of plant cell components. Whether these components are nuclear, cytoplasmic, mitochondrial, or chloroplastic in origin is an open question. At any rate, they would be responsible for the initial transcription and translation of the plasmid DNA. The functions of the resultant proteins could then be: (1) induction of the tumorous state (assuming that the mere presence of the bacterial DNA in the plant cell does not evoke tumor production), (2) cleavage or replication from the entire plasmid of the part necessary for tumor induction and maintenance, and (3) transcription of other parts of the plasmid not transcribable by host-cell components. It is also possible that the plant cell contains a site-specific DNA endonuclease analogous to bacterial restriction enzymes which would cleave from the entire plasmid the part necessary for tumor induction and maintenance.

In the second case, in which only a part of the plasmid is transferred to the plant cell, the selection of the piece to be transferred would occur in the bacterial cell in response to a signal from the plant. After transfer to the plant cell, maintenance would depend either on plant cell components or on new components synthesized from the plasmid DNA.

It seems likely that plant cell infection is merely a special type of normal bacterial conjugation. If the initiation of conjugal transfer of the Ti plasmid occurs at a specific site, as is known to occur in the conjugal transfer of the *Escherichia coli* chromosome and sex factors, and if the genes for tumor induction and maintenance are among the first portions of the plasmid transferred to the plant cell, then one might expect that different tumor lines would all contain a common piece of the Ti plasmid which would include the transfer initiation site, the tumor induction–maintenance genes, and perhaps varying amounts of the adjoining portion of the plasmid. This situation indeed seems to occur in tobacco tumors incited by octopine-utilizing bacterial strains. Results obtained in our laboratory (D. J. Merlo and R. C. Nutter, unpublished observations) reveal that four independently derived tumor lines all contain a common portion of the plasmid, and varying amounts of adjacent portions of the plasmid. This evidence seems to speak against the transposon model of gene insertion as proposed by Schell and Van Montagu (1977). I will discuss this model further in Section III,B and Section IV.

B. Replication of Plasmid DNA in the Tumor Cell

In most tumor cell lines examined in our laboratory the plasmid DNA sequences are present in multiple copies per cell. If one assumes that the initial infection involves the transfer of only a single copy of the plasmid sequences to the plant cell, it is obvious this DNA must be increased or amplified in the tumor cell. The exact intracellular location of these sequences is not known, and there are several mechanisms which can be proposed to explain their replication and amplification after infection.

Schell and Van Montagu (1977) have proposed that the entire region of the Ti plasmid which is transferred to and maintained in the tumor cell is flanked by insertion sequences (IS elements) and that this region therefore constitutes a transposon (Cohen, 1976; Kleckner, 1977). Insertion sequences are specific sequences of DNA bases that are recognized by specific endonucleases. The length of DNA between adjacent insertion sequences can thus be removed intact and be inserted in another DNA molecule in which similar cuts have been made. Such a sequence bounded at either end by an insertion sequence is called a transposon or transposable element. In this model, the transposable element would either integrate into plant cell DNA (either chromosomal or organellar) and then become amplified by a series of replications and reinsertions at different sites in the plant DNA, or the element would first be replicated several times as an autonomous unit, then integrate into several positions in the plant DNA. Another possibility is that tandemly repeated multiple copies of the element are found at a single site in the plant genome.

An alternative model suggests that the plasmid DNA is excised from the Ti plasmid as a transposon but does not become integrated into plant DNA. It would then exist as an autonomously replicating unit in the nucleoplasm, cytoplasm, or organellar cytoplasms.

A feature of both models is that they predict that the portion of the Ti plasmid which constitutes the transposon will exist as a defined unit of DNA of more or less invariant size. Since we know that not all tumor lines contain the same total amount of Ti plasmid, it is clear that the mechanism of plasmid transfer involves more than a simple transposon nature. However, since transposons are flanked by IS elements, and these elements are known to generate deletions in DNA, it is possible that a large transposon is transferred to the plant cell, but during subsequent growth of the tumor cell deletions in the part of the transposon not required for tumor maintenance could result in varied amounts of plasmid DNA being maintained in various tumor lines.

Whatever the mechanisms of replication and amplification, if the plasmid DNA is not integrated into the plant DNA it will be important to determine whether the multiple copies of the plasmid DNA exist as independent units (e.g., linear pieces or circles) or as large, tandemly repeated structures. It is also important to determine whether the total amount of plasmid DNA sequences present in the tumor exists as a continuous piece, or whether it is segmented and dispersed in the cell.

Much light could be shed upon the maintenance of the tumorous state if the intracellular location of the plasmid DNA in the tumor cell was known. The observation that shoots from teratoma could be seemingly cured of their tumor traits (Braun and Wood, 1976) suggests that the foreign DNA is not integrated into the plant DNA. The foreign DNA could presumably be lost through dilution during a series of rapid cell divisions. Additional support for the nonintegration hypothesis is provided by the observation that tumor traits are not retained during meiotic cell divisions (Turgeon et al., 1976). Since it is known that normal-appearing plants regenerated from teratoma shoots not only contain nopaline but plasmid DNA as well, it is likely that reversion merely reflects repression of tumor-maintenance genes which are present as extrachromosomal elements in the tumor cell. Experiments are underway in our laboratory to attempt to answer these important questions.

C. Expression of Plasmid DNA in the Tumor Cell

The transcription of plasmid DNA and the translation of the resultant RNA must occur either by the "eukaryotic" components of the plant, or by the "prokaryotic" components of the chloroplasts and/or mitochondria, depending on its intracellular location. Since, in the minimal case, the amount of plasmid DNA maintained in the tumor cell is sufficient to contain only about five genes, it is not likely that extensive numbers of Agrobacterium transcription or translation components are involved in these processes in the tumor cell. It is known that RNA is copied from the plasmid DNA in several types of octopine tumors (Drummond et al., 1977) and nopaline tumors (F. Yang, unpublished observation). Whether or not any proteins are synthesized from these RNA's has yet to be determined. Likely candidates to be synthesized are the octopine and nopaline dehydrogenases involved in synthesis of these compounds, pectin methylating enzymes, and enzymes involved in synthesis of plant hormones. To determine the identity of these proteins, the techniques of molecular cloning can be employed to amplify the particular segments of plasmid DNA known to be maintained in tumor lines. These genes can then be transcribed and translated *in vivo* (e.g., in *Escherichia coli* minicells) or *in vitro* to provide the specific gene products coded by these DNA segments. Antibodies raised against authentic octopine or nopaline dehydrogenase or other proteins could then be used to establish the identity of the proteins synthesized from the cloned genes.

IV. UNCONTROLLED GROWTH OF THE TRANSFORMED PLANT CELL

So far in this discussion we have assumed that a specific gene product of the plasmid DNA is responsible for the loss of normal control in the tumor cell. It is conceivable, however, that no transcription of the DNA is required to induce tumorous characteristics, if the plasmid DNA is integrated into crucial control positions in the plant genome. The mere physical act of integration of this foreign DNA into the plant DNA would disrupt the functioning of adjacent genes. To further define the model, I suggest that the inserted DNA might contain a special type of insertion sequence (IS), with the capacity to act as a gene promoter, and would therefore activate adjacent genes if present in the proper orientation. Flipped into the reverse orientation, it would inactivate adjacent genes.

The model demands that the plasmid DNA containing the IS element insert into the plant genome at crucial positions, for example, adjacent and "upstream" from the genes for plant hormone synthesis. Transcription originating in the IS element would result in expression of the adjacent plant genes coding for plant hormones. Similar insertions throughout the plant genome would activate (or inactivate) whatever genes are responsible for the tumorous state. Reversion of tumor traits would result in those rare cases where the IS element flips to the reverse orientation, thus restoring the normal control of plant growth. Certain features of this model are the same as the model proposed by Nevers and Saedler (1977) to explain transposable controlling elements of maize. Its distinguishing feature is that no gene product of the plasmid DNA is required for expression of tumor traits.

Since the amount of plasmid DNA maintained in tumor cells is considerably greater than the size of known IS elements, and RNA is transcribed from the plasmid DNA in the tumor cell, it seems more likely that the induction and maintenance of the tumorous state is the result of gene products coded by the plasmid DNA. In this context, it is important to note that the tumor-inducing gene products are not octopine or nopaline, since some tumor lines do not contain these compounds. In an effort to establish the involvement of certain gene products in tumor maintenance, it may be possible to incite tumor lines using mutagenized bacterial cells. Some of these mutations may confer temperature sensitivity to the specific gene products, so that one could observe reversion to normal phenotype when tumor lines are grown at elevated temperatures. Results of this type have been obtained with animal cells transformed by simian virus 40 (Renger and Basilico, 1972) and avian sarcoma virus (Graf and Friis, 1973). As a first approach to this type of experiment, Schilde-Rentschler *et al.* (1977) have isolated mutants of *Agrobacterium* which are temperature sensitive in their ability to induce tumors.

A final point to be made is the potential of the crown gall system for use as a tool for genetic engineering of plants. By means of *Agrobacterium* and the Ti plasmid, we have available a vehicle for the introduction, maintenance, and expression (at least via transcription) of foreign DNA in plant cells. The future use of this system to insert selected genes into plants remains an area of much speculation and excitement.

Acknowledgments

The author was supported by a postdoctoral fellowship from the National Science Foundation (SMI77–12352) during preparation of this manuscript. I thank M.-D. Chilton, M. H. Drummond, M. P. Gordon, A. Montoya, R. C. Nutter, and E. W. Nester for discussions and for reading the manuscript.

References

Beardsley, R. E. (1972). The inception phase in the crown gall disease. *Prog. Exp. Tumor Res.* **15**, 1–75.

Bhuvaneswari, T. V., Pueppke, S. G., and Bauer, W. D. (1977). Role of lectins in plant-microorganism interactions. I. Binding of soybean lectin to *Rhizobia. Plant Physiol.* **60**, 486–491.

Braun, A. C. (1952). Conditioning of the host cell as a factor in the transformation process in crown gall. *Growth* **16**, 65–74.

Braun, A. C. (1972). The relevance of plant tumor systems to an understanding of the basic cellular mechanisms underlying tumorigenesis. *Prog. Exp. Tumor Res.* **15**, 165–187.

Braun, A. C., and Wood, H. N. (1976). Suppression of the neoplastic state with the acquisition of specialized functions in cells, tissues, and organs of crown gall teratomas of tobacco. *Proc. Natl. Acad. Sci. U.S.A.* **73**, 496–500.

Chilton, M.-D., Farrand, S. K., Levin, R., and Nester, E. W. (1976). RP4 promotion of transfer of a large *Agrobacterium* plasmid which confers virulence. *Genetics* **83**, 609–618.

Chilton, M.-D., Drummond, M. H., Merlo, D. J., Sciaky, D., Montoya, A. L., Gordon, M. P., and Nester, E. W. (1977). Stable incorporation of plasmid DNA into higher plant cells: the molecular basis of crown gall tumorigenesis. *Cell* **11**, 263–271.

Cohen, S. N. (1976). Transposable genetic elements and plasmid evolution. *Nature* (*London*) **263**, 731–738.

Drummond, M. H., Gordon, M. P., Nester, E. W., and Chilton, M.-D. (1977). Foreign DNA of bacterial plasmid origin is transcribed in crown gall tumors. *Nature* (*London*) **269**, 535–536.

Genetello, C., Van Larebeke, N., Holsters, M., DePicker, A., Van Montagu, M., and Schell, J. (1977). Ti plasmids of *Agrobacterium* as conjugative plasmids. *Nature* (*London*) **265**, 561–563.

Goldmann, A., Tempé, J., and Morel, G. (1968). Quelques particularités de diverses souches d'*Agrobacterium tumefaciens*. *C.R. Seances Soc. Biol. Ses Fil.* **162**, 630–631.

Goldmann, A., Thomas, D. W., and Morel, G. (1969). Sur la structure de la nopaline, métabolite anormal de certaines tumeurs de crown-gall. *C.R. Hebd. Seances Acad. Sci., Ser. B* **268**, 852–854.

Graf, T., and Friis, R. R. (1973). Differential expression of transformation in rat and chicken cells infected with an avian sarcoma virus ts mutant. *Virology* **56**, 369–374.

Hack, E., and Kemp, J. D. (1977). Comparison of octopine, histopine, lysopine, and octopinic acid synthesizing activities in sunflower crown gall tissues. *Biochem. Biophys. Res. Commun.* **78**, 785–791.

Jensen, C. O. (1910). Von echten Geschwülsten bei Pflanzen. 2e Conf. Int. pour l'Étude du Cancer. *Arsskr., K. Vet.- Landbohoejsk., Copenhagen* pp. 243–254.

Jensen, C. O. (1918). Undersøgelser verdrørende nogle svulstlignende dannelser hos planter. *Arsskr. K. Vet.- Landbohoejsk., Copenhagen* pp. 91–143.

Kado, C. I. (1976). The tumor-inducing substance of *Agrobacterium tumefaciens*. *Annu. Rev. Phytopathol.* **14**, 265–308.

Kemp, J. D. (1976). Octopine as a marker for the induction of tumorous growth by *Agrobacterium tumefaciens* strain B6. *Biochem. Biophys. Res. Commun.* **69**, 816–822.

Kerr, A., Manigault, P., and Tempé, J. (1977). Transfer of virulence *in vivo* and *in vitro* in *Agrobacterium*. *Nature* (*London*) **265**, 560–561.

Kleckner, N. (1977). Translocatable elements in procaryotes. *Cell* **11**, 11–23.

Klein, R. M. (1965). The physiology of bacterial tumors in plants and of habituation. *In* "Handbuch der Pflanzenphysiologie" (W. Ruhland, ed.), Vol. 15, Part 2, pp. 209–235. Springer-Verlag, Berlin and New York.

Lipetz, J. (1966). Crown gall tumorigenesis. II. Relations between wound healing and the tumorigenic response. *Cancer Res.* **26**, 1597–1605.

Lippincott, B. B., and Lippincott, J. A. (1969). Bacterial attachment to a specific wound site as an essential stage in tumor initiation by *Agrobacterium tumefaciens*. *J. Bacteriol.* **97**, 620–628.

Lippincott, J. A., and Lippincott, B. B. (1975). The genus *Agrobacterium* and plant tumorigenesis. *Annu. Rev. Microbiol.* **29**, 377–405.

Lippincott, J. A., and Lippincott, B. B. (1977). Nature and specificity of the bacterium-host attachment in *Agrobacterium* infection. *In* "Cell Wall Biochemistry Related to Specificity in Host–Plant Pathogen Interactions" (B. Solheim and J. Raa, eds.), pp. 439–451. Norway Universitetsforlaget, Oslo.

Montoya, A. L., Chilton, M.-D., Gordon, M. P., Sciaky, D., and Nester, E. W. (1977). Octopine and nopaline metabolism in *Agrobacterium tumefaciens* and crown gall tumor cells: Role of plasmid genes. *J. Bacteriol.* **129**, 101–107.

Nester, E. W., and Chilton, M.-D., Drummond, M., Merlo, D., Montoya, A., Sciaky, D., and Gordon, M. P. (1977). Search for bacterial DNA in crown-gall tumors. *In* "Recombinant Molecules: Impact on Science and Society" (R. F. Beers and E. G. Bassett, eds.), pp. 179–188. Raven Press, New York.

Nevers, P., and Saedler, H. (1977). Transposable genetic elements as agents of gene instability and chromosomal rearrangements. *Nature* (*London*) **268**, 109–115.

Parsons, C. L., and Beardsley, R. E. (1968). Bacteriophage activity in homogenates of crown gall tissue. *J. Virol.* **2**, 651.

Petit, A., Delhaye, L., Tempé, J., and Morel, G. (1970). Recherches sur les guanidines des tissus de crown gall. Mise en évidence d'une relation biochimique spécifique entre les souches d'*Agrobacterium tumefaciens* et les tumeurs qu'elles induisent. *Physiol. Veg.* **8**, 205–213.

Renger, H. C., and Basilico, C. (1972). Mutation causing temperature-sensitive expression of cell transformation by a tumor virus. *Proc. Natl. Acad. Sci. U.S.A.* **69**, 109–114.

Schell, J., and Van Montagu, M. (1977). The Ti-plasmid of *Agrobacterium tumefaciens*, a natural vector for the introduction of *nif* genes in plants? *In* "Genetic Engineering for Nitrogen Fixation" (A. Hollander, ed.), pp. 159–179. Plenum, New York.

Schilde-Rentschler, L., Gordon, M. P., Saiki, R., and Melchers, G. (1977). Mutant of *Agrobacterium tumefaciens* with temperature sensitivity in respect to their tumor inducing ability. *Mol. Gen. Genet.* **155**, 235–239.

Sciaky, D., Montoya, A. L., and Chilton, M.-D. (1978). Fingerprints of *Agrobacterium* Ti plasmids. *Plasmid* **1**, 238–253.

Smith, E. F., and Townsend, C. O. (1907). A plant tumor of bacterial origin. *Science* **25**, 671–673.

Tempé, J., Petit, A., Holtsers, M., Van Montagu, M., and Schell, J. (1977). Thermosensitive step associated with transfer of the Ti plasmid during conjugation: Possible relation to transformations in crown gall. *Proc. Natl. Acad. Sci. U.S.A.* **74**, 2848–2849.

Tourneur, J., and Morel, G. (1970). Sur la présence de phages dans les tissus de 'crown-gall' cultivés *in vitro*. *C.R. Hebd. Seances Acad. Sci., Ser. B* **270**, 2810–2812.

Turgeon, R., Wood, H. N., and Braun, A. C. (1976). Studies on the recovery of crown gall tumor cells. *Proc. Natl. Acad. Sci. U.S.A.* **73**, 3562–3564.

Van Larebeke, N., Genetello, C., Schell, J., Schilperoort, R. A., Hermans, A. K., Hernalsteens, J. P., and Van Montagu, M. (1975). Acquisition of tumor-inducing ability by non-oncogenic agrobacteria as a result of plasmid transfer. *Nature* (*London*) *New Biol.* **255**, 742–743.

Watson, B., Currier, T. C., Gordon, M. P., Chilton, M.-D., and Nester, E. W. (1975). Plasmid required for virulence of *Agrobacterium tumefaciens*. *J. Bacteriol.* **123**, 255–264.

White, P. R., and Braun, A. C. (1941). Crown gall production by bacteria-free tumor tissues. *Science* **94**, 239–241.

Wood, H. N. (1972). The development of a capacity for autonomous growth of the crown gall tumor cell. *Prog. Exp. Tumor Res.* **15**, 76–92.

Zaenen, I., Van Larebeke, N., Teuchy, H., Van Montagu, M., and Schell, J. (1974). Supercoiled circular DNA in crown-gall inducing *Agrobacterium* strains. *J. Mol. Biol.* **86**, 109–127.

Chapter 10

Plant Teratomas—Who's in Control of Them?

CHARLES L. WILSON

I. INTRODUCTION

Man is a controlling animal. He prides himself on being at the top of the evolutionary ladder because he can control his environment. Man is also a controlled animal. The cells that make up his body impose the greatest form of control over him. He does not realize this control when his cells are functioning normally. But when his cells go awry he becomes terribly threatened. We are perhaps never more threatened than when we experience or witness cancerous growth ("teratomas" = monstrous growth). The realization that forces other than our own are controlling our cells is very unsettling. Thomas (1974) presents even the more humbling possibility that all our cells are acting out their own potential and that we are only rationalizing (not controlling) their behavior.

Because of the threatening nature of cancer, man has spent considerable resources trying to understand it. "Plant cancers" have been and will continue to be a valuable tool in this quest. Braun (1959, 1969b) has outlined the array of abnormal growths in plants and their terminology.

II. OBJECTIVE

Teratosis is the disease we are discussing; teratoma is the symptom. In this chapter I propose to deal with teratomas and how they develop. I shall not deal, except peripherally, with the effects of the disease as a disruptive influence on, e.g., photosynthesis, reproduction, or water and food transport.

In this essay my intent is to examine some of the more salient teratomal overgrowths and look for underlying similarities. I also hope to relate recent research that may point to new understanding. Since Merlo dealt with crown gall in Chapter 9 of this volume, I shall consider it only peripherally. Your understanding is needed, as I will undoubtedly fall into the trap of all those who look for simple answers to complex problems. Also, please accept the bias of my cells that are telling me what to say.

III. TYPES OF PLANT TERATOMAS

My thinking about teratomas has been clarified by arranging them in groups to answer the query, "Who's in control?" Others have classified teratomas as genetic teratomas; pathogen-induced teratomas; teratomas that result from interference of normal growth and development (e.g., fasciations, witches' brooms, phylloidy); and teratomas resulting from chemical, physical, or mechanical treatments.

Braun (1969b) has made a useful distinction between nonself-limiting (autonomous) teratomas (e.g., crown gall) and self-limiting teratomas (e.g., insect galls). Autonomous teratomas produce growth over which the host has no control, whereas self-limiting tumors are controlled in varying degrees by the host.

I shall propose three major groups: (1) teratomas under host control, (2) teratomas under pathogen control, and (3) teratomas under "foreign control."

Although we often think of teratomas and certainly of cancer as cells running wild and out of control, they are in fact under what we shall call foreign control. As in all teratomas, there are factors that dominate over time and set the direction of disease development. These dominant forces will be used as bases for the following classification.

A. Teratomas under Host Control

Teratomas that are under host control appear as exaggerated forms of normal growth such as fasciations, strap leaves, phylloidy, and hairy root. When undifferentiated tissue develops, it is self-limiting. A few teratomas are under host control even though they may have been initiated by certain pathogens or insects. Others are induced genetically or by chemical, physical, and mechanical means (Kehr, 1965).

B. Teratomas under Pathogen Control

Many teratomas such as nematode-induced root knots (Bird, 1974), *Rhizobium*-induced nodules (Allen and Allen, 1954), and certain insect-induced galls (Maresquelle and Meyer, 1965) are under the control of the pathogen. Structure and development of the tumor are dictated by the pathogen for the benefit of the pathogen and when the needs of the pathogen are satisfied, the tumor stops growing. We say that these are self-limiting teratomas. The host may benefit in the case of *Rhizobium* nodules or suffer in the case of nematode root knots. Nevertheless, the teratoma is primarily under pathogen control.

C. Teratomas under Foreign Control

Many teratomas are really "new organisms" synthesized with a new genome from the genomes of both host and pathogen. I suggest that these teratomas are under foreign control. Such teratomas are not self-limiting overgrowths. They exhibit autonomous growth. Examples are crown gall and the wound tumor virus disease. The new genome in crown gall and wound tumor virus disease results from the incorporation of extranuclear elements (ENEs) into the host (Reanney, 1976).

D. The Classification Is Functional

The advantage of a classification like the one presented is that it focuses on the nature of the host–pathogen interaction. Other classifications have emphasized the nature of the overgrowths themselves. The three proposed classes of teratomas represent varying degrees of host–pathogen involvement. The first (teratomas under host control) is characterized by an incidental relationship between the causal agent (pathogen) and the host, although in the case of genetic tumors no identifiable pathogen is known. Teratomas under pathogen control have a close pathogen–host association and interaction over the life of the teratoma. The last class (teratomas under foreign control) involves teratomas in which there is an incorporation of some host and pathogen genomes to form a new genome that directs disease development. The nature of this incorporation is not as clearly understood in plant teratomas as it is in animal teratomas, however.

IV. SIMILARITIES BETWEEN ANIMAL AND PLANT TERATOMAS

Similarities between plant and animal teratomas have been recognized since the original description of crown gall by Smith and Townsend (1907). Plant and animal teratomas are similar in that they have a wide range of environmental and genetic factors that are related to their initiation. They vary from highly self-limiting tumors to tumors that run wild and are out of control of the host (autonomous tumors).

Autonomous teratomas in both plants and animals generally appear to involve: (1) chemical or physical "wounds" as part of the initiation of the cancerous process, (2) a change to a more energy-requiring metabolism that favors protein synthesis and cell division, (3) more independence from hormonal requirements, (4) less control of cellular division, and (5) often the introduction of extrachromosomal elements (ECEs) (Reanney, 1976) into the cell.

Autonomous plant teratomas differ from animal tumors in their lack of invasiveness and metastases. Black (1965) thinks that this relates more to the properties of plant cell walls than to the agents that cause plant and animal teratomas. He also suggests that the greater malignancy of animal teratomas may be due to the presence of vital organs in animals that do not occur in plants.

V. CONTROL AND CELLULAR COMMUNICATION

Communication is the key to control in cells as it is in societies. Plant cells have intricate control mechanisms that govern their development and relationship to neighboring cells. Plant hormones play an important role in communication by providing messages that profoundly regulate the metabolism, growth, and differentiation of cells. Development of many plant teratomas can be explained as an interference with normal hormonal communication in the plant. Some pathogens themselves produce hormones that appear to contribute directly to the overgrowths. Helgeson discusses the role of hormones in plant disease in Chapter 8, this volume.

Plant teratomas in which either the host or pathogen is in control appear to involve primarily a change in intercellular communication. The garbling of messages between cells results in a distortion, dedifferentiation, or redifferentiation of growth. Autonomous teratomas involve profound changes in intracellular communication (new message from a new genome). The result is autonomous, generally undifferentiated growth.

A. Communication and the Autonomous Cell

Autonomous teratomas in plants have been of special interest because they may hold the key to understanding cancerous growth. The acquisition of autonomy in crown gall cells involves the acquisition of a new capability to produce certain hormones plus a break from the influence of extracellular hormones (Braun, 1969a). This independence is apparently made possible by the incorporation of pathogen ECEs into the genome of the autonomous cell (Lippincott, 1977).

Inter- and intracellular communication are both important in the conversion of cells to the autonomous state in crown gall. Wounding is necessary, plus the presence of *Agrobacterium tumefaciens* at the wound site for galls to be produced. Exactly why wounding is necessary (other than exposing "raw" tissue for infection) has not been fully explained (Lippincott and Lippincott, 1976).

Differentiated cells do not have as extensive cell-to-cell connections (hence communication) as meristematic cells. Perhaps wounding, by reinstating meristematic growth through the healing process, affords a better communication system for the multiplication and dissemination of the ECEs that induce tumors.

Black's wound tumor teratoma (Black, 1965) is another example in which wounding is necessary for teratomal development. The causal virus is present in the tissue but does not incite tumors unless there is a wound-healing process (with meristematic cells). The virus is present in differentiated cells, but multiplies 100-fold in the new meristematic cells that form in response to wounding. Physical cell-to-cell communication may be important during this period for dissemination of viral replicative units.

One of the outstanding characteristics of both plant and animal autonomous teratomas is a loss of cell-to-cell communication. Normal animal cells exhibit "contact paralysis" when they collide (Armstrong, 1977). This results in the establishment of intercellular connections and contributes to tissue stability and coordinated cellular division. Autonomous animal teratomal cells lose this capability and slide unrecognized by one another and fail to establish cell-to-cell communication.

Similar characteristics in plant teratomas would probably be displayed in wall-to-wall interactions since there is no direct plasmalemmal contact as there is in animals. The cell wall in plants has been found to be important in the initiation of certain plant teratomas.

B. Cell Wall Communication and Plant Teratomas

The cell wall was long regarded as a passive cellulosic sponge. Recent research has shown that this is an erroneous concept. The plant cell wall is now known to contain very specific recognition sites and it can respond to chemical, physical, and biological stimuli in a variety of ways. Some cell wall reactions are capable of either turning on or turning off future communication with pathogens.

Lippincott and Lippincott (1978) have shown that crown gall tumor initiation is inhibited by cell walls from normal dicotyledonous plants but not by cell walls from crown gall tumors. Bacterial adherence to normal walls and nonadherence to tumor walls is the apparent explanation.

Different investigators (Daub and Hagedorn, 1977; Goodman *et al.,* 1976; Sing and Schroth, 1977) have recently shown that cell walls are capable of immobilizing saprophytic bacteria while allowing parasitic ones to multiply freely. The immobilized bacteria are encapsulated by fibrillar extensions or exudations from the wall. The plant cell wall appears to act in some aspects as the plasmalemma of animals does as a site for immunological reactions.

Plant cell walls may also react in such a way as to promote future communication with beneficial symbionts. Roots of hosts infected with *Rhizobium* bacteria appear to be dependent on compatible immunological reactions in the wall (Bohlool and Schmidt, 1974; Dazzo and Hubbell, 1975).

Lalonde (1977) has found an interesting association between the walls of the root cells of *Alnus crispa* and a nodule-forming endophyte (presumably an actinomycete). The endophyte is a nitrogen fixer but it has not been grown in pure culture. Endophyte cells excrete compounds that act on growing roots of *Alnus* by causing the deformation of root hairs and the external excretion of blebs from the root cells. The blebs are excreted by the outer cell wall into the rhizosphere and migrate specifically toward the bacterial stage of the endophyte, fusing with it to form an encapsulation matrix. The matrix fuses with a deformed root hair and the endophyte penetrates the host by hyphal growth. The specificity of cell wall materials removed from the cell wall and functioning in the rhizosphere is interesting and raises the question of what controls this interaction.

VI. MECHANISMS OF TERATOMAL CONTROL

The classes of plant teratomas outlined have varying types of regulation, ranging from complete host control (e.g., genetic tumors) to teratomas under control of a foreign genome separate from the host and pathogen (e.g., crown gall). Understanding plant teratomas is understanding how these different genomes react with various self- and imposed regulatory systems of cells. We are only beginning to understand plant cellular regulatory systems themselves. Therefore obvious limitations are placed on our understanding of teratomal mechanisms.

Plant cells have an underlying potential for unlimited cell division. This characteristic is expressed during embryonic development. In fact, embryonic cells have been found to have many characteristics of autonomous teratomal cells—a concept that is better established in animals than in plants (Lippincott and Lippincott, 1976). The underlying host potential for teratomal development becomes evident in certain genetic lines that develop spontaneous overgrowths (Smith, 1972). Skok (1967) has shown that tobacco plants can be made to produce teratomas by repeated debudding. In such cases it appears that sufficient regulatory factors have been removed to deregulate the underlying genetic potential for unlimited growth.

The initiation of autonomous growth in cells is in one sense giving back to cells some characteristics which were lost. Differentiation appears to involve the gradual loss of certain metabolic capabilities. Cells gradually reacquire new metabolic capabilities while becoming autonomous (Meins and Binns, 1977). In some cases the acquisition of new ENE information is associated with the acquisition of new metabolic potential (Braun, 1969a).

How do autonomous cells acquire these new metabolic capabilities? Current arguments center around two main theories: (1) The somatic mutation theory which implies either a change in a structural gene (a point mutation) or a loss of genetic information (a deletion); and (2) the other derepression theory which predicates that genes for renewed growth are normally repressed in differentiated tissues and are "turned on" in autonomous cells. The reversion of tumor cells to the non-tumerous state weights the argument toward the derepression theory (Smith, 1972).

Finding that the introduction of new genetic information (extrachromosomal elements, ECEs) into the cell will incite tumors (Lippincott, 1977; Sambrook *et al.*, 1977) does not disprove either the mutation or derepression theory of cancer. Such new genetic information might incite a mutation or could serve to derepress the expression of existing genes. Since the insertion of ECEs may reinstate certain characteristics lost during differentiation (notably synthesis of growth regulators), this points to the loss of genetic information as a mechanism for differentiation. Extrachromosomal elements are suspected of controlling some differentiation processes in normal cells (Reanney, 1976). It is possible that autonomous growth could be turned off or on by the loss or acquisition of extrachromosomal elements.

A "restoration theory" for autonomous growth, which supposes that autonomous growth is activated by the restoration of certain genetic potential that is lost during embryogenesis and differentiation, is conceivable. Genetic tumors would be difficult to explain with such a theory unless a hybridization process involved transmission of ECEs which are active only in the hybrid. Also, latent tumor-inducing ECEs could possibly be present in the tobacco cells and become activated by the hybridization process. Meins and Binns (1977), however, recently presented evidence that neither permanent changes in the genetic constitution nor the presence of foreign self-replicating entities is necessary for tumor progression in crown gall.

A. Tumors under Host Control

When *Nicotiana langsdorfii* ($2n = 18$) is crossed with *Nicotiana glauca* ($2n = 24$) and the hybrid seed ($2n = 21$) grown into mature plants, spontaneous autonomous tumors develop. Transplanted callus tissue from the hybrid grows indefinitely on a simple, chemically defined medium that will not support tissue from either parent. Obviously it is a new organism with its own metabolism. Both parents appear to contribute to the new metabolic capability of the hybrid. *Nicotiana glauca* appears to provide the capacity for synthesis of thiamin as a factor for promoting a cell division factor, while *N. langedorfi* provides a capacity for synthesis of myoinositol (Hagen, 1969).

In nontumorous plants, one mechanism for the suppression of autonomous growth appears to be the loss of certain synthetic capabilities. The induction of tumors is through the acquisition of new synthetic capabilities.

Braun (1975) has indicated that genetic tumors of plants may provide a model for the so-called chromosomal imbalance theory of cancer, which postulates that tumors arise because of an upset in the balance between those genes that determine growth and those that are concerned with the regulation of growth. Reports of recovery to normal of tumor tissue of genetic origin somewhat confuse the present interpretation and suggest the possibility of the involvement of extrachromosomal elements (Braun, 1969a; White, 1939).

B. Tumors under Pathogen Control

Rhizobium root nodules, nematode root knots, and insect galls are caused by three very different organisms, and the effects on the host vary from beneficial to destructive. However, all three associations are similar in that they cause teratomas that are primarily under pathogen control. That the pathogen is in control is evidenced by: (1) the development of organized structures that the host cannot form independently, and (2) the removal of the pathogen stops further organized development of the teratoma.

1. Rhizobium Nodules

Rhizobium nodules are not merely sacs of bacteria that are capable of fixing nitrogen. They are highly organized, new plant organs that are capable of carrying on some very sophisticated new biochemistry. *Rhizobium* nodule development will cease if the *Rhizobium* bacteria die. Interestingly, some *Rhizobium* strains have been found that can cause tumorlike structures on legumes but cannot form N-fixing nodules (MacGregor and Alexander, 1971).

Phillips and Torrey (1972) indicate that *R. japonicum* and *R. leguminosarum* release sufficient cytokinin during their logarithmic phase of growth to form root nodules. The elaborate structure of *Rhizobium* nodules and the intricate interactions that are involved in host invasion and establishment presupposes numerous complex interactions between the host and parasite leading to gall formation.

2. Nematode Root Knots

Nematodes of the genus *Meloidogyne* induce teratomas in a wide range of crop and weed hosts. They cause the production within the teratoma of a unique, specialized tissue called a "syncytium." These are greatly enlarged, multinucleate cells packed with cytoplasm. Syncytia are formed from cell fusions and cell wall breakdown. The nematode draws food from the syncytium by inserting its stylet and withdrawing the contents.

No more syncytia are formed after the nematode dies. This suggests that excretions from the stylet of the nematode initiate the development of the syncytium. Indeed, Bird (1974) has found a histonelike protein exuded from the buccal stylet of the adult female *Meloidogyne*. He speculates that such proteins may account for the first steps in syncytium formation by stopping the differentiation of the normal root cell.

3. *Insect Galls*

Almost every order of insects has gall inducers. Of the 2,000 gall formers in the United States, 1,500 are species of gall wasps and gall midges which occur on roots, buds, stems, leaves, flowers, or fruits. Some relationships are very specialized, e.g., where a particular insect is adapted to a specific organ of one species of plant and induces the growth of a characteristic type of gall. Sometimes one can identify a plant more easily by its insect gall than by its own normal morphology.

The large number and variety of insect galls would indicate different mechanisms of gall formation. Unfortunately, the mechanism of the formation of insect galls is understood only in a few cases (Maresquelle and Meyer, 1965). That the insect pathogen is generally in control is evidenced from some of these studies since: (1) different insects produce very different galls on the same host, (2) galls on homologous flower and leaf structures have similar morphology, (3) the same insect may produce the same gall on different species of plants, (4) death of the insect causes cessation of gall development, and (5) tissue cultures from insect galls do not exhibit autonomous growth.

Insect gall tissue (prosoplasm) displays a distinctive pattern which the host is incapable of differentiating alone. The most common explanation for the production of a specific pattern is that it is controlled by a specific secretion (cecidogen) from a specific insect (Maresquelle and Meyer, 1965). Wound responses to the tunneling of the insect may also contribute to the pattern of overgrowth.

Lewis and Walton (1958) present an intriguing explanation of gall formation on *Hammelis virginiana* resulting from injections by the aphis *Hormaphis hammaelidis*. A crystalline cecidogen was found in the salivary glands, X-cells, mycetomes, and germ cells of the insect. The insect injects this material into the plant through a definite pattern of stinging punctures. Since the material multiplies in the host cells in relation to gall formation, it is probably a virus.

C. Tumors under Foreign Control

Reanney (1976) has proposed a catholic term, extrachromosomal elements (ECEs), to embrace all extracellular genetic elements, whether DNA or RNA, that control the formation of teratomas. The term includes all viruses, viroids, transfer factors, insertion sequences, and plasmid DNA. New types of ECEs will undoubtedly be discovered.

1. Viruses

Viruses are classic examples of extrachromosomal elements. When viral genes are combined with host genes in teratomal development and the new host–virus genome is perpetuated through cellular division, the teratoma is essentially under foreign genetic control. This is in contrast to host-controlled and pathogen-controlled teratomas where the two genomes keep their integrity. The genetic makeup and variability of the wound tumor virus genome has been studied (Reddy and Black, 1977) but the nature of its incorporation into host cells is not clearly understood.

The nature of the incorporation of ECEs into host chromosomes is not well understood in plant teratomas in general. Examples exist where the host has "grown out of" the association (Braun, 1969a), indicating that a reversible integration occurs. This could also result from the concomitant survival and multiplication in teratomal tissue of autonomous and nonautonomous cells. However, Braun (1969a) demonstrated that a single autonomous crown gall cell could revert to normal growth.

Naturally occurring tumors of animals are caused by both DNA and RNA viruses. Sambrook et al. (1968) have shown that during the process of teratomal transformation by adenoviruses and SV 40, rodent cells acquire new genetic information. Viral DNA sequences recombine with those of the host and are then passed on to the cell's descendants like any other part of the cellular genome. In studying the mechanism of integration Sambrook et al. (1977) concluded: "the possibility now becomes strong that mammalian cells may have the ability to integrate any piece of foreign DNA in a nonspecific way. In some cases this event would be lethal; in others, presumably the majority, no alteration in cellular phenotype would occur; in a small minority the integration event may become manifest as a consequence of a change in cellular behavior resulting from expression of the newly installed genes." They conclude that "what superficially may appear to be a masterpiece of genetic design may turn out to be an empirical policy of muddling through."

Originally it was thought that the mechanisms of control of RNA- and DNA-induced teratomas in animals would have to be different. However, the discovery that RNA-directed DNA polymerase (reverse transcriptase) exists in viral-infected tumor cells has established the concept of "reverse" flow of genetic information from RNA to DNA. Ultimately, similar DNA genes could control teratomal development whether from DNA or RNA viruses.

Reanney (1976) suggests that viroids in plants, like RNA tumor viruses in animals, are derived from cellular host sequences whose products are active in growth and morphogenesis, and adversely affect the regulatory system of the hosts. The old adage that viruses may be "chromosomes gone astray" is alive and well. Black (1965), however, believed that the wound tumor virus evolved outside its host and not as a deletion of the host nucleus since such an intricate relationship exists outside the plant between the virus and its insect vector.

2. Plasmids

Merlo discusses our current understanding of plasmids and the mechanism of gall induction by *Agrobacterium tumefaciens* in Chapter 9, this volume. Again we need information on the incorporation of plasmids into plant cells and the nature of expression of newly formed foreign genomes.

Viruses and plasmids have been involved in a number of plant teratomas. They also have a suspected involvement in others. As previously mentioned, viruses are thought to play a role in aphid-induced galls of *H. virginiana* and possibly in certain genetic tumors. Viruslike particles have also been reported incidentally in host–parasite associations such as those that exist in club-root of cabbage (Aist and Williams, 1971) and spot anthracnose of *Desmodium* (Mason and Wilson, 1978). Both diseases have associated overgrowths.

VII. TERATOMAS AND CELLULAR DIFFERENTIATION

In learning how cells acquire autonomy we may be gaining insights into how cells control themselves during differentiation. Autonomous teratomal growth is considered by some to be a reversion to embryonic growth by cells targeted to be differentiated (Lippincott and Lippincott, 1976). The most common difference between differentiated cells and autonomous teratomal cells is the greater dependence of the former on exogenous sources of growth substances. Perhaps this is a key to the mechanism whereby a control mechanism for differentiation operates in higher organisms.

As various cells become associated in tissues, it is reasonable that some of them will lose certain metabolic capabilities (perhaps under genetic control). In some instances, such cells could not compete and would cease to divide (a form of control). Others could continue growth and division if supplied with an exogenous source of the metabolite(s) that they no longer synthesize. Under these circumstances these cells give up their potential for autonomous growth and fall under the control of the cell or cells supplying the exogenous metabolite(s) (another form of control). The metabolite supplier is also partially controlled in that there is now a greater demand (hence less of a supply) of certain of its metabolites. Hormones would be particularly effective in such a scheme since slight amounts have profound effects on metabolites and differentiation.

VIII. BENEFICIAL AUTONOMOUS TERATOMAS

We often fail to consider the beneficial use of disease (Wilson, 1977). Situations exist in plant culture in which teratomal growth would be desirable (i.e., wound healing and the rooting of cuttings).

Blanchette and Sharon (1975) treated artificial wounds on *Betula alleghaniensis* with *A. tumefaciens*. They found a four- to fivefold increase in the rate of wound closure as opposed to untreated wounds. An added benefit occurred behind the treated wounds where there was better compartmentalization. Compartmentalization enhances the tree's defense against microorganisms. Further research of this nature seems warranted in the light of the failure of present wound dressings to offer much benefit (Wilson and Shigo, 1973). *Agrobacterium tumefaciens* may also hold promise as a root-promoting agent for rooted cuttings since callus formation is a prerequisite to rooting.

Differentiated animal tissues do not have the regenerative ability of plant tissues. Many animal diseases result from inefficient cellular repair systems. If our knowledge gets to the point at which we can control cellular differentiation following the artificial induction of cancer, a new approach to animal tissue and organ repair may be possible. Regenerative growth could be artificially initiated in nonregenerative tissues with a carcinogen and the new cells "shaped" (by controlled differentiation) into desirable new tissue. What a wonderful ending for the cancer story! Man would be back in full control of his own cells.

IX. SUMMARY

Plant teratomas can be classified according to the nature of genetic control that is exercised. At one extreme are genetic tumors and those induced by nonbiological pathogens (radiation, humidity, etc.) that are completely under host control. The other extreme includes teratomas (e.g., crown gall) under foreign control by a genome formed from ECEs from the pathogen and the host genome. Intermediate are teratomas induced by biological pathogens that cause the development of tissues foreign to the host and controlled by the pathogen.

Intercellular communication appears to be important in the initiation of autonomous teratomas (perhaps for the dissemination of ECEs). Intercellular communication is later abandoned by these cells when a new form of intracellular communication is established. Self-limiting teratomas appear to be dependent mainly on intercellular communication which is maintained throughout the life of the teratoma.

Plant cell walls (like the plasmalemma of animal cells) have immunological sites that are important in teratomal initiation.

Extrachromosomal elements have been implicated in a number of different plant teratomas including crown gall, wound tumor virus, insect galls, and possibly genetic tumors. How an ECE is incorporated with plant genomes is not known. It is suggested that one form of differentiation control in plant tissue may come about through the loss of synthetic capabilities by certain cells. Such cells would thereby fall under the control of other cells capable of producing the deficient metabolite. The beneficial use of autonomous teratomas for wound healing and rooting in plants and tissue regeneration in animals is a possibility.

Acknowledgments

Appreciation is expressed to Dr. James A. Lippincott and Mr. Charles Semer IV for supplying invaluable literature and Dr. Curt Leben for helpful suggestions and discussions.

References

Aist, J. R., and Williams, P. H. (1971). The cytology and kinetics of cabbage root hair penetration by *Plasmodiophora brassicae. Can. J. Bot.* **49**, 2023–2034.

Allen, E. K., and Allen, O. N. (1954). Morphogenesis of the leguminous root nodule. *Brookhaven Symp. Biol.* **6**, 209–234.

Armstrong, P. B. (1977). Cellular positional stability and intercellular invasion. *BioScience* **27**, 803–809.

Bird, A. F. (1974). Plant response to root-knot nematode. *Annu. Rev. Phytopathol.* **12**, 69–85.

Black, L. M. (1965). Physiology of virus-induced tumors in plants. *In* "Handbuch der Pflanzen physiologie" (W. Ruhland, ed.), Vol. 15, pp. 236–266. Springer-Verlag, Berlin and New York.

Blanchette, R. A., and Sharon, E. M. (1975). *Agrobacterium tumefaciens,* a promoter of wound healing in *Betula alleghaniensis. Can. J. For. Res.* **5**, 722–730.

Bohlool, B. B., and Schmidt, E. L. (1974). Lectin: A possible basis of specificity in the *Rhizobium*-legume root nodule symbiosis. *Science* **185**, 269–271.

Braun, A. C. (1959). Growth is affected. *In* "Plant Pathology" (J. G. Horsfall and

A. E. Dimond, eds.), Vol. 1, pp. 189–248. Academic Press, New York.

Braun, A. C. (1969a). "The Cancer Problem. A Critical Analysis and Modern Synthesis." Columbia Univ. Press, London and New York.

Braun, A. C. (1969b). Abnormal growth in plants. In "Plant Physiology: A Treatise" (F. C. Steward, ed.), Vol. 5B, pp. 379–420. Academic Press, New York.

Braun, A. C. (1975). The cell cycle and tumorigenesis in plants. In "Cell Cycle and Cell Differentiation" (J. Reinert and J. Holtyer, eds.), pp. 177–196. Springer-Verlag, Berlin and New York.

Daub, M. E., and Hagedorn, D. J. (1977). Reaction of resistant and susceptible beans to Pseudomonas syringae and Pseudomonas coronafaciens. Proc. Meet. Am. Phytopathol. Soc., 1977 p. 158.

Dazzo, F. B., and Hubbell, D. H. (1975). Cross-reaction antigens and lectin as determinants of symbiotic specificity in Rhizobium-clover associations. Appl. Microbiol. 30, 1017–1033.

Goodman, R. N., Huang, P. Y., and White, J. A. (1976). Ultrastructural evidence of immobilization of an incompatible bacterium Pseudomonas pisi in tobacco leaf tissue. Phytopathology 66, 754–764.

Hagen, G. L. (1969). Tumor growth in hybrid tobacco: The parental contributions. Proc. Int. Bot. Congr., 11th, 1969 Abstract, p. 82.

Kehr, A. E. (1965). The growth and development of spontaneous plant tumors. In "Handbuch der Pflanzenphysiologie" (W. Ruhland, ed.), Vol. 15, pp. 184–196. Springer-Verlag, Berlin and New York.

Lalonde, M. (1977). The infection process of the Alnus root nodule symbiosis. In "Recent Developments in Nitrogen Fixation" (W. E. Neuton, J. R. Postgate, and C. Rodriquez-Barrbeco, eds.), pp. 569–589. Academic Press, New York.

Lewis, I. F., and Walton, L. (1958). Gall formation of Hamammelis virginiana resulting from material injected by the aphid Hormaphis hamammelidis. Am. Microsc. Soc. 77, 146–200.

Lippincott, J. A. (1977). Molecular basis of plant tumor induction. Nature (London) 269, 465–466.

Lippincott, J. A., and Lippincott, B. B. (1976). Morphogenic determinants as exemplified by the crown-gall disease. Physiol. Plant Pathol. 4, 357–388.

Lippincott, J. A., and Lippincott, B. B. (1978). Cell walls of crown-gall tumors and embryonic plant tissues lack Agrobacterium adherence sites. Science 199, 1075–1078.

MacGregor, A. N., and Alexander, M. (1971). Formation of tumor-like structures on legume roots by Rhizobium. J. Bacteriol. 105, 728–732.

Maresquelle, H. J., and Meyer, J. (1965). Physiologie et morphogenèse des galles d'origine animale (zoocedidies). In "Handbuch des Pflanzenphysiologie" (W. Ruhland, ed.), Vol. 15, pp. 280–329. Springer-Verlag, Berlin and New York.

Mason, D. L., and Wilson, C. L. (1978). Fine structure study of the host–parasite relations in the spot anthracnose of Desmodium. Phytopathology 68, 65–73.

Meins, F., and Binns, A. (1977). Epigenetic variation of cultured somatic cells: Evidence for gradual changes in the requirements for factors promoting cell division. Proc. Natl. Acad. Sci. U.S.A. 74, 2928–2932.

Phillips, D. A., and Torrey, J. G. (1972). Studies on cytokinin production by Rhizobium. Plant Physiol. 49, 11–15.

Reanney, D. (1976). Extrachromosomal elements as possible agents of adaptation and development. Bacteriol. Rev. 40, 552–590.

Reddy, D. V. R., and Black, L. M. (1977). Isolation and replication of mutant populations of wound tumor virons lacking certain genome segments. *Virology* **80,** 336–346.

Sambrook, J., Westphal, H., Srinivasan, P. R., and Dulbecco, R. (1968). The integrated state of SV 40 DNA in transformed cells. *Proc. Natl. Acad. Sci. U.S.A.* **60,** 1288–1295.

Sambrook, J., Galloway, D., Topp, W., and Botcham, M. (1977). The arrangement of viral DNA sequences in the genomes of cells transformed by SV 40 or adenovirus 2. *In* "International Cell Biology" (B. R. Brinkley and K. R. Porter, eds.), pp. 539–552. Rockefeller Univ. Press, New York.

Sing, V. O., and Schroth, M. N. (1977). Encapsulation of pathogenic bacteria in the intercellular spaces of leaves in response to bacterial lipopolysaccharide. *Proc. Am. Phytopathol. Soc., 1977* Abstract, p. 93.

Skok, J. (1967). Tumor and teratoma induction in tobacco plants by debudding. *Plant Physiol.* **42,** 767–773.

Smith, E. F., and Townsend, C. O. (1907). A plant tumor of bacterial origin. *Science* **25,** 671–673.

Smith, H. H. (1972). Plant genetic tumors. *Prog. Exp. Tumor Res.* **15,** 138–164.

Thomas, L. (1974). "The Lives of a Cell." Viking Publishers, New York.

White, P. R. (1939). Controlled differentiation in a plant tissue culture. *Bull. Torrey Bot. Club* **66,** 507–513.

Wilson, C. L. (1977). Management of beneficial plant diseases. *In* "Plant Disease: An Advanced Treatise" (J. G. Horsfall and E. B. Cowling, eds.), Vol. 1, pp. 347–362. Academic Press, New York.

Wilson, C. L., and Shigo, A. L. (1973). Dispelling myths in arboriculture today. *Am. Nurseryman* **127,** 24–28.

Chapter 11

Dysfunction and Shortfalls in Symbiotic Responses

G. D. BOWEN

I. INTRODUCTION

In contrast to the general theme of the treatise, this chapter examines associations between higher plants and microorganisms that are usually mutually beneficial, i.e., symbiosis or "mutualisms" (Starr, 1975).

Symbiotic and rhizosphere microorganisms have played a critical role in the evolution of land plants and their productivity, for where plants have adapted to grow in soils of low nutrient status or where interspecific root competition for nutrients is high, the key to success has usually been a microorganism–root association rather than plant factors alone. Therefore, the following questions can be asked: How does the symbiotic plant differ from the infected? What is the energy cost of the symbiosis? What is the impact of the symbiosis on disease and how do disease and other factors decrease the potential benefit from symbiosis? I will deal particularly with legume nitrogen-fixing symbioses, ectomycorrhizae, which occur on some eight tree families, and endomycorrhizae, especially the vesicular arbuscular mycorrhizae (VAM), which occur on most angiosperms and many other plants (Harley, 1969).

II. SYMBIOTIC FUNCTION

The symbiotic plant differs from the nonsymbiotic in a number of ways, sometimes by the microorganism's changing the metabolism of the higher plant, sometimes by the addition of the microorganism's attributes to the plant's, and sometimes by unique interactions of microorganism and higher plant.

1. Fixation of Atmospheric Nitrogen

Nitrogen fixation occurs on many legume roots via the bacterium *Rhizobium* and in many other nonlegume roots by association with actinomycetes, blue-green algae, and possibly the bacterium *Spirillum*.

2. Enhancement of Ion Uptake from Soil

Ectomycorrhizae, ericaceous endomycorrhizae, and VAM increase ion uptake from nutrient-poor soils. The action is usually by growth of hyphae into soils (thus increasing the effective rooting length of the plant), absorption of poorly mobile ions such as phosphate and zinc, and their translocation back to the plant (Bowen, 1968, 1973; Sanders *et al.*, 1975). In mixed communities, infection also enables plants with lower rooting densities to compete effectively for nutrients (and water) with species with higher rooting densities, e.g., trees versus herbs, and legumes versus grasses (Bowen *et al.*, 1975).

3. Production of Growth Factors

In common with some plant diseases, most types of symbioses are accompanied by the production of growth factors. Mycorrhizal fungi provide germinating orchid seeds with sugars and sometimes vitamins (Smith, 1974). Auxin production is important in the nodule development of legumes (Dart, 1975), and in the dichotomous branching of ectomycorrhizae and in their elongation of radial walls of the cortex (Slankis, 1973). Cytokinin production has been recorded by some ectomycorrhizal fungi in culture, but not in others (Crafts and Miller, 1974) and may be implicated in the much-increased longevity of ectomycorrhiza cortical cells and in the enlarged nucleus of ecto-, endo-, and orchidaceous mycorrhizae. Cytokinins are also produced by rhizobia and by the meristematic cells of nodules (Syōno *et al.*, 1976) and green "islands" similar to those produced by kinins are associated with the symbiotic leaf nodules of *Psychotria* sp. (Becking, 1971). Enhanced auxin and kinin production by symbiotic associations could affect the transfer of ions to plant tops by affecting the permeability of cell membranes and, if they are also translocated to the plant shoots, may influence properties such as assimilation rate, translocation of assimilates, and longevity of tissues. The interesting possibilities of this have hardly been examined. It is important that we recognize symbiotic and nonsymbiotic roots as producers of translocatable growth factors as well as absorbers of water and nutrients.

4. Disease Resistance

The literature provides few data on the resistance of symbiotic plants to disease. Some ectomycorrhizae can provide protection against *Phytophthora cinnamomi* by their physical nature and production of antibiotics (Marx, 1973). Schönbeck and Schlösser (1976) have shown that endomycorrhizae increase the arginine content in the roots of tobacco plants and this suppresses chlamydospore formation in *Thielaviopsis basicola*. However, as with all symbiosis studies, it is important to distinguish "unique" effects due to the symbiosis from those caused merely by the changed nutritional status of the plant.

Similarly, possible production of "defense compounds" by symbiotic and diseased plants, such as phytoalexins and other compounds (Deverall, 1977), could have large implications in pathogen–symbiont interactions and symbiont–symbiont interactions. This line of investigation is worthy of far more study. Documented changes occurring with symbiotic systems include: (1) production of the phytoalexins, orchinol, and hircinol in mycorrhizal orchids (see Smith, 1974); (2) enhanced polyphenol production in the root cortex of mycorrhizal *Pinus radiata* (Foster and Marks, 1967) [the class of stilbenes in *P. radiata* extractives has particularly strong antifungal activity (Hillis and Ishikura, 1969)]; and (3) up to eightfold increases in volatile terpenes and sesquiterpenes (often fungistatic) with mycorrhizal infection of *P. sylvestris* by *Boletus variegatus* (Krupa and Fries, 1971).

III. ENERGY COST TO THE HIGHER PLANT

What are the costs of symbioses? Do costs ever outweigh the benefits?

A. Legume–*Rhizobium* Nitrogen Fixation

Nodules of legumes and nonlegumes require energy for growth of the nodule and its symbiont, for maintenance, and for nitrogen fixation which in turn is closely related to translocation of recent photosynthate to the nodule (Pate, 1976). Minchin and Pate (1973) found that for the period 12–30 days after sowing, the nodules of *Pisum sativum* commanded 32% of the net assimilate and of this, 16% (5% total) was used in nodule growth and 12% for respiration, the rest being transported out from the nodule in combination with nitrogen. The carbon cost/mg nitrogen fixed or /mg nitrogen assimilated by roots from fertilizer is usually considered similar, i.e., 0.3–7 mg carbon/mg nitrogen fixed or assimilated (Gibson, 1966; Pate, 1976). Such estimates usually arise from calculations of nitrogen increase per gram net carbon increase of plants over various periods. A more detailed diurnal analysis, however, of respiration of nodulated and of nitrate-fed *Trifolium subterraneum* growing at the same rate (Silsbury, 1977) suggests a considerably greater energy requirement for symbiotic nitrogen fixation (810 mg CO_2g^{-1} dry weight plant) than for nitrate uptake (510 mg CO_2g^{-1}). A probable reconciliation of these opposing views is that nodulated plants use more energy but also photosynthesize more actively (J. H. Silsbury, unpublished). If so, the reasons for this need further study; is it due to relaxation of feedback inhibition of photosynthesis, or to growth factors produced by nodules? Under conditions where fixation is impaired by addition of fertilizer nitrogen and many nodules still persist (and respire), there appears to be no plant yield reduction in the field (data of Ham *et al.*, 1976). Thus, if nodules are a significant energy drain, the plant appears to compensate for this adequately under normal circumstances.

There are instances, however, where nodulation is a drain on the plant. With ineffective or parasitic strains of rhizobia, little or no nitrogen is fixed but the carbon cost is increased, e.g., with ineffective rhizobia for *T. subterraneum*, nodule volume per plant may be 1.8–2.4 times that for effective rhizobia (Nutman, 1967). Plant growth reductions of up to 61% have occurred with parasitic rhizobia (Jordan, 1974). Little is known of the nature of parasitic rhizobia but researchers should be wary of this possible occurrence in genetic manipulation of rhizobia. Gibson (1966) has suggested that competition from developing nodules for photosynthates may be involved in significant reduction of inoculated *T. subterraneum* plants up to 12 days, a reduction which would affect subsequent plant growth.

B. Ectomycorrhizae

Data on partitioning of assimilates between tops and roots of ectomycorrhizal and nonmycorrhizal plants are conflicting. It has been concluded, on very slender evidence, that translocation to roots of mycorrhizal plants is greater than to roots of nonmycorrhizal plants (Nelson, 1964). By contrast, Ahrens and Reid (1973) demonstrated no difference in translocation of ^{14}C to roots of mycorrhizal and nonmycorrhizal *P. contorta* of the same size but their mycorrhizal plants were very poorly infected. The ectomycorrhizal system could be a large sink for assimilate. Harley (1975) calculated that, even omitting respiration, the carbohydrate in the sheaths and fruit bodies of ectomycorrhizae of a temperate forest approximate 10% of the timber production (500 kg ha^{-1}). However, even if translocation to the roots is increased with the mycorrhizae it may not be at the expense of top growth. Because of their proximity to the leaves, assimilate demands of shoot apices are likely to be satisfied first and increased translocation to the root could be achieved by extending the photosynthesis period during the day or increasing the photosynthesis rate. Under low nutrient conditions the benefits of mycorrhizae far outweigh any assimilate cost. If mycorrhizae are a significant sink for limited assimilates there may be situations where mycorrhizae are still found in highly fertile soils but are not needed by the plant and thus become an assimilate drain and reduce growth. There are some indications of this special situation with VAM.

Conflicting data exist also on the distribution of assimilates within mycorrhizal root systems. Bevege *et al.* (1975) found that mycorrhizae of *P. radiata* have 15 times the ^{14}C content of adjacent nonmycorrhizal roots, 24 hr after feeding $^{14}CO_2$. On the other hand, the data of Reid (1974) indicate that mycorrhizae of *P. ponderosa* have only one-third the specific activity of nonmycorrhizal short roots 12 days after giving a pulse of $^{14}CO_2$. Somewhat similar results have been obtained recently with *P. radiata* exposed continuously to $^{14}CO_2$ over 3 weeks (C. P. P. Reid and G. D. Bowen, unpublished). This lower specific activity of ectomycorrhizae could occur if the recently labeled assimilate is dissipated largely in respiration by mycorrhizae rather than by growth, for they respire much more than uninfected roots (Bevege *et al.*, 1975). However, the conflicting results can be reconciled more attractively if the growing elongating roots and mycorrhizae are regarded as sinks competing for available assimilate in a two-phase system. When root growth is active, less assimilate will reach the mycorrhizae and most will go to growing root tips but when this slows (and sink strength declines), much of the assimilate will go to mycorrhizal development and mycelial strand growth, thus exploring the soil further for nutrients.

C. Vesicular Arbuscular Mycorrhizae

Data on energy use by VAM are fragmentary. Bevege *et al.* (1975) found that mycorrhizae and uninfected roots of *Araucaria cunninghamii* are equally labeled following a $^{14}CO_2$ pulse but there was little development of external hyphae, which could be an appreciable sink. Fungal development within and outside the root varies considerably with host, endophyte, and environment (Mosse, 1973; Bevege and Bowen, 1975), but few quantitative studies exist. Bevege *et al.* (1975) found hyphae external to heavily infected clover roots to be 1% of the root weight of clover and this approximates that found by Sanders *et al.* (1977) with three endophytes of onion. However, growth within the root can be considerable and Hepper (1977) has found some 17% by weight of a heavily infected root can be fungal material. Similar heavy infections have been noted (S. E. Smith and G. D. Bowen, unpublished) on 8–10 day medic plants which may be a drain on the assimilate in the young plant; in such cases the fungus may compete with developing nodules for assimilates and early reduction in nodule numbers is sometimes noted. There is a need for much more detailed study.

Although high nutrition suppresses mycorrhizal formation, this appears to occur at supraoptimal levels of phosphate (Mosse, 1973). Below these levels, situations could occur in which the infection is present but is unnecessary for nutrition. It could thus be an assimilate sink and depress plant growth. Crush (1976) recorded decreases in growth of mycorrhizal *Medicago sativa* and *T. hybridum* of 3.5–16.2%, the decreases being more severe with phosphate addition. In other instances growth depressions of up to 40% were due to increased uptake by the mycorrhizae causing phosphate toxicity (Mosse, 1973). This aspect of VAM needs careful attention because attempts to maximize productivity by using considerable quantities of fertilizer may sometimes be counterproductive.

IV. THE DISEASE–SYMBIOSIS COMPLEX

A. A Whole Plant Approach to Disease and Symbiosis

A disease, malfunction of a symbiosis, or a symbiosis will be relevant only if it affects some rate-limiting factor, e.g., in a nutrient-rich soil a 10% loss of root by disease may be inconsequential, but serious crop reduction may occur with the same root loss in a soil low in nutrients. Therefore a whole plant quantitative approach must be taken to understand disease and symbiosis and to interpret these in terms which are meaningful to the plant, e.g., root growth, ion uptake, and water uptake. Bowen and Cartwright (1977) have discussed nutrition variables in crop growth in terms of a systems analysis framework in which crop growth, as shown in Fig. 1, is driven by leaf area, net assimilation rate, distribution of assimilates to roots, growth of roots, ion and water uptake by roots, and distribution of nutrients to stems, leaves, ears, etc., which in

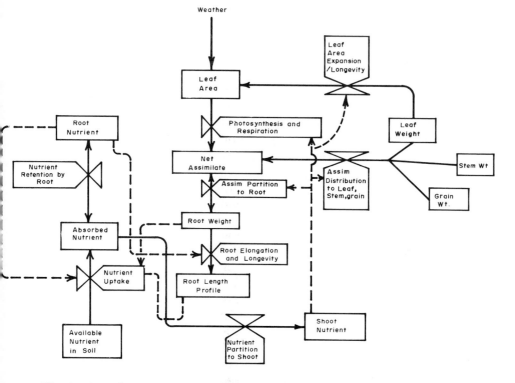

Fig. 1. Assimilation processes in plants. A scheme to represent the main compartments for assimilate and nutrient distribution in a simple model (after Bowen and Cartwright, 1977). Rectangular boxes are used for measurable variables whose values at any time describe the state of the system at that instant. These boxes are connected by continuous lines to represent material (or energy) flows. The "valve" symbols represent processes which affect the use and distribution of dry matter and nutrients. Process rates may be output values of the state variables on which they operate, or they may be affected by variables elsewhere in the system. Such controls are shown by broken lines.

turn affects leaf development and photosynthesis. Some of the salient aspects of disease and symbiosis in this framework are indicated below.

1. Leaf Area and Longevity

Leaf expansion and effective leaf area may be reduced by disease whereas relief of nutrient deficiencies by root symbioses leads to normal leaf expansion. Premature death of all or part of the leaf may result from leaf diseases but cytokinin production by nodules, mycorrhizae, increased root growth, and leaf symbionts may increase leaf longevity.

2. Net Assimilation

The rate of photosynthesis may be affected directly by leaf disease or reduced indirectly by infections acting as nutrient sinks. It will be increased by symbionts that relieve nutrient deficiencies. Respiration can be increased by leaf disease, thus lowering net assimilation.

3. Assimilate Movement to Roots

Partition of assimilate to roots is little understood except in general terms of source–sink relations and proximity to the source. Lowered assimilation caused by leaf disease or accumulation of assimilate in lesions may lead to lowered supply of assimilate to the root; e.g., a reduction of root systems occurs in cereals affected by mildews (Last, 1962). Endomycorrhizal plants usually have a similar root/shoot ratio to uninfected plants of the same size and phosphate supply, but a lower ratio than phosphate-deficient, uninoculated plants.

4. Root Length and Longevity

Infections of roots by disease organisms, symbionts, and soil pests will be determined largely by rooting density, distance over which the infecting propagule can move to the root, and the populations of microorganisms and pests (Bowen and Rovira, 1976). Epidemiology models interface with nutrition models here. Loss of roots by disease means loss of potential sites for nodulation and mycorrhiza formation.

Growth of mycelial strands and hyphae into soil from mycorrhizae increases "rooting density" at less energy cost than production of new roots and, furthermore, the hyphae are not susceptible to the same diseases as the root. Bowen *et al.* (1975) calculated that 1 mg of hyphae of 10 μm diameter had the same length of approximately 1600 mg of root of 400 μm diameter. Hyphal growth into soils is a highly effective mode of compensating for roots lost by disease, quite apart from any biological control of root disease exerted by some mycorrhizal fungi. V. Bumbieris (personal communication) found that mycorrhizal infection of *P. radiata* by *Rhizopogon luteolus* compensated for damage which would have occurred from *Phytophthora cryptogea* had the plant been nonmycorrhizal. However, mycorrhizal infection in the absence of *P. cryptogea* resulted in an 80% increase in growth.

Root longevity and nutrient uptake may be decreased by disease, but is extended, often considerably, by mycorrhizal infection (Bowen, 1973; Bowen *et al.*, 1975).

5. Use of Absorbed Nutrient

Transport of nutrients from symbiotic organs differs from that of un-infected roots. Before senescence, nodules can have twice the nitrogen content of roots (Minchin and Pate, 1973) and a much higher phosphate content. Storage of phosphate by mycorrhizae can lead to mycorrhizal roots having much higher concentrations than nonmycorrhizal roots (Harley, 1969; Sanders, 1975), especially under phosphate-rich conditions.

A potential deleterious aspect of increased nutrient absorption by my-corrhizae has been indicated by Gadgil and Gadgil (1975) who found that ectomycorrhizae of *P. radiata* decreased litter decomposition pos-sibly by competition with decomposer organisms for available nutrients. In an ecosystem depending on rapid mineralization of nutrients in litter this could decrease turnover rate, but in potentially highly leaching soils it may be beneficial.

Changes in cell composition with symbioses, and losses of substances from external hyphae of mycorrhizae as well as from the root, can lead to changes in root exudates so that the rhizosphere composition with symbiotic systems is different quantitatively and qualitatively from that of nonsymbiotic roots.

B. Compensation Strategies

Loss of some symbiotic organs does not necessarily mean decline in plant response. In the "game" of evolution and survival, plants have invested in insurances against such losses by a combination of two strate-gies described below.

1. Profuse development of the functional unit so that loss of indi-viduals is only a small fraction of the whole. Nutman (1967) showed highly effective associations of *T. subterraneum* achieved high bacteroid volume either by production of few large nodules or many small nodules —associations differing by several hundred percent in nodule number differed by only 4% in the product of bacteroid volume and longevity. It is ecologically significant that the inheritance of many nodules is dominant over few nodules—a less disastrous situation with loss of some nodules than the same loss from a sparsely nodulating clover line.

2. Increased activity of fewer units. This method of compensation has been little studied with mycorrhizae and may not occur with them, but is common with legume nodules. Increased size of fewer nodules occurs in response to adverse physical environmental factors such as

adverse soil acidity (Vincent, 1965) and in response to biological attack, e.g., by nematodes (Hussey and Barker, 1976). However, in the latter case compensation does not always occur and what happens in any particular instance is probably determined by the severity of the attack and its effect on plant growth, specifically net assimilation which "drives" root growth and nodule formation. Unfortunately most studies fail to include uninoculated fertilized controls to define effects on nodulation per se and effects due to decreased plant growth.

Compensation toward environmental fluctuation within existing nodules also occurs. Thus plants may respond to depressed nitrogen fixation caused by lower root temperatures by producing a greater volume of bacteroid tissue (Gibson, 1976). This increased growth of nodules may occur by a lessening of competing sinks for carbohydrates, e.g., lower soil temperature decreases root growth and leads to a lessening of the root apex sink for translocated assimilates (Rovira and Bowen, 1973). Similarly, following adverse conditions, degenerating nodule tissue can sometimes recover its nitrogen-fixing ability. This "resilience" is an efficient way of accommodating environmental fluctuations that affect fixation such as moisture stress, nitrogen fertilizing, and insect attack.

V. IMPEDIMENTS TO SYMBIOTIC RESPONSE

A. Dysfunction of the Legume–*Rhizobium* Symbiosis

1. Ineffective and Parasitic Associations

Little is known of the biochemical bases for ineffective and parasitic rhizobia–legume associations. (Jordan, 1974). Ineffectiveness is associated variously with the failure of bacteria to be released from the infection thread; a failure of bacteroids to form or to "mature," i.e., rapid breakdown of the bacteroids; or production of bacteroids which do not produce nitrogenase. In other instances, there is a rapid selective autolysis of host cell contents accompanying degeneration of the bacteria (Bassett et al., 1977). Most explanations are little more than descriptions of associated events or speculation. However, a biochemical basis of ineffectiveness or lowered effectiveness has been defined in some instances, e.g., the riboflavin requirement of an *R. trifolii* strain on red clover (Pankhurst et al., 1972) and an accumulation of nonprotein nitrogen in the nodules of subterranean clover which leads to less translocation of nitrogen to the rest of the plant (Freney and Gibson, 1974). Adverse environments such as high or low temperatures (presumably causing biochemical lesions) or molybdenum deficiency (Dart, 1975) can also cause ineffectiveness. It is enticing to think that if we could eliminate such dysfunctions so that all infections were highly effective, problems of shortfall in responses due to competition from indigenous, poorly effective strains would also be eliminated.

2. Competition between Symbionts

Reduction or elimination of inoculated strains by domination by poorly efficient indigenous symbionts (Ireland and Vincent, 1969) is a serious problem; e.g., Caldwell and Vest (1970), on testing 28 strains of R. japonicum on soybean in soils with indigenous strains, found that the inoculated strains produced only 5–10% of the nodules. Displacement of the native rhizobia can sometimes be achieved by increasing inoculum density considerably (Ireland and Vincent, 1969) but this does not always work (Johnson et al., 1965).

Antagonism between Rhizobium strains can occur in the laboratory either by antibiotic production (Schwinghamer and Belkengren, 1968) or by production of bacteriocins (Schwinghamer et al., 1973). The importance of such antagonistic rhizobia is unknown but they could have a large influence on the rhizobial composition of the rhizosphere. The implication of bacteriocins in control of crown gall by Agrobacterium radiobacter (Kerr and Htay, 1974) suggests it as an area worth studying with respect to rhizobial interactions.

Very large differences between Rhizobium strains can occur in rhizosphere growth and with different Rhizobium–plant combinations (Chatel and Greenwood, 1973). However, growth in the rhizosphere does not by any means fully account for competitive success; e.g., Labandera and Vincent (1975) found that the R. trifolii strain, TA1, dominated a Uruguayan strain, U73, in nodulation of T. repens and T. subterraneum by 18–94:1 when both strains were equally represented on the root surface. Success of TA1 had little relation to its relative abundance in the rhizosphere but may have been related to its consistently earlier production of nodules because although a few bacteria suffice to produce the first nodules, further infection becomes increasingly difficult.

The dynamic nature of success in nodulation in mixtures of rhizobia is illustrated by a few examples: (1) the distribution of serotypes recovered from nodules can depend on soil temperature during growth (Weber et al., 1971); (2) the relative success of two strains in competition with naturally occurring rhizobia will vary with the soil. Roughley et al. (1976) showed that two R. trifolii strains in a mixture formed 82 and 6% of the nodules of T. subterraneum, respectively, in one soil but 16 and 29%, respectively, in another; and (3) successful competition against native strains can be determined at the cultivar level. Roughley *et al.* (1976) found that introduced *R. trifolii* WU 295 formed 66% of the nodules on *T. subterraneum* cultivar Woogenellup but only 25% on cultivar Mt. Barker. In most of these cases the roles of multiplication in the rhizosphere and infectivity per se have not been separated. However, some strains do consistently perform better than others in a range of soils in the year of inoculation and in subsequent persistence (Vincent, 1965).

The soil population, including symbionts and nonsymbionts, is in a dynamic equilibrium peculiar to the soil–plant system present. Naturally occurring rhizobia, although possibly less effective than other strains, are compatible with the existing soil microflora and conditions. To introduce another organism, the system must usually be disturbed after which a new equilibrium position will evolve. If the perturbation allowing the initial introduction of an organism is maintained, that organism may persist as a significant microbial component of that system, but if the inputs are not sustained the equilibrium will tend to revert to the original or to an intermediate position. Even where no naturally occurring effective strains of rhizobia have been detected in a soil in the first year, these may occur in very small numbers, multiply under favorable conditions, and finally become a significant component of the soil population. Although it is possible to dominate nodulation with an introduced strain in the first year of an annually seeded legume, persistence in competition varies considerably with the *Rhizobium*–plant–soil combination and an inoculated strain may be almost completely eliminated by the second season (Gibson *et al.*, 1976).

3. Biological Control of Symbionts

Biological control of symbionts is just as real as biological control of pathogens, e.g., eight of nine *Rhizobium* strains studied by Chatel *et al.* (1968) multiplied in the rhizosphere of *T. subterraneum* in unsterilized soil but the other did so only if the soil had been sterilized. Large strain differences occur between *Rhizobium* strains in movement through nonsterile soil ("saprophytic competence," Chatel *et al.*, 1968) and in their ability to establish an "adequate" population in the root zone of the susceptible host ("incursion," Harris, 1954), and strains must be selected on these bases as well as on effectiveness.

A wide range of bacteria, actinomycetes, and fungi depress rhizobial growth in laboratory media but such laboratory tests are almost useless in extrapolating to interaction of microorganisms in the rhizosphere (Bowen and Rovira, 1976). However, reduction of nodulation by some microorganisms has been shown in soil, e.g., Damirgi and Johnson (1966) found a 35% reduction in soybean nodules in inoculated, sterilized soil following the introduction of an actinomycete.

Keya and Alexander (1975) recorded strains of the bacterium *Bdellovibrio bacteriovorus* parasitizing rhizobia in 32 of 90 soils examined. However, populations of rhizobia were affected experimentally only when the rhizobia were in numbers rarely found in natural soils ($> 10^6$ rhizobia g^{-1}). Similarly, although a number of protozoa can be predators on rhizobia, Danso *et al.* (1975) showed that when rhizobia were added to soil with protozoa the *Rhizobium* population rarely declined below $10^{8.5}$ g^{-1}. They suggested the size of the prey population diminishes until a density is reached at which the energy used by protozoa in hunting equals that obtained by feeding, and populations stabilize. It is thus unlikely that effects of *Bdellovibrio* or protozoa will be of major importance in the establishment of rhizobia in soil. It is important, however, to study further their effects in the rhizosphere where the spatial distribution of rhizobia is different from that in soil.

Bacteriophages that attack *Rhizobium* are extremely common and can be isolated regularly from soil and nodules. Although Kowalski *et al.* (1974) found phage for soybean rhizobia in all nine soils studied, the highest numbers were 1.4×10^3 particles g^{-1} soil. If these were uniformly distributed in soil, using the model of McCoy and Powelson (1974), the mean distance between them would be 0.9–1.0 mm, too large to intercept many rhizobia. Despite earlier claims that "rhizobiophages" are economically important, more evaluation is needed. Bruch and Allen (1955) found no growth decline in *Lotus corniculatus* when phage particles outnumbered rhizobia in inocula by three to one.

Diffusates from seed can stimulate microbial growth. On the other hand, toxins to rhizobia and a wide range of other bacteria can come from the testa of many legumes (e.g., *Centrosema pubescens*, *T. subterraneum*, *Arachis hypogaea*, and *G. max.*) (Bowen, 1961; Wahab, 1977). Wahab (1977) has shown considerable differences between the soybean variety–*Rhizobium* strain combinations in sensitivity to seed coat toxins in agar plate assays. In practice, physical separation of rhizobia from the seed coat has combated the seed coat toxin factor (Thompson, 1961).

4. Environmental Stress

Differences between symbionts in growth in soil occur with such nonbiological factors as soil acidity, high and low soil temperatures, and probably such factors as high salinity and water stress (Vincent, 1965). Within limits, physical and chemical environmental problems can be overcome by selection of strains of symbionts resistant to these conditions or by agronomic practice, e.g., changing soil acidity or other treatments. For example, desiccation and death of rhizobia in soil in hot, dry summer conditions can be prevented by montmorillonite, illite, or hematite (but not kaolinite) which form a clay "envelope" around the bacteria, reducing their internal water hydration status to a level where most enzyme activity ceases (Marshall, 1976).

Combined fertilizer and soil inorganic nitrogen can depress nodulation and nitrogen fixation (Gibson, 1976); considerable *Rhizobium* strain variation exists in this and many glasshouse studies that rank rhizobial strains in order of effectiveness in low nitrogen substrates may be of little relevance to field response. Similarly, toxicities of manganese, cadmium, cobalt, copper, and zinc have been variously recorded to depress nodule production and nitrogen fixation (Holding and Lowe, 1971; McIlveen and Cole, 1974). Döbereiner (1966) advocated the selection of manganese-tolerant *Rhizobium* strains for acid tropical soils.

Effects of high salinity on nodulation and nitrogen fixation have also received little study and it is often difficult to distinguish salinity effects from water stress. Differences in sensitivity occur between plant–*Rhizobium* associations, for Bernstein and Ogata (1966) found nodulated and nitrogen fertilized alfalfa to respond similarly to salt stress; but in soybean, *Rhizobium*-dependent plants were more severely affected by salt than were nitrogen-fertilized plants. Impairment of nitrogen fixation occurred over 0–5.4 atm and especially below 3.6 atm.

Legume nitrogen fixation is extremely sensitive to water stress; e.g., acetylene reduction in a watered grass–clover sward at soil moisture 19.2% (0 bar suction) has been recorded to be 15 times that in a non-irrigated sward under mild water stress at 16% soil moisture (-1.5 bars suction) (Sprent, 1976). Waterlogging also causes a reduction of nitrogen fixation (Sprent, 1976). Recovery from mild water stress (below 20% loss of nodule weight) involves regrowth of the nodule meristem and thus elongate nodules such as those of clover may be more resistant to stress than are the "spherical" type soybean nodules which are shed following periods as short as 24 hr stress (Wilson, 1931) and replaced subsequently by new nodules.

B. Effects of Pests and Diseases on Legume–*Rhizobium* Nitrogen Fixation

Plant diseases and pests may affect nodulation directly, or indirectly by retarding photosynthesis and root growth. In many instances the plants compensate for loss of nodules but while plant weight, nodule numbers, and nodule weight are often recorded, frequently no estimate of nitrogen fixed per gram of nodule is made. Similarly in most studies matched unnodulated plants (fed with nitrogenous fertilizers) have rarely been used as controls. This makes it impossible to distinguish the effects of the disease or pest on the plant generally from those arising uniquely from effects on the nitrogen-fixing system.

1. Fungi

The impact of fungus diseases on nodulation has been little studied. Mew and Howard (1969) found that R. japonicum reduced Fusarium root rot of soybean at pH 7, but at pH 5.2 severe root cortex necrosis and reduction or elimination of nodulation occurred. Severe root necrosis and root hair decay caused by Rhizoctonia solani attack on soybean led to a halving of nodulation (by weight) and a 20% reduction in nitrogen fixed per gram of nodule (Orellana et al., 1976). Although the fungus occurred only in the nodule outer cortex, nuclear breakdown occurred in the Rhizobium-containing cells of the central tissue, apparently by diffusion of toxin from the fungus. This raises the question of the extent to which toxins produced in other parts of a plant, and systemic pesticides may be translocated to the nodule (or mycorrhiza) and affect its functioning, for the large amount of photosynthate that reaches the nodule suggests a very efficient translocation system to it.

2. Viruses

Reductions in nodule weight per plant of up to 80% have been recorded with soybean mosaic virus, bean pod mottle virus (Tu et al., 1970a), broad bean true mosaic, pea enation mosaic, and yellow bean mosaic (Blaszczak et al., 1974) probably via their effect on plant growth, e.g., by reducing photosynthesis. Virus levels can be higher in nodules than in roots (Tu, 1973) and this may be the reason for high nitrogen levels in mosaic-infected nodules, lowered translocation of nitrogen from nodules, decreased hemoglobin, and shortened nodule life (Tu et al., 1970b). In contrast, Rajagopalan and Raju (1972) found early increases in the leghemoglobin of nodules and increased nitrogen fixation of Dolichos lablab infected with Dolichos enation mosaic virus. Nodulation was increased in sand culture but decreased in unsterilized soil.

Infection of T. repens with clover phyllody virus (thought to be a mycoplasma, Tu et al., 1970a) has led to nodules containing a much less effective Rhizobium strain than the original (Joshi et al., 1967). Isolates from these nodules were not virus-infected but were poorly effective; the virus appears to have induced genetic change to a type which, if multiplying and persisting in soil, could conceivably reduce plant response to *Rhizobium* even in subsequent clover plantings not infected by the virus.

3. Nematodes

A wide range of nodulation reactions to nematode attack on the plant has been recorded, depending partly on the severity of the attack and the ability of the plant to compensate: (1) reduction in nodule number due to reduced root growth with no effect on nitrogen-fixing efficiency but reduced longevity of the nodule (*Meloidogyne javanica* and *Heterodera trifolii* on white clover, Taha and Raski, 1969), (2) decreases in nodule number per plant (54–84%) and nodule weight per gram of root (42–67%) and in nitrogen-fixing efficiency with one of three races of *Heterodera glycines* on soybean (Lehman *et al.*, 1971), (3) increased nodule number but reduced nodule size with *Meloidogyne hapla* and *Pratylenchus penetrans* on soybean (Hussey and Barker, 1976), and (4) decreases in nodule number and size (Hussey and Barker, 1976).

With some nematodes only the outer cortex of the nodule is infested. In other cases the nematodes are restricted to the vascular bundle of the nodule but there are associated effects such as failure of bacteroid development adjacent to nematodes (Barker and Hussey, 1976). In other instances, e.g., *Acrobeloides buetschlii* on *Pisum sativum*, the nematode feeds and reproduces in the nodule, and in this case nitrogen fixation is depressed by 80–90% (Wescott and Barker, 1976).

4. Insects

Despite many records of attack on nodules and root hairs by larvae of insects, e.g., pea and bean weevil (*Ofiorrhyschus ligustici*), clover root weevil (*Amnemus quadrituberculatus*), and *Rivella* sp. (Chowdhury, 1977), there is little quantitative study of their effect on nitrogen fixation. Masefield (1955) doubled the weight of nodules per plant with partial control of *Sitona lineata* but plant growth did not increase. The severity of the attack, the dependence of the plant on nitrogen fixation, and the timing of the attack, e.g., on young or on senescing nodules, are key factors in whether plant growth is affected.

C. Shortfalls in Mycorrhizal Response

1. Competition and Establishment of Inoculated Fungi

As for the rhizobia, mycorrhizal strains vary greatly in their stimulation of the plant (Mosse, 1972; Bowen, 1973) and naturally occurring organisms may not be the most efficient. Two problems emerge—establishing an inoculant in the face of competition and antagonism from native soil flora and from indigenous mycorrhizal fungi. Introduction of selected fungi is usually much easier if the soil has been sterilized. Bowen and Theodorou (1973, also unpublished) have shown that bacteria may inhibit, depress, stimulate, or have no effect on growth of ectomycorrhizal fungi in the rhizosphere of *Pinus radiata*. Of seven fungi investigated, *Thelephora terrestris* was one of the least sensitive to several bacteria tested and this may be one reason why this fungus is one of the most common ectomycorrhizal fungi in *Pinus* forests in Australia and North America.

2. Fungal Attack

Fungal attack on ectomycorrhizae has been little studied, but a well-developed mantle gives the root some physical protection from attack as does production of antibiotics by some mycorrhizal fungi, e.g., *Leucopaxillus cerealis* var. *piceina* activity against *P. cinnamomi* (Marx, 1973). Many antibiotics have been produced by ectomycorrhizal fungi in laboratory media, but the ecological reality of these needs confirmation. Where the mantle is breached by some other agent, e.g., nematodes, the physical protection no longer exists and Barham (1972) found that entry of *P. cinnamomi* into ectomycorrhizae of *Pinus echinata* occurred following attack by the nematode *Helicotylenchus dihystera*, but not into ectomycorrhiza penetrated by *Tylenchorhynchus claytoni*.

Possible hyperparasitism of ectomycorrhizae appears not to have been studied. Growth of fungi from ectomycorrhizae is greater in sterile than in nonsterile soils (Bowen, 1973); competition and antagonism from other microorganisms limiting mycelial growth into soils could decrease mycorrhizal response considerably. Furthermore, some plants, e.g., *Calluna*, produce toxins toward mycorrhizal fungi (Robinson, 1972).

Ross and Ruttencutter (1977) identified two types of hyperparasitism of VAM: (1) growth within *Glomus macrocarpus* both inside and outside of soybean roots by a "*Pythium*-like" fungus, possibly *P. acanthicum*; and (2) parasitism of spores of *G. macrocarpus* in soil by a chytrid, a species of *Phlyctochytrium*. *Glomus gigantea* was not as susceptible and this could be important in the relative infection by a mixture of VAM fungi in soil. Hyperparasitism could seriously reduce the inoculum potential of VAM in soils.

3. Nematodes

Several root-feeding nematodes can penetrate the fungus sheath of ectomycorrhizae and reproduce in the root cortex (Ruehle, 1973). Most reports deal with histological aspects and speculate on probable impairment of the mycorrhizal response. A much more quantitative study is required. With VAM, Fox and Spasoff (1972) found *G. gigantea* and *Heterodera solanacearum* to be mutually suppressive and low populations of VAM fungi have been associated with high populations of nematodes in soybean fields (Schenck and Kinloch, 1974). This may be related to root growth reduction caused by the nematode. However, in plant–VAM–nematode studies the increase in nematode juveniles can be less than expected on a root growth basis in mycorrhizal plants (Schenck *et al.*, 1975), thus indicating a suppressive effect of the VAM on the nematodes. Ruehle (1973) found that *Pratylenchus brachyurus* attack on mycorrhizal roots caused deterioration of vesicles.

Mycophagous nematodes may play a more important role in controlling mycorrhizal development than is usually suspected. These are extremely common in soils and may depress formation of new mycorrhizae and destroy existing ectomycorrhizae (Ruehle, 1973). They would be expected to reduce mycelial growth into soils also, but quantitative data are disturbingly sparse. Similarly, their impact is potentially great on VAM, especially as mycophagous nematodes have been demonstrated experimentally to depress infection with root pathogens (e.g., Klink and Barker, 1968). Clark (1964) suggested that a mycophagous nematode *Deleadenus* sp., eating the external mycorrhizal hyphae was a prime cause of unthrifty rooted *Rhododendron* cuttings. External hyphae were few despite good VAM infection. It may well be that mycorrhizae increases that sometimes follow the use of nematicides are related more to control of mycophagous nematodes than to control of root-feeding nematodes.

4. Environmental Stress

Considerable variation exists between rhizosphere growth of mycorrhizal fungi in their reaction to environmental factors such as soil temperature, pH, and moisture (Bowen and Theodorou, 1973).

Environmental stress on mycorrhizal formation can be natural or man-made via soil-applied pesticides and fungicides or systematic fungicides (Trappe *et al.*, 1973; Nesheim and Linn, 1969; Jalali and Domsch, 1975). Fumigation of soils may eliminate mycorrhizal fungi and thus necessitate the reintroduction of the symbionts for normal growth (Kleinschmidt and Gerdemann, 1972). Thus, plant pathologists should be acutely aware of the dangers of inadvertently "killing the goose that lays the golden egg" with pesticide programs. However, properly managed, this may be an asset in allowing easier introduction of highly selected symbionts subsequently.

The factors which operate to enhance essential ion uptake by mycorrhizae will also enhance uptake of possible toxic ions such as cadmium, heavy metals, and residues of organic pesticides. This is an area requiring much further study. Inhibition or depression of mycorrhizal formation by toxic levels of heavy metals (e.g., Trappe *et al.*, 1973; arsenic) will also prevent the uptake of essential elements such as phosphate in low supply. However, in other instances the mycorrhizal fungus may possibly detoxify some toxins or store them and keep them out of the plant. The usefulness of this method of buffering the plant depends on the capacity of the fungus to store heavy metals and other toxins and is probably higher with well-developed sheaths of certain mycorrhizal fungi than with endomycorrhizae. R. C. Foster (personal communication) has shown that the ectomycorrhizal fungus *Rhizopogon luteolus* can store lead and chloride.

Any soil factor preventing growth of hyphae into soil could seriously decrease the mycorrhizal response. Although this is a key point in mycorrhizal function, factors affecting fungal growth in soil have been little studied experimentally in conditions approaching ecological reality. Skinner and Bowen (1974) found that the growth of mycelial strands into different soils could vary by as much as 70-fold. In the one soil, compacting from 1.2 g cm^{-3} to 1.6 g cm^{-3} reduced mycelial growth by 90%. Reduction of aeration with higher compaction may have been the operative factor rather than soil bulk density.

VI. THE FUTURE

The challenge facing plant scientists today is to optimize crop production in the face of dwindling resources and increasing costs. It is important, therefore, that we manage symbiotic plant systems as efficiently as possible and eliminate shortfalls in response, whatever their origin. Often the lack of symbiotic response is obvious, but where plants are apparently growing well, economically important shortfalls may be occurring very commonly without being recognized and the recognition and remedy of such shortfalls or "unseen diseases" will challenge plant scientists more and more. Plant scientists will, hopefully, continue the quest for higher yielding plants and for plants particularly suited to "difficult" environments. The first thing they should realize is that they have to manage not the plant alone but the plant–microorganism associations, for these could change the assessment of plant potential tremendously. Furthermore, there is now an increasing body of evidence to show considerable genotypic variation in susceptibility and response to nodule bacteria and to mycorrhizal fungi and this must be considered in any breeding or selection program.

For some time there will be a need to study the fundamentals of nitrogen fixation, mycorrhiza function, and the dynamics of interactions of microorganisms, not just out of curiosity but with the goal of elucidating our scope for manipulating the system. Great gaps exist in our knowledge. It is sad that factors affecting mycelial growth in soil, the very basis of mycorrhizal response, are receiving almost no study. Assimilate flow in symbiotic systems also requires more study. One must be impressed by the fine physiological balance between mutualism and parasitism and this needs more understanding, especially in these days of genetic engineering. Similarly, there will also be a continuing need for microbial symbiont strain selection programs, but with important differences. It is time to get out of the irrelevant environment of sand culture in glasshouses and to use real soils in conditions which approach those in the field, e.g., realistic soil temperatures, not the optimum 20°–25°C which plant scientists are so fond of. There is a great variation between symbiotic microorganisms and their performance in relation to potential deleterious environmental factors such as pH, heavy metal toxicity, salinity, high or low temperatures, and in their reaction to added nitrogen, and these strain variations should be taken advantage of. Progress in developing strains for particular stress situations could be relatively rapid with the short generation times of microorganisms.

The challenge, however, is not only one of selecting highly efficient organisms for a particular soil–plant combination but of selecting for ability to dominate less efficient, indigenous microorganisms and to persist. Large programs of strain selection in the laboratory are futile if the organism cannot be introduced successfully. Certain strains appear to be superior in competitive establishment and persistence, and strain selection should be based on these characteristics as well as on efficiency. Unfortunately, field behavior is not predictable from the behavior in laboratory tests. There is much to be said for selection, if possible, from the existing indigenous population (for this is already compatible with soil conditions and soil microflora) and improving such symbionts by genetic manipulation and selection. The "success" of an introduced organism will depend on its compatibility with the new equilibrium position following planting and fertilizing a new crop. Rhizobia and mycorrhizal fungi suitable for a high input agricultural system may be quite different from the successful ones in a low-input agricultural system. The successful plants of early agriculture tended to be those that formed symbioses readily with a wide range of organisms, and this was a large component of their success. However, if we are to eliminate competition from inefficient, indigenous symbiotic microorganisms, perhaps we should be breeding legumes and other plants that are extremely specific in their symbiont requirements and that are resistant to infection by indigenous organisms but susceptible to highly selected inoculated strains. Alternatively, one might select for plants which symbiose highly effectively with all indigenous strains infecting that plant.

As well as correcting for shortfalls in response, there is a need to recognize the situations in which symbioses are a liability. In most instances the net energy cost is slight compared with the benefit from response, but VAM in reasonably fertile soil may decrease plant growth, either by assimilate demands or by accumulating elements, such as phosphorus, to toxic levels. There is an increasing number of instances of this. Equally important is the possible efficient accumulation in some situations of ions detrimental not only to plant health but to animal health, for example, cadmium, and organic pesticide residues. This aspect has received almost no study.

References

Ahrens, J. R., and Reid, C. P. P. (1973). Distribution of ^{14}C-labelled metabolites in mycorrhizal and nonmycorrhizal lodgepole pine seedlings. *Can. J. Bot.* **51**, 1029–1035.

Barham, R. O. (1972). Effect of nematodes and *Phytophthora cinnamomi* on mycorrhizae of shortleaf pine seedlings. *Phytopathology* **62**, 801.

Barker, K. R., and Hussey, R. S. (1976). Histopathology of nodular tissues of legumes infected with certain nematodes. *Phytopathology* **66**, 851–855.

Bassett, B., Goodman, R. N., and Novacky, A. (1977). Ultrastructure of soybean nodules. II. Deterioration of the symbiosis in ineffective nodules. *Can. J. Microbiol.* **23**, 873–883.

Becking, J. H. (1971). The physiological significance of the leaf nodules of *Psychotria*. *Plant Soil, Spec. Vol.* pp. 361–374.

Bernstein, L., and Ogata, G. (1966). Effects of salinity on nodulation, nitrogen fixation and growth of soybeans and alfalfa. *Agron. J.* **58**, 201–203.

Bevege, D. I., and Bowen, G. D. (1975). *Endogone* strain and host plant differences in development of vesicular-arbuscular mycorrhizas. *In* "Endomycorrhizas" (F. E. Sanders, B. Mosse, and P. B. Tinker, eds.), pp. 77–86. Academic Press, New York.

Bevege, D. I., Bowen, G. D., and Skinner, M. F. (1975). Comparative carbohydrate physiology of ecto- and endomycorrhizas. *In* "Endomycorrhizas" (F. E. Sanders, B. Mosse, and P. B. Tinker, eds.), pp. 149–174. Academic Press, New York.

Blaszczak, W., Golebniak, B., and Czeszyńska, J. (1974). Effects of some viruses on the growth and nodulation of horse beans and lupins. *Zesz. Probl. Postepow Nauk Roln.* **156**, 107–119; in *Soils Fert.* **38**, 381, Item 4788 (1975).

Bowen, G. D. (1961). The toxicity of legume seed diffusates toward rhizobia and other bacteria. *Plant Soil* **15**, 155–165.

Bowen, G. D. (1968). Phosphate uptake by mycorrhizas and uninfected roots of *Pinus radiata* in relation to root distribution. *Proc. Int. Congr. Soil Sci., 9th, 1968* Vol. 2, pp. 219–228.

Bowen, G. D. (1973). Mineral nutrition of ectomycorrhizae. *In* "Ectomycorrhizae—Their Ecology and Physiology" (G. C. Marks and T. T. Kozlowski, eds.), pp. 151–205. Academic Press, New York.

Bowen, G. D., and Cartwright, B. (1977). Mechanisms and models of plant nutrition. *In* "Soil Factors in Crop Production in a Semi-arid Environment" (J. S. Russell and E. L. Greacen, eds.), pp. 197–223. Queensland Univ. Press, St. Lucia, Queensland, Australia.

Bowen, G. D., and Rovira, A. D. (1976). Microbial colonization of plant roots. *Annu. Rev. Phytopathol.* **14,** 121–144.

Bowen, G. D., and Theodorou, C. (1973). Growth of ectomycorrhizal fungi around seeds and roots. *In* "Ectomycorrhizae—Their Ecology and Physiology" (G. C. Marks and T. T. Kozlowski, eds.), pp. 107–150. Academic Press, New York.

Bowen, G. D., Bevege, D. I., and Mosse, B. (1975). Phosphate physiology of vesicular-arbuscular mycorrhizas. *In* "Endomycorrhizas" (F. E. Sanders, B. Mosse, and P. B. Tinker, eds.), pp. 241–260. Academic Press, New York.

Bruch, C. W., and Allen, O. N. (1955). Description of two bacteriophages active against *Lotus* rhizobia. *Soil Sci. Soc. Am., Proc.* **19,** 175–179.

Caldwell, B. E., and Vest, G. (1970). Effects of *Rhizobium japonicum* on soybean yields. *Crop Sci.* **10,** 19–21.

Chatel, D. L., and Greenwood, R. M. (1973). Differences between strains of *Rhizobium trifolii* in ability to colonize soil and plant roots in the absence of their specific host plants. *Soil Biol. & Biochem.* **5,** 809–813.

Chatel, D. L., Greenwood, R. M., and Parker, C. A. (1968). Saprophytic competence as an important character in the selection of *Rhizobium* for inoculation. *Proc. Int. Congr. Soil Sci., 9th, 1968* Vol. 2, pp. 65–73.

Chowdhury, M. S. (1977). Effects of soil antagonists on symbiosis. *In* "Exploiting the Legume–Rhizobium Symbiosis in Tropical Agriculture" (J. M. Vincent, A. S. Whitney and J. Bose, eds.), pp. 385–411. College of Tropical Agriculture, Univ. Hawaii, Misc. Publ. 145.

Clark, W. C. (1964). Fungal-feeding nematodes as possible plant pathogens. *N.Z. J. Agric. Res.* **7,** 441–443.

Crafts, C. B., and Miller, C. O. (1974). Detection and identification of cytokinins produced by mycorrhizal fungi. *Plant Physiol.* **54,** 586–588.

Crush, J. R. (1976). Endomycorrhizas and legume growth in some soils of the Mackenzie Basin, Canterbury, New Zealand. *N.Z. J. Agric. Res.* **19,** 473–476.

Damirgi, S. M., and Johnson, H. W. (1966). Effect of soil actinomycetes on strains of *Rhizobium japonicum. Agron. J.* **58,** 223–224.

Danso, S. K. A., Keya, S. O., and Alexander, M. (1975). Protozoa and the decline of *Rhizobium* populations added to soil. *Can. J. Microbiol.* **21,** 884–894.

Dart, P. J. (1975). Legume root nodule initiation and development. *In* "The Development and Function of Roots" (J. G. Torrey and D. T. Clarkson, eds.), pp. 467–506. Academic Press, New York.

Deverall, B. J. (1977). "Defence Mechanisms of Plants." Cambridge Univ. Press, London and New York.

Döbereiner, J. (1966). Manganese toxicity effects on nodulation and nitrogen fixation of beans (*Phaseolus vulgaris* L.), in acid soils. *Plant Soil* **24,** 153–166.

Foster, R. C., and Marks, G. C. (1967). Observations on the mycorrhizas of forest trees. II. The rhizosphere of *Pinus radiata* D. Don. *Aust. J. Biol. Sci.* **20,** 915–926.

Fox, J. A., and Spasoff, L. (1972). Interaction of *Heterodera solanacearum* and *Endogone gigantea* on tobacco. *J. Nematol.* **4**, 224–225.

Freney, J. K., and Gibson, A. H. (1974). The nature of non-protein nitrogen accumulation in *Trifolium subterraneum* root nodules. *Soil Biol. & Biochem.* **6**, 313–318.

Gadgil, R. L., and Gadgil, P. D. (1975). Suppression of litter decomposition by mycorrhizal roots of *Pinus radiata*. *N.Z. J. For. Sci.* **5**, 33–41.

Gibson, A. H. (1966). The carbohydrate requirements for symbiotic nitrogen fixation: A "whole-plant" growth analysis approach. *Aust. J. Biol. Sci.* **19**, 499–515.

Gibson, A. H. (1976). Recovery and compensation by nodulated legumes to environmental stress. *In* "Symbiotic Nitrogen Fixation in Plants" (P. S. Nutman, ed.), pp. 385–403. Cambridge Univ. Press, London and New York.

Gibson, A. H., Date, R. A., Ireland, J. A., and Brockwell, J. (1976). A comparison of competitiveness and persistence amongst five strains of *Rhizobium trifolii*. *Soil Biol. & Biochem.* **8**, 395–401.

Ham, G. E., Lawn, R. J., and Brun, W. A. (1976). Influence of inoculation, nitrogen fertilizers and photosynthetic source-sink manipulations on field-grown soybeans. *In* "Symbiotic Nitrogen Fixation in Plants" (P. S. Nutman, ed.), pp. 239–253. Cambridge Univ. Press, London and New York.

Harley, J. L. (1969). "The Biology of Mycorrhiza." Leonard Hill, London.

Harley, J. L. (1975). Problems of mycotrophy. *In* "Endomycorrhizas" (F. E. Sanders, B. Mosse, and P. B. Tinker, eds.), pp. 1–24. Academic Press, New York.

Harris, J. R. (1954). Rhizosphere relations of subterranean clover. I. Interactions between strains of *Rhizobium trifolii*. *Aust. J. Agric. Res.* **5**, 247–270.

Hepper, C. M. (1977). A colorimetric method for estimating vesicular-arbuscular mycorrhizal infection in roots. *Soil Biol. & Biochem.* **9**, 15–18.

Hillis, W. E., and Ishikura, N. (1969). The extractives of the mycorrhizas and roots of *Pinus radiata* and *Pseudotsuga menziesii*. *Aust. J. Biol. Sci.* **22**, 1425–1436.

Holding, A. J., and Lowe, J. F. (1971). Some effects of acidity and heavy metals on the *Rhizobium*-leguminous plant association. *Plant Soil, Spec. Vol.* pp. 153–166.

Hussey, R. S., and Barker, K. R. (1976). Influence of nematodes and light sources on growth and nodulation of soybean. *J. Nematol.* **8**, 48–52.

Ireland, J. A., and Vincent, J. M. (1969). A quantitative study of competition for nodule formation. *Proc. Int. Congr. Soil Sci., 9th, 1968* Vol. 2, pp. 85–93.

Jalali, B. L., and Domsch, K. H. (1975). Effect of systemic fungitoxicants on the development of endotrophic mycorrhiza. *In* "Endomycorrhizas" (F. E. Sanders, B. Mosse, and P. B. Tinker, eds.), pp. 619–626. Academic Press, New York.

Johnson, H. W., Means, U. M., and Weber, C. R. (1965). Competition for nodule sites between strains of *Rhizobium japonicum* applied as inoculum and strains in the soil. *Agron. J.* **57**, 179–185.

Jordan, D. C. (1974). Ineffectiveness in the *Rhizobium*-leguminous plant association. *Proc. Indian Natl. Sci. Acad., Part B* **40**, 713–740.

Joshi, H. U., Carr, A. J. H., and Jones, D. G. (1967). Effect of clover phyllody virus on nodulation of white clover by *Rhizobium trifolii*. *J. Gen. Microbiol.* **47**, 139–159.

Kerr, A., and Htay, K. (1974). Biological control of crown gall through bacteriocin production. *Physiol. Plant Pathol.* **4**, 37–44.

Keya, S. O., and Alexander, M. (1975). Regulation of parasitism by host density: The *Bdellovibrio-Rhizobium* interrelationship. *Soil Biol. & Biochem.* **7**, 231–237.

Kleinschmidt, G. D., and Gerdemann, J. W. (1972). Stunting of citrus seedlings in fumigated nursery soils related to the absence of mycorrhizae. *Phytopathology* **62,** 1447–1453.

Klink, J. W., and Barker, K. R. (1968). Effect of *Aphelenchus avenae* on the survival and pathogenic activity of root-rotting fungi. *Phytopathology* **58,** 228–232.

Kowalski, M., Ham, G. E., Frederick, L. R., and Anderson, I. C. (1974). Relationship between strains of *Rhizobium japonicum* and their bacteriophages from soil and nodules of field-grown soybeans. *Soil Sci.* **118,** 221–226.

Krupa, S., and Fries, N. (1971). Studies on ectomycorrhizae of pine. I. Production of volatile organic compounds. *Can. J. Bot.* **49,** 1425–1431.

Labandera, C. A., and Vincent, J. M. (1975). Competition between an introduced strain and native Uruguayan strains of *Rhizobium trifolii*. *Plant Soil* **42,** 327–347.

Last, F. T. (1962). Analysis of the effects of *Erysiphe graminis* D.C. on the growth of barley. *Ann. Bot. (London)* [N.S.] **26,** 279–289.

Lehman, P. S., Huisingh, D., and Barker, K. R. (1971). The influence of races of *Heterodera glycines* on nodulation and nitrogen-fixing capacity of soybean. *Phytopathology* **61,** 1239–1244.

McCoy, M. L., and Powelson, R. L. (1974). A model for determining spatial distribution of soil-borne propagules. *Phytopathology* **64,** 145–147.

McIlveen, W. D., and Cole, H. (1974). Influence of heavy metals on nodulation of red clover. *Phytopathology* **64,** 583.

Marshall, K. C. (1976). "Interfaces in Microbial Ecology." Harvard Univ. Press, Cambridge, Massachusetts.

Marx, D. H. (1973). Mycorrhizae and feeder root diseases. *In* "Ectomycorrhizae—Their Ecology and Physiology" (G. C. Marks and T. T. Kozlowski, eds.), pp. 351–382. Academic Press, New York.

Masefield, G. B. (1955). Conditions affecting the nodulation of leguminous crops in the field. *Emp. J. Exp. Agric.* **23,** 17–24.

Mew, T., and Howard, F. L. (1969). Root rot of soybean (*Glycine max*) in relation to antagonism of *Rhizobium japonicum* and *Fusarium oxysporum*. *Phytopathology* **59,** 401.

Minchin, F. R., and Pate, J. S. (1973). The carbon balance of a legume and the functional economy of its root nodules. *J. Exp. Bot.* **24,** 259–271.

Mosse, B. (1972). The influence of soil type and *Endogone* strain on the growth of mycorrhizal plants in phosphate deficient soils. *Rev. Ecol. Biol. Sol.* **9,** 529–537.

Mosse, B. (1973). Plant growth responses to vesicular-arbuscular mycorrhiza. IV. In soil given additional phosphate. *New Phytol.* **72,** 127–136.

Nelson, C. D. (1964). The production and translocation of photosynthate-C^{14} in conifers. *In* "The Formation of Wood in Forest Trees" (M. H. Zimmerman, ed.), pp. 243–257. Academic Press, New York.

Nesheim, O. N., and Linn, M. B. (1969). Deleterious effects of certain fungi-toxicants on the formation of mycorrhiza on corn by *Endogone fasciculata* and on corn root development. *Phytopathology* **59,** 297–300.

Nutman, P. S. (1967). Varietal differences in the nodulation of subterranean clover. *Aust. J. Agric. Res.* **18,** 381–425.

Orellana, R. G., Sloger, C., and Miller, V. L. (1976). *Rhizoctonia-Rhizobium* interactions in relation to yield parameters of soybean. *Phytopathology* **66,** 464–467.

Pankhurst, C. E., Schwinghamer, E. A., and Bergersen, F. J. (1972). The structure and acetylene-reducing activity of root nodules formed by a riboflavin-requiring mutant of *Rhizobium trifolii*. *J. Gen. Microbiol.* **70,** 161–177.

Pate, J. S. (1976). Physiology of the reaction of nodulated legumes to environment. *In* "Symbiotic Nitrogen Fixation in Plants" (P. S. Nutman, ed.), pp. 335–360. Cambridge Univ. Press, London and New York.

Rajagopalan, N., and Raju, P. N. (1972). The influence of infection by *Dolichos* enation mosaic on nodulation and nitrogen fixation by field bean (*Dolichos lablab* L). *Phytopathol. Z.* **73,** 285–309.

Reid, C. P. P. (1974). Assimilation, distribution, and root exudation of ^{14}C by ponderosa pine seedlings under induced water stress. *Plant Physiol.* **54,** 44–49.

Robinson, R. K. (1972). The production by roots of *Calluna vulgaris* of a factor inhibitory to growth of some mycorrhizal fungi. *J. Ecol.* **60,** 219–224.

Ross, J. P., and Ruttencutter, R. (1977). Population dynamics of two vesicular-arbuscular endomycorrhizal fungi and the role of hyperparasitic fungi. *Phytopathology* **67,** 490–496.

Roughley, R. J., Blowes, W. M., and Herridge, D. F. (1976). Nodulation of *Trifolium subterraneum* by introduced rhizobia in competition with naturalized strains. *Soil Biol. & Biochem.* **8,** 403–407.

Rovira, A. D., and Bowen, G. D. (1973). The influence of root temperature on ^{14}C assimilate profiles in wheat roots. *Planta* **114,** 101–107.

Ruehle, J. L. (1973). Nematodes and forest trees—Types of damage to tree roots. *Annu. Rev. Phytopathol.* **11,** 99–118.

Sanders, F. E. (1975). The effect of foliar-applied phosphate on the mycorrhizal infections of onion roots. *In* "Endomycorrhizas" (F. E. Sanders, B. Mosse, and P. B. Tinker, eds.), pp. 261–276. Academic Press, New York.

Sanders, F. E., Mosse, B., and Tinker, P. B., eds. (1975). "Endomycorrhizas." Academic Press, New York.

Sanders, F. E., Tinker, P. B., Black, R. L. B., and Palmerly, S. M. (1977). The development of endomycorrhizal root systems. I. Spread of infection and growth-promoting effects with four species of a vesicular-arbuscular endophyte. *New Phytol.* **78,** 257–268.

Schenck, N. C., and Kinloch, R. A. (1974). Pathogenic fungi, parasitic nematodes and endomycorrhizal fungi associated with soybean roots in Florida. *Plant Dis. Rep.* **58,** 169–173.

Schenck, N. C., Kinloch, R. A., and Dickson, D. W. (1975). Interaction of endomycorrhizal fungi and root-knot nematode on soybean. *In* "Endomycorrhizas" (F. E. Sanders, B. Mosse, and P. B. Tinker, eds.), pp. 607–617. Academic Press, New York.

Schönbeck, F., and Schlösser, E. (1976). Preformed substances as potential protectants. *Physiol. Plant Pathol.* **4,** 653–678.

Schwinghamer, E. A., and Belkengren, R. P. (1968). Inhibition of rhizobia by a strain of *Rhizobium trifolii*: Some properties of the antibiotic and of the strain. *Arch. Mikrobiol.* **64,** 130–145.

Schwinghamer, E. A., Pankhurst, C. E., and Whitfeld, P. K. (1973). A phage-like bacteriocin of *Rhizobium trifolii*. *Can. J. Microbiol.* **19,** 359–368.

Silsbury, J. H. (1977). Energy requirement for symbiotic nitrogen fixation. *Nature (London)* **267,** 149–150.

Skinner, M. F., and Bowen, G. D. (1974). The penetration of soil by mycelial strands of ectomycorrhizal fungi. *Soil Biol. & Biochem.* **6,** 57–61.

Slankis, V. (1973). Hormonal relationships in mycorrhizal development. *In* "Ecto-mycorrhizae—Their Ecology and Physiology" (G. C. Marks and T. T. Kozlowski, eds.), pp. 231–298. Academic Press, New York.

Smith, S. E. (1974). Mycorrhizal fungi. *Crit. Rev. Microbiol.* **2**, 275–313.

Sprent, J. I. (1976). Nitrogen fixation by legumes subjected to water and light stresses. *In* "Symbiotic Nitrogen Fixation in Plants" (P. S. Nutman, ed.), pp. 405–421. Cambridge Univ. Press, London and New York.

Starr, M. P. (1975). A generalized scheme for classifying organismic associations. *In* "Symbiosis" (D. H. Jennings and D. L. Lee, eds.), pp. 1–20. Cambridge Univ. Press, London and New York.

Syōno, K., Newcomb, W., and Torrey, J. G. (1976). Cytokinin production in relation to the development of pea root nodules. *Can. J. Bot.* **54**, 2155–2162.

Taha, A. H. Y., and Raski, D. J. (1969). Interrelationships between root-nodule bacteria, plant-parasitic nematodes and their leguminous hosts. *J. Nematol.* **1**, 201–211.

Thompson, J. A. (1961). Studies on nodulation responses to pelleting of subterranean clover seed. *Aust. J. Agric. Res.* **12**, 578–582.

Trappe, J. M., Stahly, E. A., Benson, N. R., and Duff, D. M. (1973). Mycorrhizal deficiency of apple trees in high arsenic soils. *Hortic. Sci.* **8**, 52–53.

Tu, J. C. (1973). Electron microscopy of soybean root nodules infected with soybean mosaic virus. *Phytopathology* **63**, 1011–1017.

Tu, J. C., Ford, R. E., and Quiniones, S. (1970a). Effects of soybean mosaic virus and/or bean pod mottle virus infection on soybean nodulation. *Phytopathology* **60**, 518–523.

Tu, J. C., Ford, R. E., and Grau, C. R. (1970b). Some factors affecting the nodulation and nodule efficiency in soybeans infected by soybean mosaic virus. *Phytopathology* **60**, 1653–1656.

Vincent, J. M. (1965). Environmental factors in the fixation of nitrogen by the legume. *In* "Soil Nitrogen" (W. B. Bartholomew and F. E. Clark, eds.), pp. 384–435. Am. Soc. Agron., Madison, Wisconsin.

Wahab, F. A. (1977). Factors affecting the commercial production and dissemination of rhizobia. M.Sc. Thesis, University of Malaya, Kuala Lumpur.

Weber, D. F., Caldwell, B. E., Sloger, C., and Vest, H. G. (1971). Some USDA studies on the soybean–*Rhizobium* symbiosis. *Plant Soil, Spec. Vol.* pp. 293–304.

Westcott, S. W., and Barker, K. R. (1976). Interaction of *Acrobeloides buetschlii* and *Rhizobium leguminosarum* on Wando pea. *Phytopathology* **66**, 468–472.

Wilson, J. K. (1931). The shedding of nodules by beans. *J. Am. Soc. Agron.* **23**, 670–674.

Chapter 12

Disrupted Reproduction

D. E. MATHRE

I. INTRODUCTION

Reproduction is the sine qua non of life. Living things reproduce. Nonliving things do not. Therefore, diseases that strike at the reproductive system strike at the very heart of life.

To the plant, reproduction is critical to the survival of the species. Reproduction and reseeding must occur every growing season for annual plants, while for perennials this process may not be necessary for many years. How well plants are able to reproduce may play a large role in whether they can move into and occupy an ecological niche. The development of plant communities through a series of successions is usually

257

dependent on how well the reproductive units, e.g., seeds, can be transported into and develop within a new site. In cultivated situations, many crops are established each year from seeds, the viability and health of which become critical factors in establishing a productive crop. Flowering also provides a mechanism for the genetic recombination of traits which aid in the evolution of plant species, either naturally or man guided.

Aside from their role in the maintenance of the species, the reproductive portions of a plant are often the portion that man uses for food. As such, crop productivity is often directly related to a plant's reproductive capacity.

Mangelsdorf (1966) said that over the long reaches of history man has used about 3,000 species of plants for food, that about 150 are in world commerce, but that only 15 really feed the people on this globe. Of Mangelsdorf's 15 species, the reproductive bodies of 10 are eaten by man including rice, wheat, maize, sorghum, barley, common bean, soybean, peanut (groundnut), banana, and coconut.

The reproductive disease on wheat known as bunt or stinking smut has plagued man ever since wheat was domesticated. Rice blast is a frightening word in whatever language used by the rice farmers of the world who feed more people than growers of any single other crop.

Last, since "man shall not live by bread alone," his spirit is nourished by the beauty provided by the flowers of many plants. Who among us has not been affected by the beauty of a rose, an orchid, or a tulip? The classic disease of a flower is the famous and infamous virus disease of tulips known as "breaking." The petals show streaks of color rather than being a solid color. A painting of it in 1576 in the Netherlands is probably the earliest published illustration of a plant disease. Moreover, it created the famous "tulipomania" in Holland in the 1630's. At this time, Dutch overseas trade had increased to the point that the country was ripe for an atmosphere of speculation. Tulips were relatively new to Holland, having been imported from Turkey. Ordinary tulips were easy to obtain, but the prize ones were those showing the "breaking" pattern. By 1634, the mania for tulips had increased to the point that prices of single broken tulips were bid up to fabulous levels. Large fortunes were invested in them until the bubble burst, leaving individuals and the country as a whole on the verge of bankruptcy. Today, the ornamental plant and floral industry is a thriving one, indicating man's need and quest for beauty as well as for food.

Disrupted reproduction affects, then, not only plants themselves but also man who depends on them, or is influenced by them. Much of the art and science of plant pathology is based on the prevention of this disruption.

II. PATHOGENS AND POLLEN

A. Effects on Pollen Production

Viruses sometimes induce male sterility in normally self-fertile plants. This sterility may be due to the inhibition of pollen formation, abortion of pollen after it is formed, or dysfunction of the pollen. Tomato aspermy virus in tomato interferes with normal meiotic processes, such that when tetrads are formed in pollen development they are aborted prior to maturity as pollen grains (Caldwell, 1952). An abnormal aggregation of chromosomes at pachytene may be involved in this disruption.

Pollen abortion, for whatever reason, is also involved in the male sterility of geraniums infected with tomato ringspot virus (Murdock *et al.*, 1976), and of barley infected with barley stripe mosaic virus (BSMV) (Inouye, 1962). The abortion process seems to involve a disintegration of the cytoplasm in the pollen grain followed by collapse of the wall of the pollen grain. In BSMV-infected barley, the number of pollen grains produced per anther is reduced 20–40% as compared to healthy controls.

B. Pollen Germination and Germ Tube Elongation

Pathogens may also decrease the vigor of pollen. A reduction in the number of viable pollen grains occurs in geraniums infected with tomato ringspot virus (Scarborough and Smith, 1977) and in BSMV-infected barley (Inouye, 1962). The decrease may be as much as three- to fivefold in the latter. In soybeans infected with tobacco ringspot virus, pollen production and germination problems are expressed as reduced numbers of pollen grains per flower, reduced germination, and reduced germ tube length. Similar effects on pollen vigor have been observed in tomatoes exposed to fluoride (Sulzbach and Pack, 1972). In controlled crossing experiments with raspberries, Lister and Murant (1967) noted that virus-free pollen greatly decreased the ability of pollen from infected plants to set seed, illustrating again the decreased vigor or competitive ability of virus-infected pollen.

The deleterious effect of BSMV infection in barley on pollen development and vigor is similar to that in barley with genetic male sterility (Roath and Hockett, 1971; Inouye, 1962). This suggests that the mechanisms may be similar and may involve the disruption of nucleic acid reproduction.

C. Anther Destruction

The most dramatic effect on male reproductive organs is their complete destruction. In the Netherlands, a necrosis of bud tissues in tulips has been described (deMunk and Beijer, 1971). In light cases only stamen primordia are affected, while in more severe cases the entire flower bud may be necrotic. When such bulbs are planted, some or all of the stamens are decayed but the remainder of the flower is healthy. Some fascinating detective work by deMunk (1972) has shown that this condition is caused by the invasion of mites. However, the entry site for the mites is provided by open buds. The buds open in response to increased ethylene concentrations which result from infection of the bulbs by *Fusarium oxysporum* f. *tulipae* in conjunction with poor ventilation. This is an example of the interaction between pests which can result in a more adverse effect on the plant than would be caused by either pest alone.

III. PATHOGENIC EFFECTS ON OVULES, OVARIES, AND FRUIT

A. Ovule Abortion

Abortion of flower buds and their contents occurs in some plants infected with a variety of pathogens. In some cases, the abortion occurs even prior to anthesis, while in others the ovary and its contents are affected at the time the flower opens or shortly thereafter. The mechanism of such adverse effects ranges from an interference in nuclear division at certain stages of meiosis to a wholesale rotting of the tissues. Tomato aspermy virus interferes with meiosis (Caldwell, 1952) by disrupting meiotic divisions of the megaspore mother cell. In geraniums infected with tobacco and tomato ringspot viruses the total number of potential florets is not affected, but virus infection does reduce the number of florets that complete their development (Scarborough and Smith, 1977). Infected florets show chlorosis which progresses to necrosis of the floret. By the time sepals and petals of a floret are chlorotic, the tissues interior to the petals are necrotic. Even though the ovules are oriented correctly within the ovary and development of the nucellus is normal, the integument layers fail to surround the nucellus (Murdock *et al.*, 1976).

B. Phyllody

In some diseases, sex organs are transformed into leaves, a condition known as phyllody. Smut infections are classic examples. *Sphacelotheca reiliana* in corn and sorghum often results in large, leafy structures instead of the normal smutty heads typical of this host–parasite combination. Early botanists, including Linnaeus himself, often regarded such infected plants as new species or new varieties of a species.

Smuts are not the only pathogens capable of inducing this response. In downy mildew of millet, caused by *Sclerospora graminicola*, the grain and sometimes the stamens are replaced by a short, leafy shoot. In Israel and India, Klein (1970) and Sahambi (1970) describe phyllody diseases of safflower and sesamum that are vectored by leafhoppers. Both are probably caused by mycoplasmas.

Bos (1957) has discussed the possible mechanisms of phyllody. He argues that when a shoot on a plant begins to flower, the vegetative character of the shoot is suppressed and leaves become flower parts. If, therefore, a disease counteracts the suppression during the flowering process, the vegetative character reasserts itself and leaves are formed again. Presumably the whole process is hormonally regulated.

C. Tissue Substitution

In some diseases the host tissue becomes substituted for by fungus tissue as in ergot and the smuts. The logo on the jacket of this volume depicts the substitution of the grains of wheat by the tissue of the loose smut fungus.

1. Ergot

Ergot, caused by *Claviceps purpurea*, is one of the oldest diseases known to man. It is a classic case of tissue substitution. The host tissue is transformed into an elongated purplish body called an ergot or a sclerotium. Campbell (1958) indicates that the base of the unfertilized ovary is the main site of penetration, which occurs within 24 hr of inoculation. However, others have observed penetration of the stigma. The ovary wall is colonized first, but by the fourth or fifth day, the mycelium has become intracellular and has engulfed the ovule. A hymenium develops on the surface of the ovary and produces spores in a liquid called honeydew. Mower and Hancock (1975) determined that a 10- to 50-fold difference in water potential exists between host and parasite, thus causing an increased translocation of sucrose to the infected ovary. During parasitism, the fungus converts the sucrose to a variety of honeydew sugars.

While ergot is usually thought of as a disease of rye and other cross-pollinated grasses, recent work has shown that genetically and cytoplasmically male sterile wheat and barley are also highly susceptible to ergot (Puranik and Mathre, 1971; Stoskopf and Rai, 1972). Similarly, barley made sterile by BSMV shows an enhanced susceptibility to ergot (Darlington, et al., 1976). The flowers of male sterile wheat and barley are highly susceptible to ergot because of the open flower condition which occurs in such plants. Until the flower is fertilized, it remains open and thereby is exposed to ergot inoculum for a long time. Several studies (Cunfer et al., 1975; Darlington and Mathre, 1976; Watkins and Littlefield, 1976) have shown that the fertilized ovary becomes resistant to ergot infection shortly after it has been pollinated with some indication that the resistance mechanism is induced within 30 min after pollination (Darlington and Mathre, 1976). The exact mechanism of resistance is unknown, but pollination is known to induce large changes in hormonal levels in fertilized tissue. These physiological changes may prevent germination of ergot conidia on the stigma or prevent hyphal penetration into and development within host tissue. Inoculation of diallel crosses between spring wheat lines that develop resistance quickly after pollination and those that develop resistance slowly indicate that the resistance factor is maternally controlled. In fertile spring wheat with some resistance, Platford and Bernier (1970) indicated that the ovaries are infected but further development of the fungus is inhibited and it produces only small sclerotia and reduced honeydew. If the development and use of hybrid spring wheat and barley are to be successful, the ergot problem will have to be controlled by use of resistant cultivars, resistance triggered by fertilization, or elimination of the sources of primary inoculum.

2. Smut

In addition to ergot, the other disease most often associated with the complete substitution of the female flowering parts of the plant is smut. Usually, the presence of the smut teliospores in the heads of infected grasses indicates the destruction of the floral parts, but Cherewick (1965) reports that oats infected with *Ustilago kolleri* or *U. avenae* may show "blasting" or sterility of the heads even though little or no smut develops in them.

For those smuts which attack the head and reduce the ovary to a mass of teliospores, the mycelium which produces such spores usually enters the floret as the result of a systemic infection, e.g., from mycelium which has invaded the growing point. The infective hyphae have little or no effect on the apical meristem until the plant has reached the reproductive stage when the mycelium is induced to form teliospores. According to Fischer and Holton (1957), this aspect of smut biology was one of the least understood in 1957. It still is.

D. Flower Blights

The least specific type of disease of floral parts is the complete decay or blight of flowers, as can be caused by bacteria or fungi. Perhaps the best known pathogen in this respect and one with a wide host range is *Botrytis cinerea* which can attack blossoms as well as fruit. In the "early" *Botrytis* rot of grapes, the fungus invades the stigma and style but then becomes latent in the necrotic stigmatal and stylar tissue that remains attached to the developing grape berry (McClellan and Hewitt, 1973). In this situation, as with others (Ogawa and English, 1960; Chou and Preece, 1968), the presence of pollen is stimulatory to the development of the pathogen, resulting in increased and more rapid germination of the conidia and more rapid germ tube elongation. The stimulation for these processes is in the form of nutrients in leachates or exudates from the host (Barash *et al.*, 1964; Kosuge and Hewitt, 1964). In strawberries, the presence of pollen grains in the infection droplet reduces considerably the number of conidia needed for infection (Chou and Preece, 1968). For infection of developing apricot fruit, *B. cinerea* must establish itself first on dead floral parts which adhere to the fruit. Styles that fail to dehisce are the primary avenues of infection (Ogawa and English, 1960). This suggests that a saprophytic type of growth is necessary prior to infection to allow *B. cinerea* a chance to produce pectic enzymes, particularly polygalacturonase, which breaks down or alters the host, thus facilitating entry into the ovary or developing fruit.

Other factors that affect *Botrytis* blight are related to the retention of free moisture on plant surfaces. In castorbeans, those cultivars with compact inflorescences and dwarfed internodes are most susceptible (Thomas and Orellana, 1963).

Fire blight, caused by *Erwinia amylovora*, also causes a complete destruction of flowers. While stomatalike openings in the nectaries are considered to be the main avenues of entry, numerous other natural openings on flowers are also involved, including stomata on the styles and sepals (Schroth *et al.*, 1974). In California, Miller and Schroth (1972) found *E. amylovora* existing as an epiphyte in pear flowers and

on other plant parts during the spring. Amazingly enough, during epi-
phytotics, upward of 10^4 to 10^6 cells were found in many flowers but only
a low percentage of such flowers actually became infected. Why the infection percentage of "inhabited" flowers is not greater seems to be
somewhat of a mystery, and raises questions regarding the interaction of
inoculum density and host susceptibility. Miller and Schroth (1972)
suggest that "in nature, inoculation may often be effected by a mass, or
large number of cells and not by low dosages which some workers logically contend should be used when conducting pathogenicity tests in
greenhouse and field studies to duplicate natural processes." Low dosages
of inoculum may therefore not be the norm for bacterial diseases!

An interesting bacterial–environmental interaction involving a blossom
blight has been observed in Connecticut. In years with cold, moist conditions during flowering in pears, many blossoms abort and fail to produce fruit. The usual explanation was that the abortion was due to a
physiological response of the pear tree to the adverse environment.
However, work by Sands and McIntyre (1977) has shown that the
abortion of blossoms is mainly due to infection by *Pseudomonas syringae*,
since this organism can be isolated from infected blossoms, and fruit
yields are significantly increased by use of antibiotic sprays during the
blossoming period. Such a response would not be observed if the blossom
blight were strictly a physiological response to cold temperatures.

IV. TRANSVESTISM

Webster's New Collegiate Dictionary defines transvestism as "adopting
the dress and behavior of the opposite sex." Diseases may make plants
show behavior of the opposite sex, i.e., induce stamens in the female
flower and ovaries in the male flower. For example, *Ustilago violaceae*
attacks members of the Caryophyllaceae which normally are dioecious.
In species of *Lychnis*, this fungus stimulates the production of stamens on
otherwise female flowers (see Fischer and Holton, 1957). In contrast,
this same fungus in *Melandrium album* and *M. dioecum* stimulates the
production of ovaries in the male flowers. Both responses, no doubt, are
due to altered hormonal balance in infected plants.

V. POSTFERTILIZATION EFFECTS

A. Fleshy Fruit Destruction

One of the most disease-prone plant parts is the fleshy fruit of a
number of vegetables and fruits. Latency of infection is involved in some
of these diseases. In summer bunch rot of Thompson seedless grapes in

California, Strobel and Hewitt (1964) found that spores of *Diplodia viticola* germinate on and infect the stigmatic portion of the flower. However, the mycelium in the fertilized, developing grape berry remains latent until the fruit is nearly mature. The latency mechanism is believed to be due to low pH values in the developing grape berry resulting from high levels of tartaric and malic acids. As the fruit matures, these acids are metabolized to sugars, thus raising the pH level enough to allow the pathogen to resume growth and rot the berry. An even more superficial latency of infection has been reported in avocados attacked by *Colletotrichum gloeosporioides* (Binyamini and Schiffman-Nadel, 1972). In this case the fungus forms appressoria within the wax layers on the immature fruit surface. During fruit softening at maturity, the appressoria commence growth again and hyphae penetrate on into the fruit.

The value of a food base for infection is seen in *Botrytis* rot of strawberries (Jarvis, 1962; Powelson, 1960), where infection rarely occurs in the absence of previous saprophytic growth on flower parts. Only under conditions of continuous high humidity will infections of immature fruit develop to the point that the fruit is destroyed. The role of a food base as a necessity for infection can even allow some pathogens to attack fruit which they normally would not infect. The vascular pathogen *Fusarium oxysporum* is capable of causing a pod decay of snap beans when old blossoms become infected and serve as the foci of pod infection (Goth, 1966).

The resistance of immature fruit to invasion by a variety of fungal pathogens seems to be a rather common phenomenon. In oranges, resistance to *Diplodia natalensis* is related to failure of the fungus to penetrate the cuticle and wound the periderm (Brown and Wilson, 1968). Entry can occur at the time of fruit abscission via natural openings that develop in the separation layer between the button (calyx and disc) and the fruit. The resistance of immature papaya fruit to invasion by *Colletotrichum gloeosporioides* is also related to formation of a liquefied periderm which walls off the pathogen (Stanghellini and Aragaki, 1966). In apple fruit, susceptibility occurs at or just before the climactic rise in respiration (Sitterly and Shay, 1960). This increased respiration may allow the accumulation of sugar for use by pathogens as an energy source, suggesting that the lack of such energy sources in immature fruit may be part of the reason for their resistance.

B. Nonfleshy Fruit or Seed Destruction

In those fruits where the pericarp is united to the seed coat to form a nonfleshy fruit, e.g., the caryopsis of the Graminae, or where no pericarp exists, e.g., the seeds of gymnosperms, pathogens have evolved that can injure or destroy such organs. However, few direct pathogens of gymno-sperm cones or seeds appear to exist; exceptions are the cone rusts which reduce the amount of seed produced but never destroy the entire crop (Boyce, 1961). Head, ear, and kernel rots of various cereals, however, are not uncommon and, because of the significance of cereals in world food supplies, are far from being unimportant. The scab disease of wheat and barley heads, caused by *Gibberella zeae* is particularly important in areas where corn is grown in rotation with these crops and where moisture levels at heading time tend to be high. Susceptibility to invasion of wheat heads is greatest after flowering. Strange and Smith (1971) and Strange *et al.* (1974) showed that massive fungal growth occurs on ex-truded anthers and the resulting spikelet infections are significantly greater in nonemasculated plants than in emasculated plants. Extracts of anthers stimulate fungal growth due to the presence of choline and betaine.

Kernel and ear infections in maize have received renewed interest because of the apparently high susceptibility of corn lines containing the opaque-2 gene for high lysine. These were developed because of their increased protein and nutritional value (Ullstrup, 1971). There was fear for awhile that this aspect of these new corn lines would greatly detract from their utility, but a report by Loesch *et al.* (1976) suggests that such high lysine types can be developed with good resistance to ear-rotting fungi.

C. Indirect Effects on Fruiting and Fruit Development

Pathogens need not directly invade flowers and/or fruit to have an effect on plant reproduction. The damage done is often measured by decreased numbers of seed or fruits, probably as a result of decreased photosynthesis or decreased water transport, or both, in infected plants. In plants with a determinant type of growth, e.g., wheat, the potential number of flowers that can develop is set early in the life of the plant when the spike is initiated. However, anything which affects the growth and vigor of the plant after the time the spike is differentiated will have an effect on the number of flowers that will be fertile. Early stem rust infections which greatly decrease the photosynthetic capacity of the plant and interrupt its water economy through increased transpiration can effectively sterilize a wheat plant, but later infections have a less detrimental effect on the number of fertile florets and the size of seed that develop. However, a generalization seems to be that in plants with a determinant growth habit, leaf-infecting pathogens that reduce photo-synthetic capacity probably have their biggest effect on seed or fruit size rather than on seed or fruit number. This is likely to be less true in plants with an indeterminant growth habit.

D. Effects on Floral Esthetic Qualities

While not being a direct effect on the reproduction of a plant per se, there are diseases which affect reproductive plant parts used for esthetic purposes, i.e., flowers grown as ornamentals. Such effects can range from shortened internodes in gladiolas caused by tomato ringspot virus (Bozarth and Corbett, 1958), to distortions of flower parts caused by chrysanthemum aspermy virus (Hollings, 1955), to partial or complete necrosis of flower petals as caused by *Sclerotinia sclerotiorum* (Raabe, 1971).

VI. SEED-BORNE PATHOGENS

A. Biological Implications

While in many cases pathogens disrupt reproduction to the point that the reproductive units of a plant are nonfunctional, another situation exists that is important to the host plant, its pathogens, and to man who is dependent on these units for "seed" to begin a new crop. This involves the transmission of pathogens on or within reproductive units, be they seeds, tubers, corms, or seed pieces. As Baker (1972) points out, seed transmission is a method par excellence by which pathogens (1) are introduced into new geographic areas, (2) can survive periods when the host is lacking, (3) are preferentially selected and distributed as host-specific strains, and (4) are distributed throughout the plant population as foci of primary inoculum.

The long-distance dissemination of plants has certainly had a major impact on the agriculture of the world (Crosby, 1972). Most of our major crop plants are now grown in areas where they did not evolve but have been disseminated by man, particularly as a result of what is termed "the Columbian Exchange" beginning in 1492. An important aspect of this exchange was the dissemination of plant pathogens between both the Old and the New World. A recent problem may be the dissemination of seed-borne pathogens occurring as a result of the massive movements of seed, particularly wheat and rice seed, as a part of the "Green Revolution." Quarantines and pesticide seed treatments may help eliminate the problems of pathogens borne on the outside of the seed, but they rarely have any effect on internally borne pathogens, particularly the viruses.

The survival of pathogens within seed is particularly important to those that are obligate parasites. For many viruses, seed transmission may be the single (or only) important source of virus carryover from one growing season to the next (Shepherd, 1972). This is exemplified by the case with lettuce mosaic virus in lettuce whereby the use of seed lots with no infected plants in 30,000 seeds assayed can effectively keep the crop nearly virus-free throughout the growing season (Baker, 1972).

The mechanisms by which pathogens are transferred to the seed from the mother plant are varied and include: (1) the pathogen is a passive contaminant on the exterior of the seed, (2) the pathogen spreads into the seed from the fruit, (3) the pathogen enters the seed via vascular connections, (4) the pathogen enters the embryo via infected pollen or ovarian tissue, and (5) the pathogen actively and directly penetrates the seed coat (Baker and Smith, 1966).

B. Viral Seed-Borne Diseases

1. Importance and Frequency

It is often assumed that relatively few virus diseases are seed-borne. However, Shepherd (1972) points out that at least 49 different viruses are seed-borne in at least one of their hosts. In some groups of viruses, e.g., the nematode-transmitted viruses, seed transmission is the rule rather than the exception. The degree or extent of seed transmission is highly variable and is affected by host cultivar, virus strain, stage at which the mother plant is infected, temperature after infection, and perhaps other factors. The types of viruses that are often seed transmitted include those that are readily juice transmitted, thus indicating their ability to invade parenchyma, while phloem-inhabiting, leaf-hopper vectored, and persistent-aphid vectored viruses generally are not seed transmitted (Bennett, 1969).

The importance of seed transmission in the ecology of certain viruses is illustrated by Murant and Lister's work (1967). For several of the nematode-transmitted viruses of raspberries, the persistence of such viruses through periods of fallow or fasting of the vector was found to depend on a continuing supply of infected seedlings produced by virus-containing weed seeds. Tomlinson and Carter (1970) have also documented the importance of weed seed transmission of cucumber mosaic virus in providing new sources of inoculum.

2. Mechanisms of Transmission

For many viruses, including some for which seed transmission has been reported, the seed offers a highly effective barrier to the passage of viruses from one plant generation to the next. Thus, propagation of plants through seed has therapeutic advantages to a plant species helping to ensure its survival. In view of this, the problem becomes one of determining the mechanisms by which some viruses do become seed-borne and the factors that prevent others from being propagated in this way. Many hypotheses have been presented suggesting reasons why seed transmission is somewhat rare including: (1) the presence of inhibitory materials in embryonic tissues, (2) the lack of plasmodesmatal connections between embryo and mother plant tissue, (3) the adverse effects of virus on meiotic divisions, and (4) phosphate starvation in rapidly dividing tissues which restrains or prevents virus multiplication.

a. Seed Coat Transmission. It appears that very few viruses are transmitted to new plants from inoculum existing only on the seed coat. The major exception is tobacco mosaic virus (TMV) on tomato seeds. Even in this case, if extreme care is taken to prevent wounding of emerging seedlings, little or no transmission will occur (Taylor *et al.*, 1961; Broadbent, 1965). In other cases, e.g., pea streak virus in peas, the virus may be transmitted only by immature seed but, because of its association only with the seed coat, it is inactivated as the seed matures (Ford, 1966).

b. Pollen Transmission. That infected pollen can serve to transmit viruses to the embryo, and hence to the seed, has been shown for a variety of viruses including grapevine fanleaf virus, cherry necrotic ringspot virus, bean virus one, alfalfa mosaic virus in alfalfa, and barley stripe mosaic virus (BSMV). Early studies to detect pollen transmissibility were carried out using pollen from infected plants to fertilize pistils on healthy plants. The actual mechanism of pollen transmission as observed visually is well known for BSMV. This virus is able to invade the male floral meristem with subsequent invasion of the pollen mother cells (Carroll and Mayhew, 1976a), followed by invasion of the sperm. The cell-to-cell transfer of the virus is believed to occur via microtubule transfer at the time of cell division (Mayhew and Carroll, 1974b) (Figs. 1–3). Carroll (1974) proposed two explanations for the ultimate transmission of BSMV from pollen to embryo. First, the virus could be carried by the sperm cell which infects the egg cell and thus the zygote and embryo, or second, the virus could be carried in the cytoplasm of the vegetative cell as a "contaminant" which the sperm would pass on to the egg cell. Because of the visualization of virions inside of sperm nuclei, Carroll favors the first explanation (Fig. 2). Yang and Hamilton (1974) have also studied the seed transmissibility of tobacco ringspot virus in soybeans. Virus aggregates were found consistently in the intine of pollen grains, and in the cell wall and cytoplasm of the generative cell, but not in the cytoplasm of vegetative cells nor in nuclei of vegetative or generative cells. The lack of virus in such areas probably explains why they were unable to show pollen transfer of this virus.

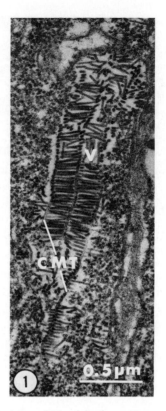

Fig. 1. Aggregate of virions (V) of barley stripe mosaic virus and cytoplasmic microtubules (CMT) in a pollen mother cell. × 35,000. (Reproduced by permission of the National Research Council of Canada from Carroll and Mayhew, 1976a.)

c. Embryonic Transmission. Transmission in the embryo implies that the embryo must become infected via one or more of three possible mechanisms: (1) through the zygote via infected sperms—a situation described above for BSMV, (2) through the zygote via infected eggs, or (3) by direct invasion of the embryo from the mother plant following embryo initiation. The first two possibilities are believed to account for the majority of cases of embryonic transmission. For BSMV in barley, early in the development of the female primary floral meristem virions can be seen in the ovular tissue often associated with microtubules (see Fig. 3) (Carroll, 1969; Carroll and Mayhew, 1976b; Mayhew and Carroll, 1974a,b). Virions are also observed in the megaspore mother cell, megaspores, and then in the embryo sac including the egg cell. These results provide strong support for Bennett's (1969) suggestion that those viruses capable of being embryonically transmitted in the seed are those which can invade the primary meristem of the host early in its develop-

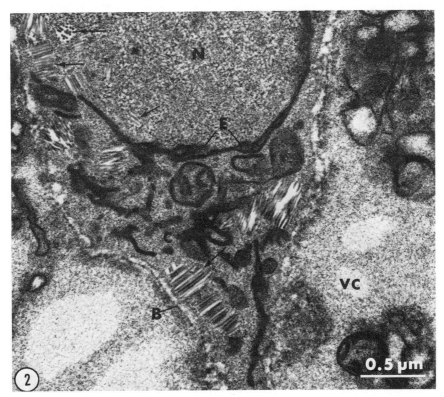

Fig. 2. Portion of BSMV-infected sperm cell. Note virions (arrows) of BSMV in the nucleus (N) and cytoplasm (C) of the sperm cell. Also shown are the nuclear envelope (E) and the boundary (B) of the sperm cell and the cytoplasm of the vegetative cell (VC). The bar represents 0.5 μm. (Reproduced by permission of Academic Press from Carroll, 1974.)

ment before the embryo is effectively "walled-off" from the mother plant. Because spherical viruses are morphologically similar to ribosomes, they are difficult to detect in the tissue with the electron microscope. However, Yang and Hamilton (1974), in a study of the seed transmission of tobacco ringspot virus in soybeans, were able to detect viruslike particles aggregated in the sieve tubes in integuments as well as in the nucleus, embryo sac wall, and megagametophytic cells.

Whether viruses can directly invade embryos from the mother plant after the embryo has been initiated is unclear. Eslick and Afanasiev (1955) reported that a low level of seed transmission of BSMV did occur in barley inoculated long after flowering, but others could not confirm this. Schippers (1963) also reported a lack of seed transmission of bean common mosaic in beans when plants were inoculated after flowering.

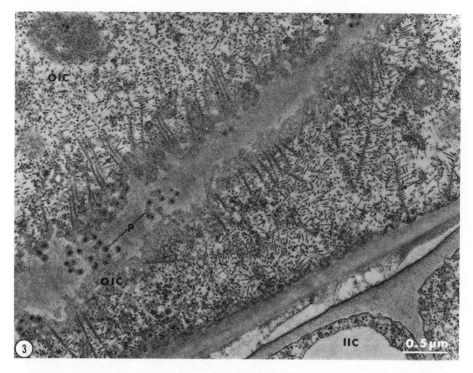

Fig. 3. Integument and nucellar cells of the ovule infected by the MI-1 isolate of barley stripe mosaic virus. Portions of two cells within the outer integument (OIC). Many virions are associated with wall microtubules (arrows). No virions are evident within plasmodesmata (P). Portions of two cells of the inner integument (IIC) are also visible. Premeiotic stage. × 23,500. (Reproduced by permission of the National Research Council of Canada from Carroll and Mayhew, 1976b).

Evidently the phenomenon of direct invasion is rare, probably because of either a barrier of disintegrating nucellar tissue around the developing embryo or the lack of plasmodesmatal or vascular connection between the mother plant and the embryo.

C. Bacterial Seed-Borne Diseases

1. Importance

The importance of seed transmission in the epidemiology of several bacterial bean and pea diseases has been a major factor in the continuing practice of producing seed in the dry irrigated areas of the western United States (Baker, 1972). In humid areas, as few as 12 infected bean seeds per acre, or 0.02% infection, can provide enough primary inoculum for the development of severe epidemics of *Pseudomonas phaseolicola*.

2. *Mechanisms of Transmission*

While seed transmission of bacterial pathogens can result from inoculum in the embryo, endosperm, or seed coat, or from contamination of the seed coat surface, it seems unlikely that embryo infections are very important. If embryos were infected, the seed would probably be nonviable. Internal infections are known, however, particularly for those diseases where the bacterium is a systemic vascular inhabitant. Zaumeyer (1932) reports that *P. phaseoli* enters bean seeds via vascular elements or the micropyle. In the latter case, the bacteria break out from systematically invaded funicular tissue or from vessel elements along the dorsal suture of the pod. The bacteria are then harbored just beneath the seed coat.

The role of the "seed contamination" in seed transmission is illustrated by the studies of Grogan and Kimble (1967) and Guthrie (1970). Development of *P. phaseolicola* in bean fields sown with western-grown seed suggested to them that such seed might be carrying low levels of inoculum. They showed that seed contamination can occur even though seed field inspections fail to detect the disease. This contamination was found to occur in small natural openings on the seed coat, including the hilum and micropyle. Similar types of contamination have also been observed for a variety of other bacterial pathogens including *Xanthomonas malvacearum*, *X. campestris*, and *X. oryzae*.

D. Fungal Seed-Borne Diseases

A wide variety of fungal pathogens are known to be seed-borne, carried either as infectious mycelium internally or as contaminants on the seed coat. The methods by which seeds become infected or contaminated include: (1) floral infection whereby the embryo becomes infected but remains viable, e.g., loose smut of cereals; (2) infection of the seed coat via vascular connections, e.g., Verticillium wilt of safflower; and (3) contamination with propagules produced on other plant parts, particularly leaves, e.g., anthracnose of beans.

1. *Mechanisms of Transmission*

With few exceptions, the loose smut fungi are unique in their ability to invade floral tissues and become established in the embryo without destroying it. Infection appears to occur directly through any part of the ovary wall (Batts, 1955) after which the mycelium becomes prevalent in the scutellar portion of the embryo. Mycelium is rarely found in the endosperm. As the seed matures, the mycelium becomes dormant and further development is restricted until the seed imbibes water and germinates. Infection can take place prior to or several days after pollination as long as the floret is open to allow entry of the teliospore inoculum. With the development of hybrid wheat and barley and the use of male sterile female plants to produce such seed, this pathogen may become more serious than it has been in the past because the flower of the male sterile female must remain open to allow for cross-pollination. This allows for a longer exposure to loose smut inoculum.

While seed-borne vascular pathogens are not very common, in those cases where there is a vascular connection into the seed coat infection has occurred. In safflower (*Carthamus tinctorius*), Klisiewicz (1963, 1975) has found *Verticillium dahliae* and *Fusarium oxysporum* f. *carthami* capable of infecting seed. In the latter case, 42% of the seed harvested from infected seed heads carried the fungus. Mycelium and spores were evident on the seed surface while hyphae were observed in the pericarp in association with xylem and sclerenchyma tissue. No embryo infections were observed.

In those plants where seeds are produced within a protective fruit wall, e.g., beans, fungal infections of the seed can occur by direct invasion of the seed coat from mycelium in the fruit or pods. Ellis *et al.* (1976), in a study of dry bean pod invasions in South America, determined that seeds become infected by a variety of fungi only in pods in contact with soil. Because of this source of inoculum, they recommended that growers in Central and South America hand pick pods not in contact with soil and save the seeds therein for the next growing season. Seed selected in this manner was found to have an emergence potential of 97% versus 52% for seeds from pods in contact with the soil.

VII. PATHOGENIC EFFECTS ON SEED QUALITY

The farmer must have seed that can germinate and develop into a healthy seedling. When seed is infected with a pathogen, the seed may be rendered nonviable or it may remain viable but produce a weak seedling. In some cases, the infected seedling may not be severely weakened, but may serve as a source of primary inoculum within a community of plants.

The extreme case where the seed is rendered nonviable occurs with those pathogens which are vigorous direct invaders of plant tissue, as illustrated by *Whetzelinia sclerotiorum* on beans (Steadman, 1975). In

Illinois, the problem in viability of soybean seed lots is correlated with environmental conditions in seed production fields (Tenne *et al.*, 1974). Seed from northern Illinois show consistently higher germination than seed from the southern part of the state. Where temperature, rainfall, and humidity are more conducive to pod infection by a variety of fungi and bacteria, there is good correlation between seed infection and lowered viability.

Corn seed pathogens are also capable of reducing seed viability. One of relatively recent importance, *Colletotrichum graminicola*, has been reported to reduce germination from 90% for healthy seed, to 37% for kernels with a few infection sites, to 6% for kernels with extensively damaged kernels (Warren and Nicholson, 1975).

References

Baker, K. F. (1972). Seed pathology. *In* "Seed Biology" (T. T. Kozlowski, ed.), Vol. 2, pp. 317–416. Academic Press, New York.

Baker, K. F., and Smith, S. H. (1966). Dynamics of seed transmission of plant pathogens. *Ann. Rev. Phytopathol.* 4, 311–334.

Barash, I., and Klisiewicz, J. M., and Kosuge, T. (1964). Biochemical factors affecting pathogenicity of *Botrytis cinerea* on safflower. *Phytopathology* 54, 923–927.

Batts, C. C. V. (1955). Infection of wheat by loose smut, *Ustilago tritici* (Pers.) Rostr. *Nature (London)* 175, 467–468.

Bennett, C. W. (1969). Seed transmission of plant viruses. *Adv. Virus Res.* 14, 221–226.

Binyamini, N., and Schiffman-Nadel, M. (1972). Latent infection in avocado fruit due to *Colletotrichum gloeosporioides*. *Phytopathology* 62, 592–594.

Bos, L. (1957). Plant teratology and plant pathology. *Tijdschr. Plantenziekten* 63, 222–231.

Boyce, J. S. (1961). "Forest Pathology," 3rd ed. McGraw-Hill, New York.

Bozarth, R. F., and Corbett, M. K. (1958). Tomato ringspot virus associated with stunt or stub head disease of gladiolus in Florida. *Plant Dis. Rep.* 42, 217–221.

Broadbent, L. (1965). The epidemiology of tomato mosaic. XI. Seed-transmisison of TMV. *Ann. Appl. Biol.* 56, 177–205.

Brown, G. E., and Wilson, W. C. (1968). Mode of entry of *Diplodia natalensis* and *Phomopsis citri* into Florida oranges. *Phytopathology* 58, 736–739.

Caldwell, J. (1952). Some effects of a plant virus on nuclear division. *Ann. Appl. Biol.* 39, 98–102.

Campbell, W. P. (1958). Infection of barley by *Claviceps purpurea*. *Can. J. Bot.* 36, 615–619.

Carroll, T. W. (1969). Electron microscopic evidence for the presence of barley stripe mosaic virus in cells of barley embryos. *Virology* 37, 649–657.

Carroll, T. W. (1974). Barley stripe mosaic virus in sperm and vegetative cells of barley pollen. *Virology* 60, 21–28.

Carroll, T. W., and Mayhew, D. E. (1976a). Anther and pollen infection in relation to the pollen and seed transmissibility of two strains of barley stripe mosaic virus in barley. *Can. J. Bot.* 54, 1604–1621.

Carroll, T. W., and Mayhew, D. E. (1976b). Occurrence of virions in developing ovules and embryo sacs of barley in relation to the seed transmissibility of barley stripe mosaic virus. *Can. J. Bot.* **54**, 2497–2512.

Cherewick, W. J. (1965). Smut, an additional cause of oat blast. *Phytopathology* **55**, 1368–1369.

Chou, M. C., and Preece, T. F. (1968). The effect of pollen grains on infections caused by *Botrytis cinerea* Fr. *Ann. Appl. Biol.* **62**, 11–22.

Crosby, A. W., Jr. (1972). "The Columbian Exchange—Biological and Cultural Consequences of 1492." Greenwood Press, Westport, Connecticut.

Cunfer, B., Mathre, D. E., and Hockett, E. A. (1975). Factors influencing the susceptibility of male-sterile barley to ergot. *Crop Sci.* **15**, 194–196.

Darlington, L. C., and Mathre, D. E. (1976). Resistance of male sterile wheat to ergot as related to pollination and host genotype. *Crop Sci.* **16**, 728–730.

Darlington, L. C., Carroll, T. W., and Mathre, D. E. (1976). Enhanced susceptibility of barley to ergot as a result of barley stripe mosaic virus infection. *Plant Dis. Rep.* **60**, 584–587.

deMunk, W. J. (1972). Bud necrosis, a storage disease of tulips. III. The influence of ethylene and mites. *Neth. J. Plant Pathol.* **78**, 168–178.

deMunk, W. J., and Beijer, J. J. (1971). Bud necrosis, a storage disease of tulips. 1. Symptoms and the influence of storage conditions. *Neth. J. Plant Pathol.* **77**, 97–105.

Ellis, M. A., Galvez, G. E., and Sinclair, J. B. (1976). Effect of pod contact with soil on fungal infection of dry bean seeds. *Plant Dis. Rep.* **60**, 974–976.

Eslick, R. F., and Afanasiev, M. M. (1955). Influence of time of infection with barley stripe mosaic on symptoms, plant yield, and seed infection of barley. *Plant Dis. Rep.* **39**, 722–724.

Fischer, G. W., and Holton, C. S. (1957). "Biology and Control of the Smut Fungi." Ronald Press, New York.

Ford, R. E. (1966). Recovery of pea streak virus from pea seed parts and its transmission by immature seed. *Phytopathology* **56**, 858–859.

Goth, R. W. (1966). A quasi-parasite of bean pods: *Fusarium oxysporum*. *Phytopathology* **56**, 442–443.

Grogan, R. G., and Kimble, K. A. (1967). The role of seed contamination in the transmission of *Pseudomonas phaseolicola* in *Phaseolus vulgaris*. *Phytopathology* **57**, 28–31.

Guthrie, J. W. (1970). Factors influencing halo blight transmission from externally contaminated *Phaseolus vulgaris* seed. *Phytopathology* **60**, 371–372.

Hollings, M. (1955). Investigation of chrysanthemum viruses. I. Aspermy flower distortion. *Ann. Appl. Biol.* **43**, 86–102.

Inouye, T. (1962). Studies on barley stripe mosaic in Japan. Sonderdruckaus Berichte des Ohara Institus fur Landwirtschaftliche Biologie, Okayama Universitat, Boncl. XI Heft 4, Marz 1962, Kuraschiki, Japan.

Jarvis, W. R. (1962). The infection of strawberry and raspberry fruits by *Botrytis cinerea* Fr. *Ann. Appl. Biol.* **50**, 569–575.

Klein, M. (1970). Safflower phyllody—a mycoplasma disease of *Carthamus tinctorius* in Israel. *Plant Dis. Rep.* **54**, 735–738.

Klisiewicz, J. M. (1963). Wilt-incitant *Fusarium oxysporum* f. *carthami* present in seed from infected safflower. *Phytopathology* **53**, 1046–1049.

Klisiewicz, J. M. (1975). Survival and dissemination of *Verticillium* in infected safflower seed. *Phytopathology* **65**, 696–698.

Kosuge, T., and Hewitt, W. B. (1964). Exudates of grape berries and their effect on germination of conidia of *Botrytis cinerea*. *Phytopathology* **54**, 167–172.

Lister, R. M., and Murant, A. F. (1967). Seed-transmission of nematode-borne viruses. *Ann. Appl. Biol.* **59**, 49–62.

Loesch, P. J., Jr., Foley, D. C., and Cox, D. F. (1976). Comparative resistance of opaque-2 and normal inbred lines of maize to ear-rotting pathogens. *Crop Sci.* **16**, 841–842.

McClellan, W. D., and Hewitt, W. B. (1973). Early *Botrytis* rot of grapes: Time of infection and latency of *Botrytis cinerea* Pers. in *Vitis vinifera* L. *Phytopathology* **63**, 1151–1157.

Mangelsdorf, P. C. (1966). Genetic potentials for increasing yields of food crops and animals. *Proc. Natl. Acad. Sci. U.S.A.* **56**, 370–375.

Mayhew, D. E., and Carroll, T. W. (1974a). Barley stripe mosaic virus in the egg cell and egg sac of infected barley. *Virology* **58**, 561–567.

Mayhew, D. E., and Carroll, T. W. (1974b). Barley stripe mosaic virions associated with spindle microtubules. *Science* **185**, 957–958.

Miller, T. D., and Schroth, M. N. (1972). Monitoring the epiphytic population of *Erwinia amylovora* on pear with a selective medium. *Phytopathology* **62**, 1175–1182.

Mower, R. L., and Hancock, J. G. (1975). Mechanism of honeydew formation by *Claviceps purpurea*. *Can. J. Bot.* **53**, 2826–2834.

Murant, A. F., and Lister, R. M. (1967). Seed-transmission in the ecology of nematode-borne viruses. *Ann. Appl. Biol.* **59**, 63–76.

Murdock, D. J., Nelson, P. E., and Smith, S. H. (1976). Histopathological examination of *Pelargonium* infected with tomato ringspot virus. *Phytopathology* **66**, 844–850.

Ogawa, J. M., and English, H. (1960). Blossom blight and green fruit rot of almond, apricot and plum caused by *Botrytis cinerea*. *Plant Dis. Rep.* **44**, 265–268.

Platford, R. G., and Bernier, C. C. (1970). Resistance to *Claviceps purpurea* in spring and durum wheat. *Nature (London)* **226**, 770.

Powelson, R. L. (1960). Initiation of strawberry fruit rot caused by *Botrytis cinerea*. *Phytopathology* **50**, 491–494.

Puranik, S. B., and Mathre, D. E. (1971). Biology and control of ergot on male sterile wheat and barley. *Phytopathology* **61**, 1075–1080.

Raabe, R. D. (1971). *Sclerotinia* blight of *Stephanotis* flowers. *Phytopathology* **61**, 1524–1525.

Roath, W. W., and Hockett, E. A. (1971). Pollen development in genetic male-sterile barley. *Barley Genet. 2, Proc. Int. Symp., 2nd, 1969* pp. 308–315.

Sahambi, H. S. (1970). Studies on Sesamum phyllody virus—virus-vector relationship and host range. *In* "Proc. First International Symposium on Plant Pathology." Indian Phytopathological Soc., New Delhi, pp. 340–351.

Sands, D. C., and McIntyre, J. L. (1977). Possible methods to control pear blast (*Pseudomonas syringae*). *Plant Dis. Rep.* **61**, 311–312.

Scarborough, B. A., and Smith, S. H. (1977). Effects of tobacco and tomato ringspot viruses on the reproductive tissues of *Pelargonium* x *hortorum*. *Phytopathology* **67**, 292–297.

Schippers, B. (1963). Transmission of bean common mosaic virus by seed of *Phaseolus vulgaris* L. cultivar Beka. *Acta Bot. Neerl.* **12**, 433–497.

Schroth, M. N., Thomson, S. V., Hildebrand, D. C., and Moller, W. J. (1974). Epidemiology and control of fire blight. *Annu. Rev. Phytopathol.* **12**, 389–412.

Shepherd, R. J. (1972). Transmission of viruses through seed and pollen. *In* "Principles and Techniques in Plant Virology" (C. I. Kado and H. O. Agrawal, eds.), pp. 267–292. Van Nostrand-Reinhold, Princeton, New Jersey.

Sitterly, W. R., and Shay, J. R. (1960). Physiological factors affecting the onset of susceptibility of apple fruit to rotting by fungus pathogens. *Phytopathology* **50**, 91–93.

Stanghellini, M. E., and Aragaki, M. (1966). Relation of periderm formation and callose deposition to anthracnose resistance in papaya fruit. *Phytopathology* **56**, 444–450.

Steadman, J. R. (1975). Nature and epidemiological significance of infection of bean seed by *Whetzelinia sclerotiorum*. *Phytopathology* **65**, 1323–1324.

Stoskopf, N. C., and Rai, R. K. (1972). Cross-pollination in male sterile wheat in Ontario. *Can. J. Plant Sci.* **52**, 387–393.

Strange, R. N., and Smith, H. (1971). A fungal growth stimulant in anthers which predisposes wheat to attack by *Fusarium graminearum*. *Physiol. Plant Pathol.* **1**, 141–150.

Strange, R. N., Majer, J. R., and Smith, H. (1974). The isolation and identification of choline and betaine as the two major components in anthers and wheat germ that stimulate *Fusarium graminearum* in vitro. *Physiol. Plant Pathol.* **4**, 277–290.

Strobel, G. A., and Hewitt, W. B. (1964). Time of infection and latency of *Diplodia viticola* in *Vitis vinifera* var. Thompson seedless. *Phytopathology* **54**, 636–639.

Sulzbach, C. W., and Pack, M. R. (1972). Effects of fluoride on pollen germination, pollen tube growth, and fruit development in tomato and cucumber. *Phytopathology* **62**, 1247–1253.

Taylor, R. H., Grogan, R. G., and Kimble, K. A. (1961). Transmission of tobacco mosaic virus in tomato seed. *Phytopathology* **51**, 837–842.

Tenne, F. D., Prasartsee, C., Machado, C. C., and Sinclair, J. B. (1974). Variation in germination and seed borne pathogens among soybean seed lots from three regions in Illinois. *Plant Dis. Rep.* **58**, 411–413.

Thomas, C. A., and Orellana, R. G. (1963). Nature of predisposition of castorbeans to *Botrytis*. II. Raceme compactness, internode length, position of staminate flowers, and bloom in relation to capsule susceptibility. *Phytopathology* **53**, 249–251.

Tomlinson, J. A., and Carter, A. L. (1970). Studies on the seed transmission of cucumber mosaic virus in chickweed (*Stellaria media*) in relation to the ecology of the virus. *Ann. Appl. Biol.* **66**, 381–386.

Ullstrup, A. J. (1971). Hyper-susceptibility of high-lysine corn to kernel and ear rots. *Plant Dis. Rep.* **55**, 1046.

Warren, H. L., and Nicholson, R. L. (1975). Kernel infection, seedling blight, and wilt of maize caused by *Colletotrichum graminicola*. *Phytopathology* **65**, 620–623.

Watkins, J. E., and Littlefield, L. J. (1976). Relationships of anthesis in Waldron wheat to infection by *Claviceps purpurea*. *Trans. Br. Mycol. Soc.* **66**, 362–363.

Yang, A. F., and Hamilton, R. I. (1974). The mechanism of seed transmission of tobacco ringspot virus in soybean. *Virology* **62**, 26–37.

Zaumeyer, W. J. (1932). Comparative pathological histology of three bacterial diseases of bean. *J. Agric. Res.* **44**, 605–632.

Chapter 13

Tissue Is Disintegrated

M. S. MOUNT

I. INTRODUCTION

Degradation and penetration of plant cell walls are major aspects of parasites and pathogenesis. Soft-rot bacteria and fungi turn fleshy tubers, fruits, and leafy vegetables into messy soups both in the field and after harvest. Root- and stem-rotting fungi and bacteria weaken or disintegrate roots and shoots of all types of plants. Many biotic pathogens, including nematodes, fungi, bacteria, and even parasitic seed plants, produce enzymes that digest or soften plant cell walls and thus aid in the penetration of the protective surfaces of plants.

The cell walls of higher plants are complex structures composed mainly of polysaccharides and lignin. These substances are laid down in a structured arrangement that is distinctive to the species or, for that matter, to the specific tissues of a given species. During elongation and maturation of plant cells, marked changes occur in their content of cellulose, pectic substances, protein, and other constituents (Meinert and Delmer,

1977). In some cases, recognition and compatibility phenomena that determine whether a given host is susceptible or resistant or whether a given pathogen is virulent or avirulent are determined by the composition of the cell walls and the induction or repression of specific degradative enzymes in the pathogen. Through the barriers they present and the enzymes they elicit, cell walls may also determine the extent of destruction by each pathogen which challenges them, ranging from penetration of a single cell to total maceration of tissue. The complexity of the enzymes involved, as well as the substrates on which they act, make experimental analysis of tissue disintegration extraordinarily difficult.

II. PLANT CELL WALL STRUCTURE

A cuticular layer, composed mainly of cutins, waxes, cellulose, and pectic substances, is found on plant epidermal cell surfaces, in substomatal areas, and in mesophyll and palisade cells. Cutin is chiefly hydroxy fatty acids, cross-linked with peroxide and ether bridges. Pectic polysaccharides, cellulose, hemicellulose, structural glycoproteins, and lignin constitute the major components of the cell wall. Minor constituents include a host of accessory substances including waxes, fats, terpenes, alkaloids, sterols, amino acids, proteins, tannins, and phenols. The cell wall may be viewed as a continuous matrix coupled with cellulose microfibrils in which there are three basic regions: middle lamella, primary cell wall, and secondary cell wall. The demarcation of the cell wall regions has been based upon the characteristic molecular architecture and chemical composition of each region. The identification of all cell wall polymers along with their network of connections should help us better understand the physiological basis of many host–pathogen interactions, and, in particular, the disintegration of plant tissues.

In general, the primary walls of various plants are probably similar in composition and structure. They consist mainly of pectic substances, celluloses, and hemicelluloses. The pectin constituents are not linearly arranged, but probably occur in a network, with the main chain of pectin being an α-1,4-linked galacturonan region interspersed with 1,2-linked rhamnosyl residues (Aspinall, 1970a; Keegstra et al., 1973; Talmadge et al., 1973). The linear portion of the galacturonan region probably contains 6–12 galacturonsyl residues (Bauer et al., 1973). The rhamnogalacturonan chain appears to be common to all pectic polysaccharides (Aspinall et al., 1967; Aspinall and Cottrell, 1971; Talmadge et al., 1973). The carboxyl groups of the uronide residues may be methylated to various degrees. Also detected, but not fully resolved, are minor amounts of 2,4- and 3,4-linked galacturonsyl residues which would cause an increased branching of the rhamnogalacturonan chain (Aspinall, 1970a; Talmadge *et al.*, 1973). Polymers of neutral sugars are also present which are probably interconnected with the rhamnogalacturonide fraction of the cell wall (Northcote, 1972).

The hemicelluloses, also a major constituent of the primary cell wall, are covalently linked by glycosidic bonds at the reducing ends of pectic polysaccharides and are linked by hydrogen bonds to the cellulose fibrils (Bauer *et al.*, 1973; Northcote, 1972). Xyloglucan consists of a β-1,4-glucan chain with frequent β-1,6-xyloside units (Bauer *et al.*, 1973). Xylans, glucomannans, galactoglucomannans, mannans, and galactomannans are also present (Aspinall 1970b; Northcote, 1972) but their arrangement in the cell wall varies in different plant species (Bateman and Basham, 1976).

A hydroxyproline-rich glycoprotein also is attached to the pectic fraction in primary walls (Keegstra, 1973; Lamport, 1970; Northcote, 1972). Its function has not been precisely determined, but it probably adds to the structural integrity of the pectic fraction. The constituents of the carbohydrate protein are arabinose and galactose (Lamport *et al.*, 1973), which are involved in the cross-linking of pectic and cellulosic segments.

Unlike the primary cell wall, secondary walls show great variation from plant to plant and tissue to tissue (Talmadge *et al.*, 1973). They are composed mainly of cellulose, pectin, hemicellulose, and lignin. a-Cellulose is the principal component of the secondary wall. It varies from 250 to about 14,000 D-glucopyranose units linked by β-1,4-bonds. These long, linear molecules are bound laterally by hydrogen bonds into microfibrillar bundles that include both crystalline and amorphous regions (Mühlethaler, 1967).

The middle lamella is the area which links cells together. In herbaceous tissues it consists mainly of pectic substances (McClendon, 1964). In lignified tissues, lignin is the major constituent of the middle lamella. It also makes up 20–30% of the primary and secondary walls (Cowling and Brown, 1969). Lignin is an aromatic compound derived from coniferyl, sinapyl, and p-hydroxycinnamyl alcohols (Kirk, 1971). Cellulose and lignin are deposited in cell walls as mutually interpenetrating polymer systems. Lignin adds to the structural integrity of the cell wall and acts as a barrier to the action of polysaccharide-degrading enzymes.

III. POLYSACCHARIDE-DEGRADING ENZYMES

In order to reach wall polysaccharides, pathogens may first employ cutin esterase or various oxidizing enzymes to penetrate the cuticular barrier. Many biotic pathogens are equipped with an arsenal of cell wall-degrading enzymes which assist in breaching the structural defenses of plants. Pathogens may produce only one type of polysaccharide-degrading enzyme or have a multienzymatic system capable of degrading most polysaccharide components of the cell wall. Although host penetration may be accomplished by both physical and chemical action, pathogen enzymatic action is probably the most important means by which organisms enter and spread through their hosts. Most of the polysaccharide-degrading enzymes are produced inductively and are extracellular; only a few constitutive enzymes are known to contribute to the degradation process. A brief summary of the types and characteristics of these enzymes is important in understanding the physiology of tissue breakdown.

A. Pectolytic Enzymes

Pectic enzymes may be classified by several criteria (Bateman and Millar, 1966): (1) the mechanism by which the 1,4-glycosidic linkage is cleaved; (2) the optimal activity the enzyme has for a substrate; and (3) whether cleavage of the substrate occurs in a random or terminal fashion. The pectic chain may be split by either a hydrolytic or transeliminative process. Pectin methylgalacturonase and polygalacturonase are hydrolytic enzymes. They usually are ineffective in the presence of Ca^{2+} ions and have an acidic pH optimum. Pectin methylgalacturonase is more reactive with pectin as a substrate rather than pectic acid, while polygalacturonase prefers pectic acid over pectin. Both of these enzymes can attack their substrates in a random or terminal manner.

The transeliminative, or lyitic enzymes degrade the pectin chain by a glycosidic split resulting in an unsaturated bond between carbons 4 and 5 of the uronide portion of the reaction products (Albersheim et al., 1960). Pectin methyltranseliminase and polygalacturonate transeliminase both require Ca^{2+} as a cofactor for optimal activity and generally exhibit their highest activity under alkaline conditions. Pectin methyltranseliminase prefers pectin as a substrate while polygalacturonate transeliminase favors pectic acid. Similar to the hydrolases, the lyitic enzymes occur as exo- and endo-types, attacking the polymer internally or removing the end linkage groups of the pectin chain.

The enzymes that hydrolyze the galactans and arabans, the neutral sugar polymers, are the galactanases and arabinases. Endo-β-1,4-galactanases and β-galactosidases are enzymes which attack the galactans (Knee and Friend, 1968; Van Etten and Bateman, 1969). The galactanases may be either exo- or endo-enzymes. The arabinases, however, are of the exo type since only monomeric arabinose is detected in a reaction mixture (Cole and Bateman, 1969; Fuchs et al., 1965; Kaji et al., 1965).

Pectin methylesterase is produced by most phytopathogens and in some higher plants. It hydrolyzes the esterified methyl group of the galacturonosyl chain producing uronic acid groups and methanol as reaction products. The pectin chain length is not affected by the action of this enzyme, but the removal of the methyl groups makes the pectin more vulnerable to other polysaccharide-degrading enzymes.

B. Cellulolytic Enzymes

Native cellulose, which is insoluble and crystalline, is cleaved by cellulolytic enzymes and peroxide produced by certain plant pathogens to soluble, oligoglucosides and glucose (Bateman, 1964, 1969; Kelman and Cowling, 1965). The enzymes responsible for cellulose degradation are C_1 and C_2 enzymes, which change the partly crystalline cellulose in an unknown manner to a more soluble cellulose or a "short-fiber" chain derivative, respectively (Halliwell and Mohammed, 1971; Reese, 1956). The resulting modified β-1,4-glucan chains are further hydrolyzed by endoglucanases, commonly known as C_x enzymes, to cellobiose, which, in turn, is converted to glucose by β-glucosidase (Reese, 1956). True cellulolytic organisms are those which are capable of degrading native cellulose to glucose, while noncellulolytic organisms may possess the ability to hydrolyze soluble cellulose and/or cellobiose by C_x enzymes and β-glucosidase (Reese, 1956).

C. Hemicellulolytic Enzymes

Our knowledge of hemicellulose degradation by plant pathogens is somewhat incomplete; however, a number of hemicellulases have been described (Bauer et al., 1973; English et al., 1971; Kaji and Yoshihara, 1970; King and Fuller, 1968; Reese and Shibata, 1965; Strobel, 1963; Van Etten and Bateman, 1969). Endoglucanase, which is a C_x cellulase, is capable of hydrolyzing β-1,4-xyloglucan linkages which make up a large proportion of the hemicelluloses (Bauer et al., 1973). As shown with cellulose degradation, complete degradation of many hemicelluloses requires multienzyme systems. The xylans can be hydrolyzed to xylose oligomers by endoxylanases (Strobel, 1963). The dimer, xylobiose, is further split by a β-xylosidase resulting in xylose monomers (King and Fuller, 1968). A two-enzyme system, endomannanase and β-mannosidase, is necessary for the complete hydrolysis of β-1,4-mannans (Reese and Shibata, 1965). Galactomannans are degraded by three enzymes: endo-1,4-β-mannanase, β-mannosidase, and α-galactosidase (Reese and Shibata, 1965; Van Etten and Bateman, 1969).

IV. PATHOGENESIS AND CELL WALL DEGRADATION

Numerous reviews have been written on the polysaccharide-degrading enzymes of plant pathogens. The latest and most constructively critical analysis is by Bateman and Basham (1976). They emphasize that the ability of a pathogen to produce a given enzyme *in vitro* or the detection of an enzyme in diseased tissue provides only circumstantial evidence of a role in pathogenesis. Evidence must be given demonstrating specific changes in the structural framework of host cells in response to specific enzymes. Light and electron microscope observations of hyphal penetration of plant cuticle and cell walls showing modifications have often been taken as evidence of enzyme action (Aist, 1976; McKeen *et al.*, 1969; Van den Ende and Linskens, 1974). Such observations are not certain proof that penetration was achieved enzymatically. Nonproteinaceous metabolic catalysts such as peroxide and iron could be involved, as was suggested by Koenigs (1974). Similarly, *in vitro* enzyme production analyses are suspect because the rate and type of enzymes synthesized by plant pathogens have been shown to be greatly influenced by the environmental conditions under which pathogens were grown (Bateman and Basham, 1976).

Demonstration of cell wall breakdown products provides the most direct evidence that enzymes play an important role in pathogenesis (Bateman *et al.*, 1969; Bateman and Basham, 1976; Cole and Wood, 1961; Hancock, 1967; Kelman and Cowling, 1965). In all cases, solubilization of certain cell wall polymers was measured and could be associated with enzymes produced by the pathogens.

Tissue susceptibility or the rate of tissue disintegration is probably greatly influenced by the sequence of enzyme synthesis in the pathogen. It is evident that endopectic hydrolases and lyases increase the susceptibility of cell walls to nonuronide degradation (Talmadge *et al.*, 1973). Therefore, any wall modifications resulting from wounding or enzymatic action could greatly influence the rate of tissue degradation by particular enzymes. The induction and repression of specific polysaccharide-degrading enzyme synthesis could play a significant role in disease expression (see Section VI).

V. MACERATION AND CELL DEATH IN HERBACEOUS TISSUES

Maceration and death of host cells is a classical problem in plant pathology (Brown, 1915). The enzymatic basis of tissue maceration is well established. The principal enzymes involved are endopolygalacturonases and endopectate lyases (Basham and Bateman, 1975a; Mount *et al.*, 1970). The rate of maceration is directly related to the log of the enzyme concentration (Bateman, 1968). Calcium either increases or decreases the rate of tissue maceration, depending upon the macerating enzyme involved (Heath and Wood, 1971). Calcium pectates are more resistant to hydrolysis by endohydrolases, but are more susceptible to lyase activity. Calcium has been postulated to increase host resistance to pathogens that produce pectin endohydrolases (Bateman and Millar, 1966).

In some cases, the pattern of maceration leading to cell death has not been observed: an endopolygalacturonase separated from an endopectin methyltranseliminase produced by *Sclerotinia fructigena* and an isozyme of endopolygalacturonate transeliminase isolated from *Erwinia chrysanthemi* did not exhibit maceration capabilities on selected host tissues (Byrde and Fielding, 1962, 1968; Garibaldi and Bateman, 1971). The reasons for these anomalous events are unknown. Some polysaccharidases contribute to the maceration process in the presence of endopectic enzymes but will not readily macerate tissues by themselves (Bateman and Basham, 1976).

Maceration of plant tissues by endohydrolases and lyases is accompanied by death of the affected, unplasmolyzed cells (Basham, 1974; Mount *et al.*, 1970). Death is associated with rapid permeability changes indicating plasma membrane damage (Basham and Bateman, 1975a; Hall and Wood, 1973). Plasmolyzing plant cells before treatment with pectic enzymes delays cell death, but maceration is not affected (Fushtey, 1957; Tribe, 1955).

Several hypotheses have been advanced to explain the killing of cells by pectic enzymes (Bateman, 1976; Bateman and Basham, 1976; Fushtey, 1957). One suggests that the products of cell wall maceration are lethal (Fushtey, 1957). However, soluble products prepared from enzyme-treated plant tissues fail to bring about any lethal effects on plant cells (Basham and Bateman, 1975a; Mussell, 1973). Another hypothesis is that certain noncatalytic properties of purified pectic enzymes, such as different net ionic charges, isoelectric points, pH optima, and substrate cleavage mechanisms, were responsible for toxic action, but no experimental evidence has been found to support this idea (Basham and Bateman, 1975a).

Pectic substances may be associated with the cell membrane of plant cells and may be subject to degradation by pectic enzymes (Mount *et al.*, 1970). When isolated cucumber protoplasts were treated with purified endopolygalacturonate transeliminase, no physiological effects were observed (Tseng and Mount, 1974), indicating that substrate sites for pectic enzymes, if present, were protected when the protoplasts were plasmolyzed. However, the results of a study of the effects of osmotic stretching and pectic enzymes on tobacco leaf protoplasts suggested that plasmolysis was not protecting the cell membrane from lyase activity (Basham and Bateman, 1975a).

Still another hypothesis suggests that the pectic enzymes alter the rhamnogalacturonide portion of the cell wall so that increased osmotic stress ruptures the plasmalemma (Hall and Wood, 1973). This hypothesis has gained the most support from experiments showing the effects of plasmolysis and deplasmolysis on electrolyte leakage from purified, lyase-treated potato tissue. Results from these studies point to a direct relationship between alteration of the cell wall and physical damage to the cell membrane (Basham and Bateman, 1975b).

VI. REGULATORY MECHANISMS OF POLYSACCHARIDE-DEGRADING ENZYMES

The majority of the regulatory mechanisms operating in tissue disintegration have not been deduced and the general picture remains unclear. In many host–parasite interactions, however, the production of cell wall-degrading enzymes in conjunction with disease expression has been shown to be subject to catabolite repression (Bateman, 1976). Moran and Starr (1969), studying *Erwinia carotovora* and *E. aroideae,* found that the enzyme was strictly constitutive and that only release from catabolite repression enabled glucose-grown cells to initiate degradation of the pectic acid molecule. Hsu and Vaughn (1969) found that maximum *Aeromonas liquifaciens* pectate lyase activity occurs under conditions of restricted growth. In an isolate of *E. carotovora,* a direct relationship was demonstrated between glucose repression of endopolygalacturonate transeliminase and adenosine-3′,5′-cyclic monophosphate (cAMP) levels, implicating cAMP as a regulatory factor of pectic enzyme synthesis (Hubbard *et al.,* 1978).

Specific inducers have been determined in some instances. Cooper and Wood (1975) established that the production of a number of cell wall-degrading enzymes from *Verticillium albo-atrum* and *Fusarium oxysporum* f. sp. *lycopersici* was specifically induced by different plant cell wall sugars. High concentrations of these sugars, however, were inhibitory to enzyme induction. The induction of cellulase in *Trichoderma viride* can be brought about by cellulose, cellobiose, lactose, and sophorose (Mandels, 1975). High concentrations of cellobiose, glucose, or glycerol repress cellulase formation, and cellobiose can inhibit cellulase activity.

Evidently a number of different mechanisms may be responsible for the repression and induction of degradative enzymes in phytopathogens.

VII. DISINTEGRATION OF WOODY TISSUES

Woody perennials make up a very large fraction of the total biomass of the earth. Thus, decay of wood is both a major aspect of pathogenesis and a source of great loss in the volume of timber in use around the world. But the degradation of woody tissues has received scant attention compared to that of herbaceous plants. Three major types of decay fungi are known: white rot, brown rot, and soft rot. In the following paragraphs an outline is presented of biochemical activities of these types.

A. Structure of Woody Tissues

The stems of woody plants consist of an inner xylem core surrounded by the bark (phloem). The xylem is usually divided into an outer, light-colored sapwood and an inner, sometimes dark-colored heartwood. The heartwood functions only in mechanical support, while the sapwood also functions in water transport. The xylem parenchyma cells function as a food storage depot. Transformation of sapwood to heartwood is an aging process in which xylem parenchyma cells are depleted of nutrients and accessory substances are deposited in the wood cells. These deposits give heartwood a wide range of colors and, in some species, make it toxic to decay fungi. The proportion of heartwood in tree stems is quite variable, even within a given tree species (Hillis, 1971). Heartwood usually contains less moisture than sapwood.

The principal constituents of wood cell walls are cellulose, hemicellulose, and lignin. The hemicelluloses are short heteropolymers of arabinose, galactose, glucose, mannose, and uronic acids. Cellulose composes about 45% of a typical wood cell and is a cohesive, three-dimensional structure. The bulk of cellulosic material is in a thick, secondary wall surrounded by a thin, primary wall. The secondary wall is divided into three layers, designated S1, S2, and S3, and can be distinguished by the orientation of the long, slender bundles of cellulose microfibrils which are bound to each other laterally by hydrogen bonds. The S1 and S3 cellulose layers are oriented in a flat helix arrangement to the fiber axis, while the S2 layer, which contains most of the cellulosic substance of the cell wall, is deposited almost parallel to the fiber axis.

Lignin is the most complex component of wood cell walls. It is a three-dimensional, branched polymer formed by oxidative polymerization of p-coumaryl, coniferyl, and sinapyl alcohols. The relative amounts of each alcohol derivative vary substantially among tree species. Lignin contains three basic types of intermonomer linkages: the arylglycerol-β-aryl ether linkage, the phenylcoumaran structure, and the biphenyl structures. Lignin and the wood polysaccharides are deposited as mutually interpenetrating polymers so that each constituent provides a protective physical barrier to the enzymatic degradation of the other constituents (Kirk, 1971).

B. Discoloration and Decay of Living Trees

Wounding of trees often leads to discoloration of xylem. This is caused by a combination of host responses to injury and to microorganisms that develop after wounding (Shigo and Hillis, 1973). The discoloration and decomposition of wood involves a sequence of events: Stage I—the events associated with wounding which include the chemical interactions of the tree with the abiotic environment; Stage II—the development of nondecay organisms in the vicinity of the wound; and Stage III—the development of decay. The degree of discoloration and decay is dependent upon the severity of the wound, the types of microorganisms present, and the vigor of the tree.

The nondecay microorganisms are usually bacteria and nonhymenomycetous fungi (Shigo, 1965, 1972). The organisms enter living cells, obtain nutrients, and detoxify chemicals. As the nondecay organisms advance and condition woody tissue, decay organisms may follow. The decay organisms are generally Hymenomycetes and degrade the cell wall constituents, mainly lignin and cellulose.

C. Decay of Wood Products

Proper use of wood in construction requires sound understanding of wood decay processes and factors that either favor or inhibit the development of decay fungi. Wood maintained below 20% moisture content by weight cannot be decayed. Liquid water must be present in wood cells, apparently to serve as a medium for the diffusion of enzymes and partial degradation products between hyphae of the fungi and the wood cell walls. If wood cannot be kept dry, decay-resistant heartwood or wood pressure-treated with preservatives should be used.

Heartwood of some timber species is more resistant to decay than sapwood; however, the supply of decay-resistant heartwood is diminishing. Many of the fast-growing trees today contain mostly sapwood which is highly susceptible to decay no matter what species of tree is involved.

The susceptibility of wood and wood products to enzymatic degradation is dependent on several physical and chemical attributes of the timber: (1) its moisture content; (2) the degree of cellulose crystallinity; (3) the size and diffusibility of cellulolytic enzymes in relation to cell wall capillary structure; (4) the unit-cell dimensions of cellulose crystal structure; (5) steric rigidity and conformation of anhydroglucose units in crystalline regions; (6) the degree of polymerization of cellulose; (7) the nature and amount of accessory substances in the wood including minerals, organic materials, phenolic compounds, and lignin; and (8) the distribution of substituent groups such as methyl, carboxymethyl, sulfate, etc. These parameters all affect the rates of enzyme attack and accessibility to cellulose while, at the same time, the types of enzymes produced by microorganisms determine the characteristics of wood degradation.

1. White Rot

Several hundred species of the Hymenomycetes are capable of attacking both lignin and cellulose and are known to produce cellulases, xylanases, mannases, and lignin-degrading enzymes (Kirk, 1975). White-rotted tissue is bleached in appearance. White-rot fungi can degrade lignin completely and sometimes do so more rapidly than the polysaccharide components. The rate of lignin degradation is variable involving a complex system of enzymes and coenzymes (Kirk, 1975). The enzymes are capable of cleaving aryl-glycerol-β-aryl ether bonds between phenylpropane units and disrupting aromatic rings.

The lignin-degrading enzyme systems of white-rot fungi have been studied by analyzing the products of model substrates in liquid cultures and by extracting partially decomposed lignin from decayed wood. Breakdown products of lignin in white-rot fungus-seeded liquid cultures are similar to those detected in decomposed wood (Kirk, 1973). Degraded lignin contains less hydrogen, carbon, and methoxyl groups but possesses a higher percentage of oxygen than native lignin (Ishikawa et al., 1963). The proportion of hydrolyzable bonds in lignin is small, probably indicating the necessity for the action of nonhydrolytic enzymes in its degradation (Kirk, 1975).

The enzymes involved in lignin degradation have not yet been isolated from decay fungi. Nevertheless, degradation of lignin is known to involve the removal of methoxyl groups, formation of α-carbonyl groups, oxidation of side chains of the polymer, oxidative cleavage of aromatic rings which are still integrated in the lignin polymer, and release of aliphatic molecules after ring cleavage (Kirk, 1975). Lignin is apparently unique among natural polymers in being degraded by enzymes that cleave the polymer within monomer units (in this case, cleavage of the aromatic ring) rather than the bonds between the monomer units.

In a model system using *Pseudomonas acidovorens* on a compound containing a β-ether linkage, it was demonstrated that an α-carbon oxidation occurred first, followed by a splitting of the β-ether (Crawford *et al.*, 1973, 1975). This shows the sequential enzyme attack pattern typical of the degradative process; the decomposition of lignin probably occurs in a similar fashion (Kirk, 1975). The number of enzymatic events which contribute to the degradation of lignin could involve demethylation–aromatic hydroxylation, hydroxylation–decarboxylation reactions, lactonizations, delactonizations, and dehydrogenations (Kirk, 1975). Laccases, peroxidases, and tyrosinases are produced by white-rot fungi, but their role in the decomposition of lignin is little understood (Kirk, 1973).

White-rot fungi degrade cellulose rapidly but apparently do so "one-molecule-at-a-time" so that the cellulose remaining in white-rotted wood at any stage of decay is not substantially less in degree of polymerization than in the original wood before decay. *Polyporus versicolor,* a typical white-rot fungus, degrades both crystalline and amorphous cellulose at rates that are constant in all stages of decay (Cowling, 1961).

2. Brown Rot

The brown-rot fungi are Holobasidiomycetes. The wood decomposed by these fungi retains a characteristic brownish discoloration and cubical checking pattern due to the shortening of cellulose chain molecules (Cowling, 1961). Brown-rotted wood is friable and brittle compared to white-rotted wood, which is more elastic. The brown-rot organisms live in the fiber lumina where they secrete polysaccharidases which result in the dissolution of cell wall constituents (Cowling and Brown, 1969). Brown-rot fungi decompose carbohydrates and alter the lignin, but do not utilize it as a source of food. The strength and elasticity of brown-rotted wood is reduced before any visible decay is evident.

The principal polysaccharide-degrading enzymes produced by brown-rot fungi are cellulases and hemicellulases. Brown-rot fungi can be readily distinguished from white-rot organisms in that brown-rotters do not exhibit an extracellular polyphenol oxidase system. In the initial stages of brown rot, polysaccharides in the amorphous regions of wood cell walls are depolymerized rapidly whereas, in the later stages, the residual crystalline fractions are depolymerized at a much slower rate. Thus polysaccharidases of brown-rot fungi are capable of reacting with microfibrils at the fine structural level. This is in contrast to the action of the white-rot fungus *Polyporus versicolor,* which appears to be confined to the gross capillary surfaces of the wood, at least until decay is well advanced (Cowling, 1961).

3. Soft Rot

Soft rot of woody tissues is a less severe decay, caused by certain Asco-mycetes and Fungi Imperfecti (Duncan, 1960). Details concerning the physiology of soft-rot degradation are lacking. Soft-rot fungi occur in wood that is extremely wet, a condition which is not optimal for the wood-rotting Basidiomycetes. The hyphae of soft rotters are usually not evident in the cell lumen, but are found in degraded cylindrical cavities which they create within the secondary cell wall. These characteristic cavities are aligned parallel to the orientation of microfibrils within the S2 layer. The enzymatically hydrolyzed cellulose causes considerable weight loss to the wood. In general, soft-rot fungi degrade only a small portion of the lignin (Kirk, 1971), although soft rot of beech wood caused by *Chaetomium globosum* results in significant losses of lignin (Levi and Preston, 1965).

D. Role of Nitrogen in Wood Decay

Many cellulolytic organisms are unable to cause wood decay, in part because they require more nitrogen for growth and development than wood-decay Basidiomycetes. In general, heartwood contains less than 0.05% total nitrogen per dry weight of tissue (Merrill and Cowling, 1966). The carbon:nitrogen (C:N) ratio of wood generally varies between 350:1 to 1,250:1 and comparable ratios for herbaceous tissues are 10:1 to 100:1. The rate of decay of woody tissues is directly related to the nitrogen content of the wood (Levi and Cowling, 1969; Reis, 1973). Decay fungi probably survive and promote decay by one or more of the following mechanisms (Cowling and Merrill, 1966): (1) fungal adapta-tions that result in preferential allocation of nitrogen to substances and pathways that are highly efficient in the utilization of wood constituents; (2) reutilization of available nitrogen by active mycelium through a dynamic and continuous process of autolysis and reuse; and (3) utiliza-tion of nitrogen from sources besides the wood itself, such as nitrogen fixation by cohabiting bacteria.

In synthetic media containing 32:1 and 1,600:1 ratios of C:N, *Polyporus versicolor* preferentially utilized the available nitrogen for the synthesis of nucleic acids (Levi and Cowling, 1969). Under conditions of nitrogen starvation, the nitrogen content of the mycelia was substantially less than that of mycelia produced when more nitrogen was available. These re-sults support the first hypothesis listed above.

Support for the second hypothesis stems from observations that a portion of the hyphae of wood-destroying fungi commonly disappear or appear fragmented in decayed wood. Levi *et al.* (1968) demonstrated that certain wood-decay fungi are capable of utilizing extracts of their own vegetative mycelium as their sole source of nitrogen when grown in the presence of glucose, and that proteins, peptides, and amino acids are more efficient in supporting growth than are nucleic acids, nucleotides, or cell wall materials. Thus the capacity to recycle nitrogen in their own mycelium contributes to the success of wood-destroying Basidiomycetes in elicting decay.

There is no evidence that wood-decay fungi can themselves fix atmospheric nitrogen; however, large populations of nitrogen-fixing bacteria have been isolated from decayed wood (Aho *et al.*, 1974). Since nitrogen-fixing bacteria can be abundant in decayed wood, they most likely play an important role in the decay of heartwood in living trees by supplying the much-needed nitrogen to the decay fungi. Conversely, the decay fungi aid the bacteria by decomposing cellulose and other polysaccharides to a usable form (Consenza *et al.*, 1970). The benefits for the organisms are mutualistic and the third hypothesis is satisfied.

VIII. CONCLUSION

The ultimate goal in disease physiology concerning the interaction of plant cell walls and pathogens and/or their enzymes is to understand the biochemical systems which lead to tissue disintegration. Knowledge of these biochemical events should help us develop new approaches to disease control. Cell wall alteration can involve a series of complex events brought about by many enzymes from diverse groups of pathogens.

There are many questions that can be asked regarding the role of cell wall-degrading enzymes in pathogenesis. It is well established that pectic enzymes (endo type) are involved in soft-rot diseases, but there is little evidence supporting the direct involvement of polysaccharidases in diseases other than those which are typified by extensive tissue decay. Are cell wall degrading enzymes in specific host–pathogen combinations a primary determinant of disease or are they merely secondary to other aggressive traits? To what extent are cellulases and hemicellulases responsible for pathogenic processes? Some pathogens produce isozymes of cell wall degrading enzymes. Are these enzymes involved in disease expression? Indirect evidence suggests that the sequence of enzyme synthesis may be important in determining the extent of cell wall alteration and, consequently, parasitism. The induction and repression of various enzymes may be determined by the accessibility of certain cell well substrates. Much of the research today is directed toward the study of cell wall degradation after extensive decay has taken place. Compatibility, however, is generally evident very soon after confrontation of host and parasite. More information is needed regarding the production and function of cell wall-degrading enzymes during the crucial initial stages of infection.

There is one point which may not be obvious to all readers and which must be stressed. In a search of the literature, it becomes evident that there is a lack of genetic studies in relation to cell wall-degrading enzyme production and pathogenesis. Much of the evidence presented within the last few years implicating cell wall-degrading enzymes in disease development is circumstantial. It seems logical that genetic comparisons must be performed before anyone can say conclusively that a particular enzyme is necessary for pathogenesis. Mutants which lack the ability to synthesize specific enzymes may not elicit the typical cell wall modifications, but may nevertheless be found to be fully capable of causing disease. There is no conclusive report that the production of polysaccharide degrading enzymes is the single determinant of virulence. It is imperative that genetic studies be performed, along with biochemical evaluations, in order to elucidate the involvement of cell wall-degrading enzymes in pathogenesis.

The challenge is before us and I, for one, have an overwhelming desire to return to the laboratory to resolve some of these problems.

References

Aho, P. E., Seidler, R. J., Evans, H. J., and Raju, P. N. (1974). Distribution, enumeration, and identification of nitrogen-fixing bacteria associated with decay in living white fir trees. *Phytopathology* **64**, 1413–1420.

Aist, J. R. (1976). Cytology of penetration and infection—fungi. *In* "Physiological Plant Pathology," Encyclopedia of Plant Physiology, Vol. 4. (R. Heitefuss and P. H. Williams, eds.), pp. 197–221. Springer-Verlag, New York.

Albersheim, P., Neukom, H., and Deuel, H. (1960). Über die Bildung von ungesättigten Abbauprodukten durch ein pektinabbauendes Enzym. *Helv. Chim. Acta* **43**, 1422–1426.

Aspinall, G. O. (1970a). Pectins, gums, and other plant polysaccharides. *In* "The Carbohydrates" (W. Pigman and D. Horton, eds.), 2nd ed., Vol. 2B, pp. 515–536. Academic Press, New York.

Aspinall, G. O. (1970b). "Polysaccharides," p. 228. Pergamon, Oxford.

Aspinall, G. O., and Cottrell, I. W. (1971). Polysaccharides of soy-beans. VI. Neutral polysaccharides from cotyledon meal. *Can. J. Chem.* **49**, 1019–1022.

Aspinall, G. O., Begbie, R., Hamilton, A., and Whyte, J. N. C. (1967). Polysaccharides of soy-bean. III. Extraction and fractionation of polysaccharides from cotyledon meal. *J. Chem. Soc. C* pp. 1065–1070.

Basham, H. G. (1974). The role of pectolytic enzymes in the death of plant cells. Ph.D. Thesis, Cornell University, Ithaca, New York.

Basham, H. G., and Bateman, D. F. (1975a). Killing of plant cells by pectic enzymes: The lack of direct injurious interaction between pectic enzymes or their soluble reaction products and plant cells. *Phytopathology* **65**, 141–153.

Basham, H. G., and Bateman, D. F. (1975b). Relationship of cell death in plant tissue treated with a homogeneous endo-pectate lyase to cell wall degradation. *Physiol. Plant Pathol.* **5**, 249–261.

Bateman, D. F. (1964). Cellulase and the *Rhizoctonia* disease of bean. *Phytopathology* **54**, 1372–1377.

Bateman, D. F. (1968). The enzymatic maceration of plant tissues. *Neth. J. Plant Pathol.* **74**, Suppl. 1, 67–80.

Bateman, D. F. (1969). Some characteristics of the cellulase system produced by *Sclerotium rolfsii* Sacc. *Phytopathology* **59**, 37–42.

Bateman, D. F. (1976). Plant cell wall hydrolysis by pathogens. *In* "Biochemical Aspects of Plant–Parasite Relationships" (J. Friend and D. R. Threlfall, eds.), pp. 79–103. Academic Press, New York.

Bateman, D. F., and Basham, H. G. (1976). Degradation of plant cell walls and membranes by microbial enzymes. *In* "Physiological Plant Pathology," Encyclopedia of Plant Physiology, Vol. 4 (R. Heitefuss and P. H. Williams, eds.), pp. 316–355, Springer-Verlag, New York.

Bateman, D. F., and Millar, R. L. (1966). Pectic enzymes in tissue degradation. *Annu. Rev. Phytopathol.* **4**, 119–146.

Bateman, D. F., Van Etten, H. D., English, P. D., Nevins, D. J., and Albersheim, P. (1969). Susceptibility to enzymatic degradation of cell walls from bean plants resistant and susceptible to *Rhizoctonia solani* Kuhn. *Plant Physiol.* **44**, 641–648.

Bauer, W. D., Talmadge, K. W., Keegstra, K., and Albersheim, P. (1973). The structure of plant cell walls. II. The hemicellulose of the walls of suspension-cultured sycamore cells. *Plant Physiol.* **51**, 174–187.

Brown, W. (1915). Studies in the physiology of parasitism. I. The action of *Botrytis cinerea. Ann. Bot. (London)* **29**, 313–348.

Byrde, R. J. W., and Fielding, A. H. (1962). Resolution of endopolygalacturonase and a macerating factor in a fungal culture filtrate. *Nature (London)* **196**, 1227–1228.

Byrde, R. J. W., and Fielding, A. H. (1968). Pectin methyl-*trans*-eliminase as the macerating factor of *Sclerotinia fructigena* and its significance in brown rot of apple. *J. Gen. Microbiol.* **52**, 287–297.

Cole, A. L. J., and Bateman, D. F. (1969). Arabinase production by *Sclerotium rolfsii. Phytopathology* **59**, 1750–1753.

Cole, M., and Wood, R. K. S. (1961). Pectic enzymes and phenolic substances in apples rotted by fungi. *Ann. Bot. (London)* [N.S.] **25**, 435–452.

Consenza, B. J., McCreary, M., Buck, J. D., and Shigo, A. L. (1970). Bacteria associated with discolored and decayed tissues in beech, birch, and maple. *Phytopathology* **60**, 1547–1551.

Cooper, R. M., and Wood, R. K. S. (1975). Regulation of synthesis of cell well degrading enzymes by *Verticillium albo-atrum* and *Fusarium oxysporum* f. sp. *lycopersici. Physiol. Plant Pathol.* **5**, 135–156.

Cowling, E. B. (1961). Comparative biochemistry of the decay of sweetgum sapwood by white-rot and brown-rot fungi. *U.S. Dep. Agric., Tech. Bull.* **1258**, 1–79.

Cowling, E. B., and Brown, W. (1969). Structural features of cellulosic materials in relation to enzymatic hydrolysis. *In* "Cellulases and Their Application. "Advances in Chemistry, Series 95, pp. 152–187. (R. F. Gould, ed.), Washington, D.C., Amer. Chem. Soc.

Cowling, E. B., and Merrill, W. (1966). Nitrogen in wood and its role in wood deterioration. *Can. J. Bot.* 44, 1539–1554.

Crawford, R. L., Kirk, T. K., Harkin, J. M., and McCoy, E. (1973). Bacterial cleavage of an arylglycerol-β-aryl ether bond. *Appl. Microbiol.* 25, 322–324.

Crawford, R. L., Kirk, T. K., and McCoy, E. (1975). Dissimilation of the lignin model compound veratrylglycerol-β-(σ-methoxyphenyl) ether by *Pseudomonas acidovorans*: Initial transformations. *Can. J. Microbiol.* 21, 577–579.

Duncan, C. G. (1960). Wood-attacking capacities and physiology of soft-rot fungi. *U.S. For. Serv. Rep.* 2173, 1–70.

English, P. D., Jurale, J. B., and Albersheim, P. (1971). Host-pathogen interactions. II. Parameters affecting polysaccharide-degrading enzyme secretion by *Colletotrichum lindemuthianum* grown in culture. *Plant Physiol.* 47, 1–6.

Fuchs, H., Jobsen, J. A., and Wouts, W. M. (1965). Arabinases in phytopathogenic fungi. *Nature (London)* 206, 714–715.

Fushtey, S. G. (1957). Studies in the physiology of parasitism. XXIV. Further experiments on the killing of plant cells by fungal and bacterial extracts. *Ann. Bot. (London)* [N.S.] 21, 273–286.

Garibaldi, A., and Bateman, D. F. (1971). Pectic enzymes produced by *Erwinia chrysanthemi* and their effects on plant tissue. *Physiol. Plant Pathol.* 1, 25–40.

Hall, J. A., and Wood, R. K. S. (1973). The killing of plant cells by pectolytic enzymes. *In* "Fungal Pathogenicity and the Plant's Response" (R. J. W. Byrde and C. V. Cutting, eds.), pp. 19–38. Academic Press, New York.

Halliwell, G., and Mohammed, R. (1971). Interactions between components of the cellulase complex of *Trichoderma koningii* on native substrates. *Arch. Mikrobiol.* 78, 295–309.

Hancock, J. G. (1967). Hemicellulose degradation in sunflower hypocotyls infected with *Sclerotinia sclerotiorum*. *Phytopathology* 57, 203–206.

Heath, M. C., and Wood, R. K. S. (1971). Role of cell-wall-degrading enzymes in the development of leaf spots caused by *Ascochyta pisi* and *Mycosphaerella pinodes*. *Ann. Bot. (London)* [N.S.] 35, 451–474.

Hillis, W. E. (1971). Distribution properties and formation of some wood extractives. *Wood Sci. Technol.* 5, 272–289.

Hsu, E. J., and Vaughn, R. H. (1969). Production and catabolite repression of the constitutive polygalacturonate *trans*-eliminase of *Aeromonas liquifaciens*. *J. Bacteriol.* 98, 172–181.

Hubbard, J. P., Williams, J. D., Niles, R. M., and Mount, M. S. (1978). The relationship between glucose repression of endo-polygalacturonate *trans*-eliminase and adenosine 3',5'-cyclic monophosphate levels in *Erwinia carotovora*. *Phytopathology* 68, 95–99.

Ishikawa, H., Schubert, W. J., and Nord, F. F. (1963). Investigations on lignin and lignification. XXVII. The enzymic degradation of soft-wood lignin by white-rot fungi. *Arch. Biochem. Biophys.* 100, 131–139.

Kaji, A., and Yoshihara, O. (1970). Production of α-L-arabinofuranosidase from various strains of *Corticium rolfsii*. *Agric. Biol. Chem.* 34, 1249–1253.

Kaji, A., Tagawa, K., and Motoyama, K. (1965). Studies on the enzymes acting on araban. VII. Properties of arabanase produced by plant pathogens. *J. Agric. Chem. Soc. Jpn.* 39, 352–357.

Keegstra, K., Talmadge, K. W., Bauer, W. D., and Albersheim, P. (1973). The structure of plant cell walls. III. A model of the walls of suspension-cultured sycamore cells based on interconnections of the macromolecular components. *Plant Physiol.* **51**, 188–197.

Kelman, A., and Cowling, E. B. (1965). Cellulase of *Pseudomonas solanacearum* in relation to pathogenesis. *Phytopathology* **55**, 148–155.

King, N. J., and Fuller, D. B. (1968). The xylanase system of *Coniophora cerebella*. *Biochem. J.* **108**, 571–576.

Kirk, T. K. (1971). Effects of microorganisms on lignin. *Annu. Rev. Phytopathol.* **9**, 185–210.

Kirk, T. K. (1973). Polysaccharide integrity as related to the degradation of lignin in wood by white-rot fungi. *Phytopathology* **63**, 1504–1507.

Kirk, T. K. (1975). Lignin-degrading enzyme system. *In* "Cellulose as a Chemical and Energy Resource" (C. R. Wilke, ed.), pp. 139–150. Wiley, New York.

Knee, M., and Friend, J. (1968). Extracellular "galactanase" activity from *Phytophthora infestans* (Mont.) de Bary. *Phytochemistry* **7**, 1289–1291.

Koenigs, J. W. (1974). Hydrogen peroxide and iron: A proposed system for decomposition of wood by brown-rot basidiomycetes. *Wood Fiber* **6**, 66–80.

Lamport, D. T. A. (1970). Cell wall metabolism. *Annu. Rev. Plant Physiol.* **21**, 235–270.

Lamport, D. T. A., Katona, L., and Roerig, S. (1973). Galactosylserine in extensin. *Biochem. J.* **133**, 125–132.

Levi, M. P., and Cowling, E. B. (1969). Role of nitrogen in wood deterioration. VII. Physiological adaptation of wood-destroying and other fungi to substrates deficient in nitrogen. *Phytopathology* **59**, 460–468.

Levi, M. P., and Preston, R. D. (1965). A chemical and microscopic examination of the action of the soft-rot fungus *Chaetomium globosum* on beechwood (*Fagus sylv.*). *Holzforschung* **19**, 183–190.

Levi, M. P., Merrill, W., and Cowling, E. B. (1968). Role of nitrogen in wood deterioration. VI. Mycelial fractions and model nitrogen compounds as substrates for growth of *Polyporus versicolor* and other wood-destroying and wood-inhabiting fungi. *Phytopathology* **58**, 626–634.

McClendon, J. H. (1964). Evidence for the pectic nature of the middle lamella of potato tuber cell walls based on chromatography of macerating enzymes. *Am. J. Bot.* **51**, 628–633.

McKeen, W. E., Smith, R., and Bhattacharya, P. K. (1969). Alterations of the host wall surrounding the infection peg of powdery mildew fungi. *Can. J. Bot.* **47**, 701–706.

Mandels, M. (1975). Microbial sources of cellulase. *In* "Cellulose as a Chemical and Energy Resource" (C. R. Wilke, ed.), pp. 81–105. Wiley, New York.

Meinert, M. C., and Delmer, D. P. (1977). Changes in biochemical composition of the cell wall of the cotton fiber during development. *Plant Physiol.* **59**, 1088–1097.

Merrill, W., and Cowling, E. B. (1966). Role of nitrogen in wood deterioration: Amounts and distribution of nitrogen in tree stems. *Can. J. Bot.* **44**, 1555–1580.

Moran, F., and Starr, M. P. (1969). Metabolic regulation of polygalacturonic acid *trans*-eliminase in *Erwinia*. *Eur. J. Biochem.* **11**, 291–295.

Mount, M. S., Bateman, D. F., and Basham, H. G. (1970). Induction of electrolyte loss, tissue maceration, and cellular death of potato tissue by an endo-polygalacturonate *trans*-eliminase. *Phytopathology* **60**, 924–931.

Mühlethaler, K. (1967). Ultrastructure and formation of plant cell walls. *Annu. Rev. Plant Physiol.* **18**, 1–24.

Mussell, H. W. (1973). Endopolygalacturonase: Evidence for involvement in *Verticillium* wilt of cotton. *Phytopathology* **63**, 62–70.

Northcote, D. H. (1972). Chemistry of the plant cell wall. *Annu. Rev. Plant Physiol.* **23**, 113–132.

Reese, E. T. (1956). A microbiological progress report. Enzymatic hydrolysis of cellulose. *Appl. Microbiol.* **4**, 39–45.

Reese, E. T., and Shibata, Y. (1965). β-Mannanases of fungi. *Can. J. Microbiol.* **11**, 167–183.

Reis, M. S. (1973). Variation in decay resistance of four wood species from southeastern Brazil. *Holzforschung* **27**, 103–111.

Shigo, A. L. (1965). The pattern of decays and discolorations in northern hardwoods. *Phytopathology* **55**, 648–652.

Shigo, A. L. (1972). Successions of microorganisms and patterns of discoloration and decay after wounding in red oak and white oak. *Phytopathology* **62**, 256–259.

Shigo, A. L., and Hillis, W. E. (1973). Heartwood, discolored wood, and microorganisms in living trees. *Annu. Rev. Phytopathol.* **11**, 197–222.

Strobel, G. A. (1963). A xylanase system produced by *Diplodia viticola*. *Phytopathology* **53**, 592–596.

Talmadge, K. W., Keegstra, K., Bauer, W. D., and Albersheim, P. (1973). The structure of plant cell walls. I. The macromolecular components of the walls of suspension cultured sycamore cells with a detailed analysis of the pectic polysaccharides. *Plant Physiol.* **51**, 158–173.

Tribe, H. T. (1955). Studies in the physiology of parasitism. XIX. On the killing of plant cells by enzymes from *Botrytis cinerea* and *Bacterium aroideae*. *Ann. Bot. (London)* [N.S.] **19**, 351–368.

Tseng, T. C., and Mount, M. S. (1974). Toxicity of endopolygalacturonate transeliminase, phosphatidase, and protease to potato and cucumber tissue. *Phytopathology* **64**, 229–236.

Van den Ende, G., and Linskens, H. F. (1974). Cutinolytic enzymes in relation to pathogenesis. *Annu. Rev. Phytopathol.* **12**, 247–258.

Van Etten, H. D., and Bateman, D. F. (1969). Enzymatic degradation of galactan, galactomannan, and xylan by *Sclerotium rolfsii*. *Phytopathology* **59**, 968–972.

Chapter 14

The Engineering Mechanics of Pathogenesis

ELLIS B. COWLING

I. INTRODUCTION

A moment's reflection makes it obvious that most plants are cantilevered beams anchored in the soil. If the anchorage is weakened by rot
in the roots or the beam is weakened by cankers or decay of the stem,

the plant is likely to break or be toppled by wind or heavy loads of snow. If a pathogen were to inhibit the synthesis of cellulose or the lignification of stems, the affected plants would become rubbery and grow like vines on the ground. Every farmer knows that grain crops that do not stand erect rarely produce well, and often do not get harvested. Farmers also know that acute-angle crotches in fruit trees are weak, and avoid them like the plague.

Since plant pathogens contribute to lodging and stem breakage in crops, it makes sense for plant pathologists to join with agronomists, horticulturalists, and foresters in the study of lodging and other mechanisms of failure in the support system of plants. Thus this chapter is designed to explore how disease affects the structural mechanics of whole plants and the integrity of their parts. Understanding both is essential to gain profitable yields of food and fiber. Researching both represents a fascinating opportunity to advance the scientific frontiers not only in plant pathology but also in agronomy, horticulture, and forestry.

As Mark Mount has shown very well in Chapter 13, this volume, present-day understanding of the role of enzymes in tearing plant tissues apart is useful, impressive, and growing year by year. But despite this impressive understanding of the enzymology of tissue disintegration, most plant pathologists have essentially ignored the effects of these molecular processes on the functional integrity of whole plants and their structural components. As a result, we know very little about how the disintegration of plant tissues by parasitic and pathogenic organisms affects the ability of plants to function and especially to hold themselves together and erect under the physical forces of wind and rain, snow and ice, and heavy loads of fruit.

Since quantitative information on the specific effects of pathogens on the strength of plants and organs is so limited, our discussion must be largely qualitative and, in part, theoretical. We shall begin by describing the various physical factors of the environment that impose mechanical or other physical stresses on plants. Then we shall analyze the engineering mechanics of growing plants and, finally, discuss how various pathogens and other microorganisms affect the strength of plants and the structural integrity of organs.

II. MECHANICAL STRESSES ON THE SUPPORT SYSTEM OF PLANTS

Many factors in the environment impose mechanical stresses on plants. The most important of these are physical forces imposed by wind, wind-blown rain and snow, the flow of water over soil, accumulation of ice and snow on plant surfaces, and the freezing of soil and of the plants themselves. In addition, impact loads are applied by falling hail and abrasive forces are applied by wind-blown soil and sand as well as by snow and ice particles.

A. Wind Stress on Whole Plants

With and without accompanying rain or snow, wind imposes very large physical forces on plants. Lodging of crops and blow-down in forests provide vivid illustrations of how severely plants can be stressed by wind. Every pathologist who has tried to rig a sail in a high wind can appreciate what the stresses must be like in the stem of a mature tree supporting a full canopy of leaves—especially during a hurricane or a typhoon!

Figure 1 shows the relationship between the velocity and the physical forces imposed by wind. The magnitude of these forces increases by the square of the velocity of the wind according to the equation $F = KAV^2$, where F = force of the wind, A = area of the plant, and V = velocity of the wind. When A is measured in cm^2 and V in m/sec, $K = 0.0066$ (Neenan and Spencer-Smith, 1975). Wind-tunnel experiments show that many plants, including large trees, respond to the force of wind by streamlining their shape to decrease the effective value of A in the equation shown above (Hillaby, 1962). Surprisingly, the forces imposed by wind-blown rain are only slightly greater than those imposed by wind without rain (Neenan and Spencer-Smith, 1975).

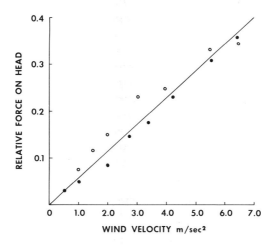

Fig. 1. Relationship between wind velocity and force exerted on the head of a barley (O) or a wheat (●) plant. (Adapted from Neenan and Spencer-Smith, 1975.)

Single plants growing alone are subject to very large forces imposed by wind. When plants grow together in stands, however, they act together to modulate the stress on any given plant. The wind velocity is frequently many times greater above or at the top of the plant canopy than at the base of plants. When root-rot pockets or other openings occur in an otherwise continuous crop or forest canopy, the flow of air becomes highly turbulent and the plants at the edge of the opening will be buffeted by erratic gusts of wind that may change direction and speed greatly from one moment to another. The intensity and variability of the forces involved depend on many factors, but especially on the velocity of the wind and the size of the opening in the plant canopy. These microclimatic relationships have been reviewed by Geiger (1965).

Plants respond to wind by swaying in light breezes and by bending to an extent that is proportional to the velocity of the wind, the size and shape of the canopy, leaves, branches, and stem, and the mechanical properties of the plant material.

The forces imposed by wind are noticeable first in the trembling or rapid oscillation of leaves, petioles, and small branches, and next in the bending of branches and stems. Toppling of plants is resisted by the development of tensile stresses in the lateral roots which, in turn, induce shear forces at their surfaces, thus calling into play the resistance of the aggregate of soil within the root mass. Complete uprooting of trees, palms, and corn plants is a common consequence of high winds, especially when the plants are shallow rooted, or are growing on wet soil or where root rot is prevalent. A gale in Scotland, in 1953, for example, blew down in a single day, five times as much timber as the annual felling quota for the whole of Great Britain (Hillaby, 1962)!

The stems of most plants are highly flexible and bend or twist in conformity with the distribution of stresses imposed by wind. Such movements often reduce the forces on plants by reducing the area exposed to the wind. Sizeable forces can be absorbed within the limits of elasticity of stems, branches, twigs, etc. If the force applied is greater than their elastic limit, however, permanent deformation or fracture will result. These deformations range from microscopic internal planes of slippage to cracks, splits, and ultimately to complete fracture of the plant structure.

B. Wind-Blown Organs

High winds can cause leaves, fruit, and branches on the same plant to rub against one another and thus cause injury by mutual abrasion. Contacts between organs of different plants are frequent whenever plants are crowded together in fields or forests. Under the influence of wind, these contacts also lead to the battering or smashing of fruits and leaves as well as abrasions which can result in natural grafting of branches (Groves, 1946), transfer of systemic pathogens, and development of specialized decay fungi such as *Hymenochaetae agglutinans* (Hepting, 1971).

High winds frequently whip leaves and floral parts violently back and forth. Hurricane-force winds can simply blow the leaves off plants or cause the leaves to tear or shatter. This can occur with healthy leaves, especially when they are immature. But such damage is even more important following attack by various pathogens.

C. Wind-Blown Sand and Snow

"Sand blasting" is the apt name given by agronomists to the abrasion of young seedlings exposed to wind-blown soil and sand particles (Skidmore, 1966). Sand storms are a serious problem in many regions of the world where vegetable or grain crops are grown on sandy soils (Fryrear *et al.*, 1973). Sand blasting also is important in establishing vegetational cover on rangelands that have been overgrazed, airport "rights-of-way," and sand dunes. Sand blasting is an important factor in the epidemiology of wheat streak-mosaic virus (Sill *et al.*, 1954). Poole (1933) reports injury by wind-blown sand in mature orchard fruit and grapes elevated well above the soil. Such damage is most frequent at wind speeds above 30 km per hour.

A similar abrasion of stem and leaf surfaces by wind-blown snow occurs in many alpine environments. This can result in plants growing only away from the direction of prevailing winds.

D. Flooding

The physical force of water flowing over soil can be disastrous for crops, especially at the seedling stage of development. The water first softens the soil and then washes it away from the roots. If flow continues, the water may push the plant over and bury it under mud, or even float it away entirely. In violent storms that combine heavy rainfall with strong wind, the anchorage of plants in soil is taxed severely. Plants with small, shallow root systems are much more susceptible than those with massive roots, especially deeply penetrating tap roots. When trees are lashed by gale-force winds they rock back and forth making a hydraulic pump which can cause water to spurt out at the periphery of the root mass in wet soil (Hillaby, 1962).

E. Hail

Hail imposes very severe impact stress on plants, especially relatively delicate leaves and floral parts, but also on sturdy branches and stems. Hail damages plants by crushing cells on the upper surfaces, and even ripping holes in the leaves of broad-leaved plants like tobacco. Hail can shatter the leaves of plants already attacked by diseases such as the shothole disease of peaches and the anthracnose diseases of sycamore and walnut.

F. Ice

Accumulations of ice and snow on plant surfaces are very common in alpine regions but tend to be even more damaging on the rare occasions in which they are observed in temperate and subtropical regions (Groves, 1946). Accumulations of even a few millimeters of ice add huge loads of weight to branches and stems, sometimes leveling whole orchards or forests. A single ice storm in February of 1972 felled 7 million m^3 of timber in South Carolina and an additional 4 million m^3 in North Carolina. Breakage of stems and branches, especially of open-grown orchard trees, can be serious indeed. Branches afflicted with cankers or decay are usually the first to break.

G. Freezing

Many herbaceous plants, palms, and ferns, and even some woody plants such as most species of *Eucalyptus* cannot survive more than a few degrees below freezing. By contrast, many woody perennials of temperate regions are adapted for survival during very cold winter seasons. The mechanisms of frost hardiness are discussed by Kozlowski in Chapter 2, this volume. Freezing sometimes creates very large physical forces which induce radial or tangential frost cracks in plant stems.

Very rapid freezing of wet soil can also apply large forces to plant stems and roots. Since water expands as it freezes, wet soil likewise expands and lifts the frozen soil above the unfrozen soil beneath. Farmers call this "frost heaving"; it can apply huge tensile forces to root systems that extend from the frozen into the nonfrozen soil.

III. THE ENGINEERING MECHANICS OF GROWING PLANTS

Agricultural engineers and timber technologists have provided much insight into the structural mechanics of plants and the strength of various organs. These subjects have been reviewed recently by Neenan and

Spencer-Smith (1975). The role of mechanics in the evolution of herbaceous plant stems is also discussed by Smith (1948).

The following definitions and concepts are useful in understanding the impact of mechanical stresses on diseased plants. They were kindly suggested by my colleague in timber technology, Ronald Pearson. "Stress" may be defined as the magnitude of force applied per unit area of plant material. Typically, stress is expressed in such units as kilograms of force per cm^2 of material but Newtons per mm^2 is now the preferred unit. "Strain" may be defined as the deformation resulting from the application of a given stress and is usually expressed as units of deformation per unit length of material.

In "elastic" organs, such as the stems, branches, or roots of a plant, the ratio of stress to strain is called the "modulus of elasticity" or "Young's modulus." If the material is stressed beyond its elastic range, some permanent damage is done and the material will not recover its original dimensions when the force is removed. The magnitude of stress beyond which the material no longer behaves elastically is called the "elastic limit" or the "proportional limit" (see Figure 2). Deformations for a given increment of stress are generally much larger beyond than within the elastic range. If the applied force is large enough, failure will occur either by excessive deformation or by rupture of the material. Plant materials which show a large deformation beyond the elastic limit are said to be "tough," whereas those that show little deformation beyond the elastic limit are said to be "brittle." This distinction is shown graphically in Figure 2. Note the tough breakage of the sound wood compared to the brittle breakage of the decayed wood.

A. Types of Applied Force

It is convenient to distinguish three types of force and associated stresses and strains: (1) tensile forces which cause elongation of plant materials parallel to the direction of the applied force; (2) compressive forces which cause contraction of the material parallel to the direction of the applied force; and (3) shear forces which produce slippage between adjacent parallel planes and thus change the shape but not the dimensions of plant materials, i.e., they produce angular deformations.

Bending by wind or heavy loads of snow or fruit involves a combination of all three types of force. The concave side of a stem or branch (assuming it was straight to begin with) is in compression and the convex side is in tension. The bending stresses in the organ vary linearly from maximum compressive stress at the concave surface through zero in the middle to maximum tensile stress at the convex surface of the member. However, the shearing stresses in a bent member are maximum at the middle of the cross-section and may contribute to the failure of branches or stems under some circumstances. For example, longitudinal splits may be observed when a bent member breaks. Maximum shear stress develops at an angle of about 45° to the applied force and rupture or gross deformation ("slip planes") may occur along planes at about this angle from the applied force.

Bending of plant materials is a result of a "bending moment" which is the product of the force times the distance from the force to the position of interest in the material. For erect plants, the maximum bending moment in a stem bent by wind would occur just below the ground line. This is rarely the position of maximum stress, however, because the stem is usually enlarged near the ground line. Thus, rupture of a plant stem in a high wind usually occurs somewhat above the ground line where the combination of bending moment and cross-sectional dimensions gives the highest stress, or where the stem has been weakened by disease, insects, or other factors.

Most plant stems, branches, and roots consist of longitudinal fibers cemented together by intercellular substances such as pectin and/or lignin. Tensile or compressive forces are transferred from one fiber to the next in shear through these matrix substances. Failure in tension may occur in two main ways. If the fibers are thin walled or otherwise weak, then the fiber walls may break with simultaneous rupture of the matrix substances to give a "brash" or "carroty" fractured surface. If the fibers are relatively strong, however, then the shear strength of the matrix substances may be exceeded and the fibers will pull out from the matrix, giving the fractured surface a "splintery" appearance. Larger deformations usually accompany the second type of failure. Usually, tough materials fail in tension with a splintery fracture.

B. The Growth Habit of Plants

From the standpoint of structural rigidity in plants, three general habits of plant growth must be recognized: (1) the erect plants such as elm trees, palms, petunias, and corn; (2) the climbing plants such as ivy and grapes; and (3) the prostrate or creeping plants such as cucumber and strawberry. Obviously, strength of stems and roots is most important in the erect plants and least critical in the prostrate ones.

Tissues forming the junctures between organs are the weakest points in the shoots of all three types of plants. These tissues usually consist of specialized, thin-walled cells which function in the abscission of leaves, needles, flowers, cones, fruits, buds, and branches. In healthy plants, abscission is inhibited during the growing season by growth-regulatory substances synthesized in various organs.

C. The Strengthening Tissues of Plants

The strengthening tissues of herbaceous monocots and dicots consist mainly of sclerenchyma cells. Vascular bundles usually are buried in the pith or form a vascular ring beneath the cortex of the stem and root. In grasses with hollow stems there is no pith and the vascular bundles form part of the vascular sheath of the stem.

In woody angiosperms and gymnosperms the strengthening tissues are concentrated in the solid cylinder of xylem inside the bark. In most woody plants, each new layer of normal xylem is laid down by the cambium and prestressed in tension parallel to the length of the branch or stem. These built-in stresses are commonly called "growth stresses." Their magnitude varies with the species and growing conditions of the plant. Generally growth stresses are much larger in fast-grown than in slow-grown plants, and in angiosperms than in gymnosperms. They act to maintain the plant in an upright position and enable it to recover quickly when bent by wind or accumulations of ice or snow. They also help branches sustain their own weight. The tensile force in the outermost layers of the xylem must be overcome before the concave side of a bent stem goes into compression. Because the compressive strength of the xylem is only about 20–25% of its tensile strength, much larger bending forces can be supported than would be the case if the stem was not prestressed. Manufacturers of concrete beams have only recently learned the benefits of prestressing.

Some plants anchor themselves in soil mainly by taproots, others by a branching network of lateral roots, sometimes by both. In some plants, large structural roots can be distinguished readily from smaller feeder roots. In other plants this distinction does not hold. Fine roots frequently penetrate small capillaries in soil and provide very large surfaces for aggregation and adhesion of soil particles. The size of this aggregated mass of soil often is increased by the hyphae of mycorrhizal fungi (Sutton and Sheppard, 1976). (See also Chapter 11 by Bowen, this volume.) Thus, the roots of healthy plants usually are intimately aggregated within a sizeable mass of soil.

D. Deformation and Breakage of Plant Parts

The cantilevered beams in erect plants can fail in three major ways: (1) the branches or stems can break, (2) the roots can break, or (3) the contact with the soil can break and the plant can be uprooted. The mode of failure will be determined by the relative strength and rigidity of the branches, stem, and roots in relation to the weight and shear strength of the mass of soil adhering to the roots (Neenan and Spencer-Smith, 1975). If the plant has a taproot like cotton, the likelihood of uprooting will be less than if only lateral roots are present, as in corn. Huge forces are imposed at the fulcrum of the cantilevered beam. That is why tree trunks and branches grow biggest at the fulcrum and why corn plants have bracing roots. Taproots below the fulcrum of the cantilevered beam are stressed by bending of the plant's stem. Lateral roots are stressed mainly in tension on the side of the plant opposite to the direction in which the stem is bent.

Breakage of plant organs from wind or snow results whenever the applied stresses exceed the elastic limits of the organ. The stems of wheat and other small grains are generally thin walled and often hollow. Buckling is the most common mode of failure in such plants. In healthy corn stems, however, the pith tissue offers some resistance to buckling.

Failure in the stems of woody plants usually is the result of excessive bending forces. The appearance of the fractured surface depends on the properties of the particular stem in question. Gross deformation usually occurs on the compressive face where wrinkles due to telescoping and the crushing of fibers often can be seen. Cracks and splits then appear on the convex surface due to breakage of fibers in tension. These may develop into a break showing numerous long splinters. Longitudinal shear and tensile stresses perpendicular to the grain often influence the type of failure. Wood is relatively weak in both these properties; the tensile strength of clear wood perpendicular to the grain is usually less than 6%, and the shear strength is usually less than 10%, of the tensile strength parallel to the grain. Sloping splits are common, particularly with relatively brittle wood, such as is found in young pine stems. These result from the effect of a combination of longitudinal shear and tension forces perpendicular to the grain.

E. Compensation for Weakening of Plant Parts

Herbaceous annuals have only a limited capacity for secondary growth and only a single growing season in which to adjust their habit of growth so as to strengthen roots or shoots weakened by pathogens. Woody perennials, on the other hand, are capable of secondary growth and thus can strengthen or even replace plant parts weakened by pathogens or injury (Steucek and Kellogg, 1972). Side branches can assume apical dominance; adventitious buds can give rise to new shoots and roots; photosynthate can be allocated to the synthesis of tissues which will strengthen the stem where it has been weakened by injury or disease or simply stressed by wind; compression wood in gymnosperms and tension wood in angiosperms can be formed in an effort to restore a leaning stem to a vertical orientation or to strengthen large branches (Larson, 1965).

IV. EFFECTS OF DISEASE ON LODGING OF GRAIN CROPS

Susceptibility to lodging is one of the most important hazards of crop production in corn, rice, wheat, oats, and barley. The high yields obtained with the so-called green-revolution grain crops are due in part to their short stature and stiff stems—traits that confer major resistance to lodging. Losses in yield caused by lodging may be as high as 60% (Mulder, 1954), with average losses more commonly in the range of 5–20%

A. Types of Lodging

Lodging of grain crops is classified as root lodging (toppling of the entire plant) and stem lodging (collapse of the stem, usually by buckling). In small grains, stem lodging is common but root lodging is rare. In corn, both types are common although the literature on stem lodging is more abundant, perhaps because it is easier to investigate. The difference in frequency of the two types of lodging in corn and small grains is due in part to the contrasting geometry of the two types of plants. Small grains have larger root systems relative to stems; they also have stems of greater flexibility than corn, partly because their stems are hollow. Also, most of the stem tissues in small grains become lignified early in the growing season. This provides early protection against many stem pathogens. In corn, by contrast, the rind tissues become lignified early but lignification of the pith tissue usually occurs late in the season and even then is not complete. A very good correlation between the crushing strength of corn stalks and the tendency for lodging has been found by Zuber et al. (1957) and by Thompson (1964).

B. Disease-Induced Lodging of Corn

In corn, the diseases most commonly associated with susceptibility to lodging are the stalk rots induced by *Diplodia zeae*, *Gibberella saubinetti*, *G. fujikuroi*, *Nigrospora oryzae*, *Fusarium moniliforme* and other *Fusarium* sp., and *Basisporium gallarum*. Lodging also has been associated with the brown spot disease induced by *Physoderma maydis*; various root rots; a complex of maize viruses (Josephson *et al.*, 1976); and various root-feeding and stem-boring insects, especially the European corn borer and the corn root worm. In many cases, these insects provide avenues of entrance for the stalk-rotting and root-rotting fungi.

The stalk-rotting fungi in corn usually attack the soft pith tissue inside the rind. When the stem of a healthy corn stalk is stressed by winds, a much larger force of bending is required to buckle a healthy stem than a stem in which the pith tissue has been destroyed by stalk-rotting fungi or bacteria. The principle involved here is like that of bending a pipe. A pipe packed with sand will bend farther without buckling than a hollow pipe because the sand tends to keep the concave (compression) side of the pipe from collapsing against the convex (tension) side.

Pappelis and Smith (1963) have shown that only dead cells in the pith of corn can be invaded by *Diplodia*. This observation explains the differential susceptibility to stalk rot by this pathogen among plants of different varieties, the gradient of decreasing susceptibility with increasing number of internodes above the soil, and changes in susceptibility to stalk rot during the growing season. The gradient mentioned above has an important bearing on lodging because corn stems usually break at the second or third internode above ground (Thompson, 1964). C. M. Nagel (personal communication) has recently demonstrated a relationship between the lignification of pith and its resistance to stalk-rotting pathogens. He has also demonstrated the heritability of resistance to these pathogens in corn. Contrary results also have been reported recently in other selections of corn (Undersander *et al.*, 1977).

C. Disease-Induced Lodging of Small Grains

In small grains, the major diseases associated with lodging are root and stem blights, although foliage diseases also have been implicated. In rice, lodging is associated with the rotting of culms induced by *Leptosphaeria salvinii* and with the "Bakanae" or "foolish rice disease" induced by *Gibberella fujikuroi*. In this case, the slender stems of infected plants elongate abnormally and thus are highly susceptible to buckling by wind. Rice blast induced by *Pyricularia oryzae* results in decay of the panicle and its branches so that the grain simply drops off. Lodging in wheat is a characteristic symptom of infection by *Cercosporella herpotrichoides* (Dickson, 1947) and it has been associated with both stem rust and leaf rust of wheat. Stem rust apparently weakens the stem directly, whereas leaf rust does so indirectly (Chester, 1946). According to T. T. Hebert (personal communication), this association between lodging and leaf rust (and possibly mildews as well) probably is related to interference with photosynthesis or distribution of photosynthate (see Chapter 4 by Kosuge in this volume). These pathogens are well known to decrease the yield of grain. In addition, the stems of rusted or mildewed plants may not develop normally (perhaps because the pathogens interfere with lignification) and may tend to buckle at weak points far removed from the site of infection.

D. The Value of Cooperative Research on Lodging

Considering the importance of disease as a contributory factor in lodging, it is remarkable how few plant pathologists have joined with their colleagues in agronomy to study the process together. The papers by Zuber *et al.* (1957); Pappelis and Smith (1963); and Nagel (1973), who worked together with Boyd Shank, show how profitable such cooperation can be.

The discovery of lodging resistance in corn inbred line C103 also illustrates how valuable studies of the engineering mechanics of pathogenesis can be. In the early days of testing for lodging resistance it was common to kick the stems of corn plants to see which ones were strong and which ones were weak. Oliver Nelson, at that time a graduate student working with Donald Jones at the Connecticut Agricultural Experiment Station, applied this test in the early 1940's and noted a relationship between the force required to break the stems and their susceptibility to stalk-rotting organisms. This observation, among others, led to the release of line C103 as a "low breakage line" which now is appreciated around the world for its resistance to stalk rot and to lodging. Once again "chance favored the prepared mind" (Singleton *et al.*, 1948; Day, 1969).

V. THE RUBBERY WOOD DISEASE OF APPLES

Knowing a little about the influence of various constituents of plant cell walls on the mechanical properties of organs makes it possible to speculate about the probable effects of pathogens that would interfere with the synthesis of lignin, cellulose, and pectin or their deposition within plant cell walls. This sort of theoretical biology (logicians call it inductive reasoning) is not common in plant pathology—at least not in our published papers. Nevertheless it is fun and can be useful. The rubbery wood disease of apples provides stimulus for such useful speculation.

Rubbery wood is a virus disease of certain varieties of apple (Prentice, 1949). The widely used English variety, Lord Lambourne, is so highly susceptible that it can be used as a biological indicator of the disease.

The disease has been reported on the Golden Delicious variety in North America, New Zealand, and Europe. The virus is latent in many other varieties and can be transmitted by budding and by natural root grafts. Thus, symptoms may show for the first time when apple trees are several years old.

The characteristic symptom of the disease is an extreme flexibility of the stem and branches—a branch 2–3 cm in diameter can be bent easily with the fingers of one hand! Severely affected trees literally lie down on the job—older trees show a "weeping" habit of growth. Trees infected when they are young grow like a vine on the ground, have very short internodes, produce fruit earlier than noninfected trees, and are frequently stunted.

The affected wood consists of cells rich in cellulose but poor in lignin. The affected cells occur in patches separated by normal cells—the amount of flexibility is a function of the relative proportions of normally lignified and abnormally lignified tissues. Prentice (1949) has made an interesting study of the mechanical properties of affected stems and concluded that rubbery wood disease is caused by a graft-transmissible virus.

Sondheimer and Simpson (1962) isolated and characterized the lignin from affected and healthy trees in an effort to understand the mechanism by which the virus inhibits lignin biosynthesis. Nitrobenzene oxidation of the lignin from rubbery wood yielded fewer methoxyl groups and less vanillin and syringaldehyde than the lignin from normal apple wood.

Sondheimer and Simpson (1962) conclude their paper with the following comment: "A detailed investigation of the biochemical difference between normal and virus-infected apple seedlings will yield information not only on this specific disease, but also on the more general aspects of the biosynthesis of lignin." Here, then, is another case where study of diseased plants can contribute to understanding of healthy ones. It also provides a fine example of the engineering mechanics of pathogenesis. Finally, this case also offers encouragement for theoretical plant pathology. What symptoms would be observed in a disease that inhibits synthesis of pectin or cellulose? What kind of pathogen would most likely have such an effect? It would be interesting to know.

VI. STEM PITTING AND GRAFT INCOMPATIBILITY IN FRUIT TREES

Stem pitting is another virus disease that weakens the stems and roots of fruit trees (Jones, 1976). It is induced by a special strain of the tomato ring spot virus and is very common on apple and peach, but also occurs on other stone fruits including apricot, nectarine, plum, and cherry. The wood and bark in the lower stem and structural roots of infected trees are disorganized. Frequently, the root system develops so poorly that growth of the whole tree is inhibited. The disorganized wood of the roots is so weak in comparison with that in healthy trees, that stem-pitted trees commonly break just below the ground line. Many trees carry the disease as a latent infection which becomes obvious when scion material from infected trees is grafted onto susceptible root stocks. In stone fruits the disease is transmitted by the nematode *Xiphinema americanum.*

Breakage of supporting stem tissue is a frequent consequence of incompatibility or lack of coordinated growth in grafted orchard, fruit, and nut trees and in seed orchards for forest trees. Typically the xylem of the scion and the rootstock fail to form a structurally sound union even though the graft may be adequate physiologically. The tissue of the graft union often is disorganized with conspicuous voids or pits developing both above and below the plane of the graft. Stem-boring insects as well as decay- and canker-inducing fungi sometimes further weaken the stems by invading the disorganized tissue of incompatible grafts.

Mechanically unstable stems also can result from diameter growth of a scion that is more rapid than that of the rootstock to which it is joined. This condition can develop to ridiculous extremes where an obviously undersized rootstock simply is not able to support the massive stem above it. This has been a particular problem with certain size-controlling rootstocks used in peach and apricot orchards (C. N. Clayton, personal communication).

VII. DAMPING-OFF OF SEEDLINGS

Plants rarely collapse under the stress of their own weight but this apparently does occur in the case of damping-off of seedlings. Numerous pathogens are involved, including *Rhizoctonia solani, Fusarium moniliforme,* and other *Fusarium* sp., as well as various species of *Pythium, Phytophthora,* and *Sclerotinia.* The cortical tissues of delicate young seedlings are neither suberized nor lignified and have only a very thin protective cuticle. Thus, seedlings of almost all plant species are highly susceptible to penetration and disintegration by pathogens with potent pectin-degrading or cellulolytic enzyme systems. Pathogens such as *Pythium* sp., which produce only pectic enzymes, attack only very young seedlings, whereas *Rhizoctonia solani* and *Fusarium moniliforme* also produce cellulolytic enzymes and thus can cause the collapse of root and stem tissues in older seedlings (Husain and Kelman, 1959).

VIII. DISEASE-INDUCED WEAKENING OF BONDS BETWEEN ORGANS

Strong bonds between organs are essential to the structural and physiological integrity of plants. These bonds remain firm in healthy plants throughout the growing season and, in perennial plants, sometimes for many seasons. In diseased plants, however, the bonds between organs are often weakened or broken entirely (Osbourne, 1968). Under the stimulus of biotic disease or water imbalance, diseased leaves or needles are separated from twigs, buds are separated from shoots, flowers and fruit are separated from their supporting stems, branches are separated from stems, lateral roots are separated from major roots, and cortical tissues or bark are separated from the underlying xylem.

Examples of most of these disease-induced changes are well known. Premature abscission of leaves is induced by many pathogens including the bacterial leaf spot of stone fruits induced by *Xanthomonas pruni* (Jones, 1976), cherry leaf spot induced by *Coccomyces heimalus*, anthracnose of walnut and oak (Berry, 1964), *Cercospora* leaf spot of many field crops, *Septoria* brown spot of soybean (Dickson, 1947), and many other foliar pathogens. Certain diseases are named "needle cast" or "leaf drop" because of their very marked tendency to induce premature abscission of leaves. Examples include *Rhabdocline pseudotsuga* needle cast on Douglas fir and *Pseudopeziza medicaginis* leaf spot of alfalfa.

Premature abscission of diseased fruit also is induced by apple scab and California blight of cherry induced by *Coryneum beijerinckiii*, and weather-induced shedding of cotton bolls. Shedding of diseased roots is observed with *Fomes annosus* root rot of pines. In the shot-hole disease of peach a structurally weak corky layer is formed around lesions on the leaf and the diseased part simply falls out and blows away. In all these cases the cohesive strength of the organ or of its parts is reduced to the point where the ordinary forces of wind or weight of fruit are sufficient to shatter the organ itself or to dislodge or otherwise separate the infected organ from the stem or root of the plant. This premature abscission of organs has been attributed to imbalance among growth-regulatory substances in the organs affected. It also involves the enzymatic digestion of cell wall material in the abscission layer (Osbourne, 1968). Here then is another fertile area for cooperation between plant pathologists and plant physiologists.

IX. DISEASE-INDUCED CHANGES IN PHYSICAL PROPERTIES OF FRUITS AND VEGETABLES

As in many other aspects of plant pathology, Horsfall was also a pioneer in studying the engineering mechanics of pathogenesis. In the early 1930's he compared the rate of change in mechanical crushing strength of healthy peas with that of peas affected by a complex of root-rotting pathogens including *Pythium* sp., *Aphanomyces euteiches, Fusarium martii pisi,* and *Rhizoctonia solani* (Horsfall, *et al.,* 1932). As reported in the paper itself, "The load necessary to crush one pea was much higher, size for size, on the same harvest date in the diseased than in the normal samples. This is another way of saying that on the same harvest date, diseased peas were poorer in quality than normal peas."

Many bacterial and fungal pathogens induce watery or soft rot of fruits and vegetables. These pathogens include various species of *Erwinia* and *Xanthomonas* and several species of fungi including *Botrytis cinerea, Penicillium digitatum, Rhizopus stolonifer, Pythium* sp., *Sclerotinia sclerotiorum,* and *Monolinia fructicola.* These organisms produce pectinolytic enzymes which can rapidly dissolve the intercellular substance of parenchyma cells and thus transform a crisp head of lettuce or a cucumber, or a firm tomato, or banana or potato into a soft, watery and sometimes foul-smelling "soup." These pathogens frequently enter their host organs through wounds induced by mechanical harvesting or processing machinery and are often destructive after harvest.

These diseases illustrate the economic value of understanding the impact of plant disease on the mechanical and other physical properties of fruits and vegetables. Pathogens that induce rapid deterioration in the handling characteristics and texture of plant parts deserve special attention.

X. DISEASE-INDUCED SPLITTING AND CRACKING OF FRUITS AND STEMS

When stems and fruits expand in size, substantial tensile forces develop as growth stresses the surface tissues of the organ. Sometimes huge forces may be involved as when shade trees split the concrete of sidewalks. During picnics, many of us have observed these stresses—the ripe watermelon splits ahead of our knife. In healthy plants, these normal growth stresses are balanced by the strength of the cells and the bonds between them in the surface layers of the organ. When pathogens weaken these cells, however, cracks and splits of various kinds develop on fruit and stems. Examples of this phenomenon are provided by the cracking of apple fruit affected with scab, the so-called collar crack of rubber trees associated with *Armillaria mellea*, and the splitting of sweet potato and tomatoes in wet weather especially after a period of drought.

Another example of failure by splitting of the support system is provided by the development of frost cracks in trees. Sometimes the splitting is so sudden and violent that it sounds like a rifle shot. Herrington *et al.* (1964) have shown that the coefficients of thermal contraction of dry wood in the tangential direction, and of ice in any direction are, respectively, 40 and 51×10^{-6} mm mm^{-1} °C^{-1}. By contrast, the corresponding coefficient for wood at 119% moisture content was about 380×10^{-6} mm mm^{-1} °C^{-1}. The latter measurements are in agreement with the size of radial splits that develop in tree stems during freezing. No mechanism is known to account for this great difference in coefficient of thermal contraction. In theory, the partial pressure of ice formed in the lumina of sap-filled cells could be sufficiently lower than that of water bound within cell walls that water vapor could be withdrawn from the cell walls during rapid freezing. Thus, shrinkage of the cell walls during freezing would be analogous to drying below the fiber saturation point. Such removal of water from cell walls could cause failure in tension around the circumference of stems and thus induce the formation of radial frost cracks.

XI. WEAKENING OF STEMS BY CANKER- AND GALL-FORMING PATHOGENS

Many pathogens form localized necrotic spots, cankers, or galls on the main stem and branches of plants. In almost all cases, these localized sites of infection become points of weakness in the plant stem, predisposing the plant to breakage by wind, heavy loads of fruit, or accumulations of ice or snow. The true cankers of tree stems provide conspicuous examples of this type of failure in the support system (Hepting, 1971). *Hypoxylon prunatum* on aspen, *Eutypella parasitica* on maple, *Nectria galligena* on birch, *Strumella coryneoides* on oaks and maples, and the bacterial canker induced by *Pseudomonas syringe* on cherry, apparently cause less structural weakening than the so-called canker-rot fungi such as *Hericium erinaceus* on oak, *Poria obliqua* on birch, *Polyporus glomeratus* on maple, and *P. hirsutus* on various angiosperms.

Branches heavily laden with apples frequently break after they are weakened by *Physalospora obtusa*. Canker diseases of herbaceous crops also can cause the structural weakening of stems. Various species of *Col-*

letotrichum and *Gloeosporium* induce such canker diseases of bean, tomato, flax, and tobacco (Husain and Kelman, 1959).

Fungi that form galls on plants also can induce structural weakening of stems. Wind breakage is a very common symptom of fusiform rust induced by *Cronartium fusiforme.* In an attempt to explain this symptom, Myron and Kelman (1975) cut small wood specimens from the galls induced by this pathogen in loblolly pine and found that the impact strength of gall wood was only about 25% of that for tissue from healthy trees. Although the greater cross-sectional dimensions of these rust galls tends to compensate somewhat for this difference in strength, the galls frequently are further weakened by insects and discoloring and decay fungi. Veal *et al.* (1974) also showed that the yield of pulp obtained from wood infected by *C. fusiforme* is less than from healthy wood.

XII. ROOT ROT IN ORCHARD, PLANTATION, AND FOREST TREES

Toppling of mature trees by wind is one of the most spectacular symptoms of structural weakness in roots (Hillaby, 1962; Hepting and Cowling, 1977). The fungi involved are legion. On a world-wide basis, *Armillaria mellea* and *Clitocybe tabescens* are among the most common fungi that attack structural roots in orchards of fruit, citrus, and nut trees or in plantations of coffee, tea, and rubber. These fungi also are important in forest trees but *Fomes annosus, Poria wierii, Polyporus schweinitzii,* many other wood-destroying Basidiomycetes, and a few Ascomycetes such as *Xylaria* sp. also attack the root systems of gymnosperms and angiosperms (Hepting, 1971). Sudden blow-down of apparently healthy trees is sometimes the first above-ground symptom of attack by these pathogens. Roots of pines sometimes are so badly decayed by *Fomes annosus* that trees 20–25 cm in diameter can be pushed over by hand even though the crown is green and diameter growth is continuing at a near-normal rate (Cowling and Johansson, 1970; Froelich *et al.,* 1977). These observations, which have been repeated in Europe as well as in North America, demonstrate the marked loss in strength which can occur in roots that remain functional in water transport despite extensive invasion and structural weakening of root tissues by pathogens. These observations for structural root pathogens stand in distinct contrast to the major effects of feeder root pathogens such as *Phytophthora cinnamomi,* various *Pythium* sp., and many root-feeding nematodes, including species of *Xiphenema, Pratylenchus, Meloidogyne,* and *Belonolaimus.* All of these organisms kill young feeder roots and inhibit uptake of water and nutrients by them. This prevents normal water and mineral transport from soil to functional xylem but does not contribute much to blow-down. In later stages of the diseases induced by *Phytophthora* and *Pythium* sp. weakening of structural roots is observed. For example, Ko (1971) has reported that *Phytophthora palmivora* can cause such extensive decay of the lateral and tap roots of papaya that older trees are weakened and frequently toppled by wind.

XIII. DECAY OF TREE STEMS AND TIMBER PRODUCTS

Internal decay of tree stems is one of the most conspicuous causes of failure of the support system of plants. The common term, heart rot, is a misnomer for two reasons: (1) decay is important in many species of trees which do not form a normal heartwood (a genetically controlled central cylinder of xylem devoid of living parenchyma cells); (2) healthy trees of such diffuse-porous timber species as maple, birch, beech, and sweetgum contain only sapwood from cambium to pith even when they are mature. Also, decay in tree species which normally produce heart-wood often is not confined to the heartwood portion of the tree stem. Discolorations and decays of tree stems are initiated by wounds. It involves a succession of bacteria and nonhymenomycetous fungi as well as the well-known wood-destroying Basidiomycetes and certain Ascomycetes (Shigo and Hillis, 1973).

About 2,000 species of fungi are known to cause decay in living trees. Some, such as *Polyporus betulinus,* attack only a single genus (*Betula*) of host trees. Others, such as *Fomes pini* and *F. igniarius,* occur on many species of forest trees (Hepting, 1971). A few species of decay fungi are known to attack both angiospermous and gymnospermous hosts.

These stem-rotting fungi cause greater loss of value in the forests of the world than any other type of damage. In the United States, these fungi weaken, destroy, or degrade approximately one-half as much timber as is harvested for use by the people of this country (Hepting and Jemison, 1958). In addition, decay fungi that attack wood in storage and in use cause substantial losses in the value of wooden boats, bridges, homes, churches, piling, railroad ties, utility poles, and other wooden structures. Albion (1926) concluded from his study of the relationship between forests and seapower in Great Britain during the period from 1652–1862 that more wooden ships were lost to decay than to enemy vessels. The wood preservation industry uses more fungicides for decay prevention in timber than the total of all other uses of fungicides. Thus, failure of the support system of trees and decay of wood in use affects the economies of many countries.

Decay in stems of living trees is initiated by wounds that expose the xylem beneath the bark. These wounds may be natural in origin, such as wind-broken branches, tops, or roots; fire scars resulting from natural fires; and injuries induced by woodpeckers and large animals, such as deer or porcupines, and small animals such as the stem-boring insects. Wounds also can be induced by man, for example, during logging, pruning, or controlled-burning operations.

Following injury to stems, a sequence of events is initiated, including development of bacteria and nonhymenomycetous fungi in the tissues near the wound and invasion of the discolored wood by decay fungi. These processes normally are confined to tissues present in the tree at the time of injury. A barrier zone of fungus-impervious tissue is laid down by the cambium in the vicinity of the wound. In addition, reaction zones rich in phytoalexins retard the radial, tangential, and longitudinal spread of discoloration and decay in tissues formed prior to wounding. The extent and rapidity of weakening of the tree stems by this succession of organisms is a complex function of the aggressiveness of the microorganisms involved, the size of the wound, and the capacity of the tree to close the wound and form effective barrier zones and reaction zones. The biochemical and microbial dynamics of these processes are described in the papers of Shain (1967), Shigo and Hillis (1973), Chen *et al.* (1976), and Shortle and Cowling (1978a,b).

Detailed studies have been made of the progressive changes in strength properties caused by various types of decay fungi (Scheffer, 1936; Scheffer *et al.*, 1941; Cartwright and Findlay, 1946). The results of one such study are shown in Fig. 2. Note the difference in maximum load and the abrupt loss of strength in the decayed wood compared to the sound wood. Much more mechanical work (area under the curve) was necessary to carry the load to its maximum value in the case of sound wood than in the case of the decayed wood.

Loss in strength by most white-rot and brown-rot fungi is much more rapid than changes in specific gravity. This is especially true of decay by brown-rot fungi for reasons that are shown in Fig. 3. Even in the incipient states of brown rot, the average length (and therefore the strength) of cellulose (plus hemicellulose) molecules is reduced to a small fraction of what it is in sound wood. The cellulose molecules remaining in white-rotted wood are not much shorter than those in sound wood. For this reason, white-rotted wood can be used to produce useful plywood, particle boards, and good-quality pulp and paper products, but brown-rotted wood is worthless for such purposes even at very early stages of decay.

It is rare that decayed tissues from herbaceous crops can be converted into useful products. This is not true of forest products, however. For

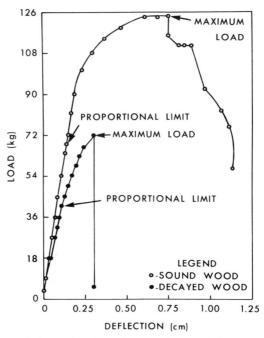

Fig. 2. Effects of decay by the white-rot fungus *Polyporus versicolor* on the bending strength of sweetgum sapwood. Weight loss due to decay was 17.6%. Note the difference in maximum load and abrupt loss of strength in the decayed compared to the sound wood. More mechanical work (area under the curve) was necessary to carry the load to its maximum value in the sound than in the decayed wood. (Adapted from Scheffer, 1936.)

example, during the past several years the technical utilization requirements for Douglas fir timber have been adjusted to permit utilization of logs partially decayed by *Fomes pini*. Plywood containing so-called white-pocket veneers is used as decorative paneling and in packaging crates, sheathing, and other products where structural integrity and/or attractiveness are desired rather than maximum strength. Use of veneer sheets containing substantial amounts of white-pocket wood is permitted; these new specifications of the American Plywood Association (1974) have provided a major stimulus for the utilization of logs infected with *Fomes pini*. Reasonable yields of good-quality particle boards and pulp and paper products can be made from logs too decayed for utilization as plywood (Waters and Cowling, 1976). In fact, paper sheets made from logs of southern pine decayed by *F. pini* are stronger in tensile strength, although somewhat weaker in tear resistance, than papers made from nondecayed logs (Reis and Libby, 1960). This is also true of timber decayed by *Fomes annosus* and many other white-rot fungi.

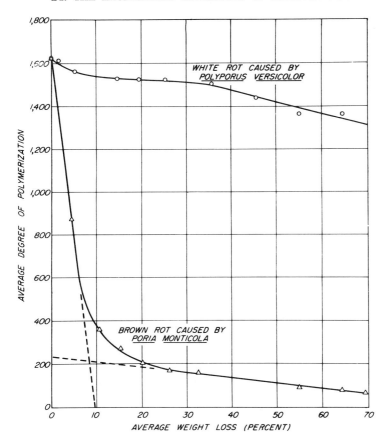

Fig. 3. Progressive changes in degree of polmerization of the cellulose plus hemicelluloses during decay of sweetgum sapwood by representative white-rot and brown-rot fungi. Note the strikingly rapid depolymerization induced by the brown-rot fungus in the initial stages of decay. This accounts for the extreme loss in strength of wood due to brown rot. (From Cowling, 1961.)

These potentials for utilization of partially decayed timber illustrate the practical value of research on the engineering mechanics of pathogenesis in forest trees (Weiner *et al.*, 1974).

XIV. CONCLUSIONS

In this chapter we have discussed various ways in which disease causes failure in the support system of plants. The amount of qualitative information available is impressive, but the amount of quantitative data is limited. In almost all cases we do not know the magnitude of physical force necessary to cause failure, and in many cases we do not know the physiological or biochemical mechanisms by which structural weakening is achieved. This is particularly true in the case of premature abscission of diseased organs and the cracking and splitting of fruits and stems.

Lodging, wind breakage, wind throw, ice damage, sand blasting, frost cracks, and premature abscission of organs all represent serious sources of loss in crop production. Plant pathologists can make very substantial contributions both to understanding these phenomena and in developing practical methods for their amelioration.

Groves (1946) had a vision of this potential when he wrote his Station Bulletin on "Weather injuries to fruits and fruit trees." Prevention of damage in orchard crops and lodging in grain crops provide particularly promising areas for investigation. The paper by Neenan and Spencer-Smith (1975) provides an excellent introduction to the engineering mechanics of the process. Many different types of pathogens affect lodging. Agronomists have devised physical testing methods which permit rapid measurements of pathogen effects on the strength of stems. Thus cooperative research together with agronomists, horticulturalists, or foresters, or with agricultural engineers, wood technologists, or meteorologists should be highly productive (Groves, 1946; Zuber et al., 1957; Herrington et al., 1964; Nagel, 1973).

As in so many other aspects of disease processes, it appears that the molecular biology of pathogenesis has preceded understanding of the effects of disease on whole plants and organs. In preparing this chapter it has been fun to explore the engineering mechanics of pathogenesis. It will be fun also to see what new insights may develop in this new aspect of our field in the years ahead.

Good luck, good hunting, and "may the wind be always at your back."

Acknowledgments

Many thanks are due the following friends and colleagues who kindly provided useful comments and suggestions regarding this chapter: R. G. Pearson, Alexa Williams, R. M. Kellog, R. A. Zabel, D. J. Thompson, J. G. Horsfall, C. N. Clayton, T. T. Hebert, L. F. Grand, T. C. Scheffer, and C. M. Nagel.

References

Albion, R. G. (1926). Forests and sea power: The timber problem of the Royal Navy 1652–1862. *Harv. Econ. Stud.* **29**, 1–485. Harvard University Press, Cambridge, Massachusetts.
American Plywood Association (1974). "U.S. Product Standard PS 1–74 for Construction and Industrial Plywood with Typical APA Grade-trademarks," Ameri-

can National Standard, A 199.1–1974. Am. Plywood Assoc., Tacoma, Washington.

Berry, F. H. (1964). Walnut anthracnose. *U.S., For. Serv., For. Pest Leafl.* **85**, 1–4.

Cartwright, K. St. G., and Findlay, W. P. K. (1946). "Decay of Timber and Its Prevention." HM Stationery Office, London.

Chen, C-L., Chang, H-m., Cowling, E. B., Huang Hsu, C-Y., and Gates, R. P. (1976). Aporphine alkaloids and lignans formed in response to injury of sapwood in *Liriodendron tulipifera. Phytochem.* **15**, 1161–1167.

Chester, K. S. (1946). "The Nature and Prevention of the Cereal Rusts as Exemplified in the Leaf Rust of Wheat." Chronica Botanica, Waltham, Massachusetts.

Cowling, E. B. (1961). Comparative biochemistry of the decay of sweetgum sapwood by white-rot and brown-rot fungi. *U.S., Dep. Agric., Tech. Bull.* **1258**, 1–78.

Cowling, E. B., and Johansson, M. (1970). Physiology of *Fomes annosus. In* "Proceedings of the Third International Conference on *Fomes annosus*" (C. S. Hodges, J. Rishbeth, and A. Yde-Anderson, eds.), pp. 192–205. Aarhus, Denmark (printed by Southeast. For. Exp. Stn., Asheville, North Carolina).

Day, P. R. (1969). New wealth from corn genes. *Front. Plant Sci.* **21**, 6–7.

Dickson, J. D. (1947). "Diseases of Field Crops." McGraw-Hill, New York.

Froelich, R. C., Cowling, E. B., Collicott, L. V., and Dell, T. R. (1977). *Fomes annosus* reduces height and diameter in growth of planted slash pines. *For. Sci.* **23**, 299–306.

Fryrear, D. W., Stubbendieck, J., and McGully, W. G. (1973). Grass seedling response to wind and wind blown sand. *Crop Sci.* **13**, 622–625.

Geiger, R. (1965). "The Climate Near the Ground." Harvard Univ. Press, Cambridge, Massachusetts.

Groves, A. B. (1946). Weather injuries to fruits and fruit trees. *Va. Agric. Exp. Stn., Bull.* **390**, 1–39, Blacksburg, Virginia.

Hepting, G. H. (1971). Diseases of forest and shade trees of the United States. *U.S., Dep. Agric., Agric. Handb.* **386**, 1–658.

Hepting, G. H., and Cowling, E. B. (1977). Forest pathology: Unique feature and prospects. *Annu. Rev. Phytopathol.* **15**, 431–450.

Hepting, G. H., and Jemison, G. M. (1958). Forest protection. *In* "Timber Resources for America's Future." *U.S., For. Serv. Res. Rep.* **14**, 184–200. Washington, D.C.

Herrington, L., Parker, J., and Cowling, E. B. (1964). The coefficient of expansion of wood in relation to frost cracks. *Phytopathology* **54**, 128.

Hillaby, J. (1962). Gales and trees. *New Sci.* **13**, 493–495.

Horsfall, J. G., Kertesz, Z. I., and Green, E. L. (1932). Some effects of root rot on the physiology of peas. *J. Agr. Res.* **44**, 833–848.

Husain, A., and Kelman, A. (1959). Tissue is disintegrated. *In* "Plant Pathology: An Advanced Treatise" (J. G. Horsfall and A. E. Dimond, eds.), Vol. 1, pp. 143–188. Academic Press, New York.

Jones, A. L. (1976). "Diseases of Tree Fruits," North Cent. Reg. Ext. Publ. No. 45. Michigan State University, East Lansing.

Josephson, L. M., Graves, C. R., and Kincer, H. C. (1976). Response of corn hybrids to the virus disease complex. Tenn. Farm and Home Sci., Tenn. Agr. Expt. Stn. Progress Report **97**, 16–17, Knoxville, Tennessee.

Ko, W. H. (1971). Biological control of seedling root rot of papaya caused by *Phytophthora palmivora. Phytopathology* **61**, 780–782.

Larson, P. R. (1965). Stem form of young Larix as influenced by wind and pruning. *For. Sci.* **11**, 412–424.

Mulder, E. A. (1954). Effect of mineral nutrition on lodging of cereals. *Plant Soil* **5**, 246–306.

Myron, D. T., and Kelman, A. (1975). Fungi inhabiting fusiform rust galls and some properties of rust infected pine wood. *Plant Dis. Rep.* **59**, 148–151.

Nagel, C. M. (1973). Techniques and methods useful in the selection of root and stalk resistance in corn. *Proc. 28th Annu. Corn and Sorghum Res. Conf., Am. Seed Trade Assoc.* pp. 51–57.

Neenan, M., and Spencer-Smith, J. L. (1975). An analysis of the problem of lodging with particular reference to wheat and barley. *J. Agric. Sci.* **85**, 495–507.

Osbourne, D. J. (1968). Defoliation and defoliants. *Nature (London)* **219**, 564–567.

Pappelis, A. J., and Smith, F. G. (1963). Relationship of water content and living cells to spread of *Diplodia zeae* in corn stalks. *Phytopathology* **53**, 1100–1105.

Poole, R. F. (1933). Wind and sand injury to leaves and fruit. *J. Elisha Mitchell Sci. Soc.* **49**, 171–175.

Prentice, I. W. (1949). Experiments on rubbery wood disease of apple trees. *1949 Annu. Rep., East Malling Res. Stn., Kent*, pp. 122–125.

Reis, C. J., and Libby, C. E. (1960). An experimental study of the effect of *Fomes pini* (Thore) Lloyd on the pulping qualities of pond pine, *Pinus serotina. Tappi* **43**, 489–499.

Scheffer, T. C. (1936). Progressive effects of *Polyporus versicolor* on the physical and chemical properties of red gum sapwood. *U.S., Dep. Agric., Tech. Bull.* **527**, 1–46.

Scheffer, T. C., Wilson, T. R. C., Luxford, R. F., and Hartley, C. (1941). The effect of certain heart rot fungi on the specific gravity and strength of Sitka spruce and Douglas fir. *U.S. Dept. Agr. Tech. Bull.* **779**, 1–24.

Shain, L. (1967). Resistance of sapwood in stems of loblolly pine to infection by *Fomes annosus. Phytopathology* **54**, 1034–1045.

Shigo, A. L., and Hillis, W. E. (1973). Heartwood, discolored wood, and microorganisms in living trees. *Annu. Rev. Phytopathol.* **11**, 197–222.

Shortle, W. C., and Cowling, E. B. (1978a). Development of discoloration, decay, and microorganisms following wounding of sweetgum and yellow-poplar trees. *Phytopathology* **68**, 609–616.

Shortle, W. C., and Cowling, E. B. (1978b). Interaction of live sapwood and fungi commonly found in discolored and decayed wood. *Phytopathology* **68**, 617–623.

Sill, W. H., Jr., Lowe, A. E., Bellingham, R. C., and Fellows, H. (1954). Transmission of wheat streak-mosaic virus by abrasive leaf contacts during strong winds. *Plant Dis. Rep.* **38**, 445–447.

Singleton, W. R., Jones, D. F., and Everett, H. L. (1948). Sweet dent silage. *Conn., Agric. Exp. Stn., New Haven, Circ.* **165**, 1–14.

Skidmore, E. L. (1966). Wind and sandblast injury to seedling green beans. *Agron. J.* **58**, 311–315.

Smith, R. B. (1948). The role of mechanics in the evaluation of the herbaceous plant stem. Ph.D. Thesis, State University College of Forestry at Syracuse University, Syracuse, New York.

Sondheimer, E., and Simpson, W. G. (1962). Lignin abnormalities of "rubbery apple wood." *Can. J. Biochem. Physiol.* **40**, 841–846.

Steucek, G. L., and Kellogg, R. M. (1972). The influence of stem discontinuity on xylem development in Norway spruce, *Picea abies. Can. J. Res.* **2**, 217–222.

Sutton, J. C., and Sheppard, B. R. (1976). Aggregation of sand-dune soil by endomycorrhizal fungi. *Can. J. Bot.* **54**, 326–333.

Thompson, D. L. (1964). Comparative strength of corn stalk internodes. *Crop Sci.* **4**, 384–386.

Undersander, D. J., Bauman, L. F., Lechtenberg, V. L., and Zuber, M. S. (1977). Effect of cyclic selection for high and low crushing strength on rind, pith, and whole stalk composition in corn. *Crop. Sci.* **17**, 732–734.

Veal, M. A., Blair, R. L., Jett, J. B., McKean, W. T., and Cowling, E. B. (1974). Impact of fusiform rust on pulping properties of young loblolly and slash pines. *Proc. Amer. Phytopath. Soc.* **1**, 61.

Waters, W. E., and Cowling, E. B. (1976). Forest pest management—a silvicultural necessity. *In* "Integrated Pest Management" (J. L. Apple and R. F. Smith, eds.), pp. 149–177. Plenum, New York.

Weiner, J., Roth, L., Cowling, E. B., and Hafley, W. L. (1974). "Changes in the Value and Utility of Pulpwood, Sawlogs, and Veneer Bolts During Harvesting, Transport and Storage," Bibliogr. Ser., Spec. No. 60. Inst. Paper Chem., Appleton, Wisconsin.

Zuber, M. S., Grogan, C. O., Michaelson, M. E., Gehrke, C. W., and Monge, J. F. (1957). Studies of the interrelationship of field stalk lodging, two stalk rotting fungi, and chemical composition of corn. *Agron. J.* **49**, 328–331.

Chapter 15

Disease Alterations in Permeability and Membranes

HARRY WHEELER

I. INTRODUCTION

The concept of permeability includes both the capacity of an agent to penetrate, diffuse into, or pass through (i.e., permeate) a substance or a mass and the properties of the substance or mass which allow permeation. In a given system, an alteration in permeability may reflect a change in the permeating agent, a change in the substance being permeated, or a change in both components of the system. If the substance or mass being permeated is alive and in good health, it will exhibit permeability properties quite different from those observed in nonliving systems. Living organisms, tissues, cells, and even subcellular organelles have the ability to regulate permeation. This property, termed selective, differential, or semipermeability, is a characteristic of life and the loss of selective permeability is often the first sign of loss of vitality.

Although altered permeability was shown to be characteristic of diseased plant tissues nearly four decades ago (Thatcher, 1939), 30 years passed before the first general review of this aspect of plant pathogenesis appeared (Wheeler and Hanchey, 1968). Current renewed interest was apparently aroused by the discovery that pathotoxins produced by certain plant pathogens cause rapid and drastic changes in the permeability of susceptible plant tissues (Wheeler and Black, 1963). Interest was further stimulated when chain-splitting pectolytic enzymes were found to cause marked changes in permeability well before tissue maceration or cell death could be detected (Mount et al., 1970).

The extensive literature on the effects of pathogen-produced toxins and enzymes on the permeability of plant cells and tissues has been, to say the least, adequately covered in recent reviews (Luke and Gracen, 1972; Page, 1972; Scheffer and Yoder, 1972; Wood, 1972; Hislop et al., 1973; Kaars Sipesteijn and Van Dijkman, 1973; Hall and Wood, 1973; Patil, 1974; Strobel, 1974; Wheeler, 1975, 1976; Bateman and Basham, 1976; Rudolph, 1976; Scheffer, 1976). In addition, a comprehensive review of the role of permeability and host membranes in disease specificity is forthcoming (Hanchey and Wheeler, 1978). This plethora makes it easy to comply with the editors' request that this chapter not be a review of the literature. At the same time, it makes their challenge—to produce fresh ideas and to synthesize new concepts—more demanding.

II. CONCEPTS OF CELL PERMEABILITY

Any valid concept or theory of the nature of cell permeability must account for the phenomenon of selective permeability mentioned in the preceding section as one characteristic of living organisms. A well-known and much-studied example of the phenomenon is the ability of cells, both plant and animal, to accumulate potassium and to exclude sodium in defiance of the laws of ordinary diffusion. Algal cells growing in seawater provide a striking example of selective permeability. Inside such cells the concentration of potassium is 40 times higher, whereas that of sodium is 5 times lower, than the concentration of these elements in the surrounding seawater (Osterhout, 1922). The ability of algae to accumulate potassium is even more striking when cells are grown in solutions which contain only trace amounts of this element. Under these conditions, accumulation ratios (ratio of the concentration of potassium in the cell sap to that in the external solution) in excess of 2,000 have been observed (Hoagland and Davis, 1923). Selective permeability of living cells is also reflected by their bioelectric properties. The potential difference (PD) between the interior of a plant cell held in a dilute salt solution and the surrounding medium usually falls in the range of -50 to -150 mV. Maintenance of one component of this PD requires metabolic energy (Higinbotham, 1973).

Three general theories of cell permeability have been proposed. The first, and by far the most widely accepted, assumes that membranes are the controlling elements. The second, supported most strongly in the USSR, holds that sorption to protoplasmic colloids and macromolecules plays a vital role. The third, supported by a few scattered iconoclasts, proposes that cytosis or vesicular transport, known to operate in certain animal cells, also functions in higher plants. Evidence and arguments pro and con for each of these theories have been reviewed and evaluated (Wheeler, 1976). This review and his own experience led the author to conclude that all three theories of cell permeability have some degree of validity. What follows is not an attempt to bolster this conclusion. Instead it is intended to serve as a warning that an overly simplistic concept of cell permeability, based solely on membrane theory, may be dangerous to the health of research in this area.

Currently, membrane theory so dominates concepts of cell permeability that any change in permeability is taken as evidence of a change in some membrane system. Cells in diseased plant tissues leak electrolytes because their plasmalemmas have been disrupted. Electrochemical potentials drop in cells exposed to pathotoxins because their membranes have been depolarized. Infected tissues become waterlogged because membranes have been altered. The refrain could continue, but the danger referred to earlier in complete and uncritical reliance on membrane theory is not that unjustified or erroneous conclusions may be drawn. The danger is that innovative, unconventional approaches to the many unsolved problems of cell permeability may be suppressed. For those who may feel that such a danger does not exist, the writer, during discussions of cell permeability, has had the temerity to suggest that sorption or cytosis might be involved. This suggestion has elicited two kinds of reactions: blank stares from those who had never heard of either phenomenon and wrathfully righteous rejections from true believers whose dogma had been challenged.

To warn of the danger of complete reliance on membrane theory is not to imply that membranes play less than a central role in cell permeability. Although questions remain about permeability phenomena which are dependent on metabolic energy (active transport), there can be no doubt that membranes serve as barriers to provide internal compartmentalization and to isolate the cell interior from its surrounding environment. This function of membranes was emphasized by Collander (1959), who also pointed out that preoccupation with membrane permeability has tended to make us forget that membranes are, in fact, highly impermeable to most substances. The increase in permeability, or more properly, the loss of resistance to permeation, observed by Thatcher (1939) and later by many others indicates that, in diseased plant tissues, membranes lose the ability to function effectively as barriers to permeation. Therefore, the remainder of this chapter will focus on membranes and the question of how changes in membrane structure and function may be involved in pathogenesis. At the same time, the possibility that sorption and cytosis may play roles in permeability phenomena in diseased plants will not be ignored.

III. MEMBRANE STRUCTURE AND FUNCTION

The flood of proposed models of membrane structure which began about 15 years ago apparently has subsided. Currently, a "fluid mosaic" model (Singer and Nicolson, 1972) appears to be gaining in popularity. In this model, as in most others recently proposed, the matrix or core of the membrane is assumed to be a bilayer of amphipathic phospholipid molecules oriented with their ionic polar groups toward the exterior and their nonpolar hydrocarbon chains toward the interior. Integral globular proteins also arranged in an amphipathic structure, with polar groups protruding from the membrane and nonpolar groups extending into the interior, are assumed to be embedded in, and in some cases to span, the lipid bilayer which is in a fluid liquid state. Thus membranes are visualized as mosaics composed of a lipid bilayer interrupted in a nonrandom manner by globular proteins.

The idea that membranes are mosaics of lipids and proteins is far from novel. Such a mosaic structure was proposed by Nathansohn (1904) to account for the ability of both lipid-soluble and lipid-insoluble substances to penetrate cell membranes. Furthermore, the concept of a lipid bilayer, arranged in an amphipathic structure, has long been regarded as the basic backbone of membranes (Gorter and Grendel, 1925). The novel feature of the current model is its fluid nature which allows lateral migration of integral proteins and dynamic changes in structure without loss of membrane integrity.

Concomitant with the rise in popularity of the fluid mosaic model of membrane structure has been the development of interest in the chemiosmotic hypothesis of membrane function in active transport processes (Mitchell, 1966). This hypothesis assumes that hydrolysis of adenosine triphosphate (ATP) results in the separation of H^+ and OH^- ions across membranes. Originally applied in the area of phosphorylation, this model proposed that in mitochondria H^+ ions are extruded (pumped out) whereas in chloroplasts the H^+ ion pump works in the opposite direction. Currently, electrogenic pumps as mechanisms by which net charges are transferred across membranes at the expense of metabolic energy are the focus of attention not only in investigations of phosphorylation but also in studies of electrochemical potentials, ion uptake, and active transport in general. Despite their popularity, electrogenic pumps have proved to be, in the words of one investigator (Poole, 1973), "remarkably elusive to experimental approach." Perhaps the best evidence for ion pumps in plant cells is the occurrence of cell membrane potentials which are dependent on metabolic energy and which exceed the Nernst potential for any of the diffusible ions involved (Higinbotham, 1973).

Current concepts of membrane structure and function have been derived from work with animal and microbial systems. Since the rather sparse data obtained with plants have been reviewed (Anderson, 1973; Higinbotham, 1973; Mooré, 1975), the remainder of this section will be devoted to a brief general description of membrane systems in plant cells and the roles these are thought to play in cell permeability.

A. Plasmalemma

This membrane, which in plants separates the protoplast from the cell wall, is considered to be not only the chief barrier to passive permeation, but also the site of the enzymatic machinery (currently conceived to be ion pumps) which function in active transport. Electron micrographs of thin sections and freeze-etch preparations indicate that the plasmalemma is asymmetric, both in the distribution of particles on or embedded in the two surfaces (Staehelin and Probine, 1970) and in that densely stained material, thought to be glycoprotein, coats the outer surface (Cook, 1971). There has been much speculation, but little evidence, that the particles on or in the membrane represent enzyme complexes involved in active transport and that the outer coat, rich in carbohydrates, provides receptor or recognition sites similar to those found on animal cells (Hughes, 1976). Marked increases in asymmetry and width of the plasmalemma have been observed in cells exposed to divalent cations (Wheeler, 1974).

In general, the plasmalemma appears to be chiefly responsible for the bioelectrical properties of plant cells. Most evidence for this view has come from work with giant cells of freshwater algae in which electrochemical potentials across the plasmalemma are large whereas those across the tonoplast are small (Spanswick and Williams, 1964). There are, however, some questions about extrapolation of these findings to cells of higher plants. In certain marine algae, the electrical resistance of the tonoplast is much higher than that of the plasmalemma and the potential across the tonoplast is large, with the vacuole positive (Higinbotham, 1973). In addition, flux analysis of ion leakage from tissue slices of storage organs has implicated the tonoplast rather than the plasmalemma as the chief barrier to ion flux (Van Steveninck, 1975).

B. Tonoplast

Other than the plasmalemma, the tonoplast which surrounds the vacuole is the membrane which has received the most attention in studies of plant cell permeability. Currently, cells are conceived as consisting of two permeability compartments: (1) the cytoplasm, bounded on the exterior by the plasmalemma, and (2) the vacuole, separated from the cytoplasm by the tonoplast. Flux analysis has been used in attempts to identify the contributions made by each of these two compartments and the extent to which active and passive processes are involved. Results currently available appear highly inconsistent (MacRobbie, 1971; Anderson, 1973). In some cases, single inflections on efflux or influx curves have been obtained and these are consistent with a simple, two-compartment model operating in series. In other cases, multiple inflections, indicative of multiple compartments, or no inflections, indicative of a single compartment, have been observed. Flux analysis has also failed to give a clear answer to the question of whether active transport processes operate at both the tonoplast and the plasmalemma or only at the latter.

Traditionally, vacuoles have been considered to function as osmoregulators and as temporary storage areas for metabolites such as organic acids and phenolics. More recently, a lysosomal role for vacuoles has been suggested (Wilson, 1973). In both views, the tonoplast serves as a protective barrier to the penetration of potentially lethal agents into the cytoplasm.

C. Membrane-Bound Organelles

The permeability characteristics of chloroplasts and mitochondria, which have clearly defined outer and inner membranes, are, in general, similar to those of whole protoplasts. These two organelles contain enzyme systems which generate energy, either through respiratory or photosynthetic phosphorylation. Loss or disruption of these energy-generating systems would be expected to lead to a loss of energy-dependent properties of other membranes, especially the plasmalemma.

Little is known about the permeability of other internal membrane systems. The double membrane which surrounds the nucleus and is perforated by numerous pores cannot be an important barrier to permeation. In secretory cells, vesicles formed by the Golgi apparatus are excreted by exocytosis (Schnepf, 1969) but the significance of this process in relation to cell permeability is not understood. Even more uncertain is the possible role of vesicular endocytotic uptake and transfer (Anderson, 1973). If lysosomal hydrolytic enzymes are sequestered in vesicles or vacuoles as some propose (Wilson, 1973), maintenance of the integrity of membranes surrounding these structures would be essential to avoid autolysis. Finally, virtually no attention has been given to the endoplasmic reticulum even though, as MacRobbie (1973) points out, there is a high probability that extensive sheets of this membrane system provide a considerable barrier to permeation.

The concept of a cell consisting of two independent permeability compartments introduced in Section III,B is almost certainly an oversimplification and may be highly inaccurate. Evidence of continuities between membranes (plasmalemma–vesicle–Golgi apparatus–endoplasmic reticulum–nuclear envelope–tonoplast) led to the concept of an endomembrane system as a structural and functional continuum (Morré and Mollenhauer, 1973). Chloroplasts and mitochondria were regarded as at least semi-autonomous systems, although membrane continuities between the latter and the endoplasmic reticulum had been reported. If the endomembrane concept is correct, permeability would be controlled by the entire system and could be altered by a change at any point in the membrane chain. Continuities between the outer chloroplast membrane and various other membrane systems have also been observed in electron micrographs. The most puzzling of these indicate that such continuities provide open channels extending from the vacuole through the perichloroplast space to the cell wall (Crotty and Ledbetter, 1973). Equally difficult to explain, at least on the basis of membrane theory, are distinct discontinuities in the plasmalemma (Wheeler, 1974). In view of the artifactual nature of all electron microscopic images, it would be premature to speculate about the possible significance of these observations in relation to cell permeability.

IV. PATHOLOGICAL CHANGES IN PERMEABILITY

A. Ubiquity in Diseased Plants

Changes in permeability appear to be universally characteristic of diseased plant tissues regardless of disease type or the nature of the pathogenic agent. The ubiquity of this phenomenon among diseases caused by fungi was established by Thatcher (1942) who studied rusts, powdery mildews, downy mildews, soft rots, dry rots, and wilts. Later work has extended the phenomenon to diseases caused by bacteria, viruses, atmospheric pollutants, environmental stresses, and pathogen-produced metabolites. The latter, in the form of enzymes and toxins, have been the focus of attention during the past decade. Results with enzymes and toxins as agents which disrupt permeability will be summarized very briefly since this work has already been the subject of many reviews (see Section I for a list of references).

B. Effects of Enzymes

The long history of research on the role of cell wall-degrading enzymes in soft-rot diseases has been well covered by Brown (1965). Comprehensive reviews of more recent work have been provided by Wood (1972) and by Bateman and Basham (1976). These should be consulted for details of findings on which the following summary is based.

Results with highly purified enzyme preparations indicate that only those with pectolytic activity are capable of attacking intact cell walls and bringing about tissue maceration. Pectic enzymes, in particular endo-polygalacturonases and lyases, loosen cell wall structure and expose non-pectic polymers to attack by other enzymes. When disks of storage tissues are exposed to purified pectic enzymes, leakage of electrolytes begins almost immediately (within 3–5 min) whereas tissue maceration and cell death are detectable only after a delay of 20–60 min. These results seem to indicate that cell death is the result of damage to the plasmalemma, either directly by the pectic enzymes or indirectly by some product of their activity.

Other observations are difficult to reconcile with the idea that damage to the plasmalemma, either by direct enzymatic activity or through other biochemical reactions, is responsible for the lethal effects of pectic enzymes. Loss of electrolytes and cell death can be prevented if tissues are held in hypertonic solutions while undergoing maceration by pectic enzymes. Furthermore, pectic enzymes have no effect when added to isolated protoplasts held in hypertonic solutions. These results and the fact that cells plasmolyzed in hypertonic solutions gradually deplasmolyze in 1–2 hr (Thatcher, 1942) make it unlikely that the protective effect of hypertonic solutions can be attributed to a physical separation of the plasmalemma from the cell wall or to failure of pectic enzymes or their reaction products to gain access to sensitive sites on the protoplast surface.

The hypothesis proposed by Wood (1972) that the plasmalemma is mechanically damaged by failure of enzyme-treated cells walls to provide adequate support for the protoplast has been evaluated by Bateman and Basham (1976). They point out that several lines of evidence tend to support this hypothesis. (1) Purified pectic lyases can release more than 50% of the total sugars from cell walls and much of this occurs before loss of cohesion in tissue disks can be detected. (2) Cell wall degradation, measured by the release of unsaturated uronides, begins within a few minutes after the loss of electrolytes is detected and the rate of release of uronides closely parallels that of the loss of electrolytes. (3) When enzyme-treated tissues are held in hypertonic solutions, cell wall degradation proceeds but little or no loss of electrolytes occurs. However, if after 1 hr the tissues are rinsed in osmoticum to remove the enzyme and then transferred to distilled water, all electrolytes are lost very rapidly at a rate expected of free-space equilibration.

Although current data are compatible with the hypothesis that lethal effects of pectic enzymes result from rupture of the membrane caused by rapid swelling in the absence of restraining intact cell walls, the prevailing assumption that rupture or damage to the plasmalemma is of primary importance may be in error. Brown (1965) states that cells killed by pectic enzymes do not burst under turgor pressure; instead, the protoplasts collapse. This description would also apply to cells in injured or diseased tissues which undergo false plasmolysis. Evidence from light and electron microscopy suggests that false plasmolysis results from disruption of the tonoplast, which allows contents of the vacuole to mix with the cytoplasm (Hanchey and Wheeler, 1969). Electron microscopy of tissues during maceration by pectic enzymes might provide useful information on the site and nature of changes in permeability and eventual lethal effects.

C. Effects of Toxins

Permeability changes caused by pathogen-produced toxins, in particular those which are selectively toxic to plants susceptible to the pathogen, have been the subjects of intense investigation during the past two decades. Selectively toxic products of plant pathogens, termed selective pathotoxins (Wheeler and Luke, 1963) or host-specific toxins (Pringle and Scheffer, 1964), have been used in lieu of living pathogens to study the sequence of events in pathogenesis. Such model systems offer several advantages; host responses can be studied without interference by metabolic activities of the pathogen; pathological events which occur over a period of days in inoculated plants can be compressed to a few hours, and quantitative data are more readily obtained. On the other hand, results obtained with such model systems must be interpreted with caution. Even though some selective pathotoxins appear to produce all the visible and physiological symptoms found in naturally infected plants, it is unlikely that any single metabolic product of a pathogen will be found to be solely responsible for all aspects of any plant disease.

Permeability changes induced by selective pathotoxins have been investigated most extensively with the model system of *Helminthosporium victoriae*–victorin–Victoria blight of oats. The permeability characteristics of susceptible tissues treated with victorin are very similar to those of plants infected by *H. victoriae* or various other pathogens. Water permeability, estimated from the time required for plasmolysis and deplasmolysis, is increased. Cells become leaky and lose large quantities of electrolytes and other materials. Uptake and accumulation of mineral salts and other materials are inhibited. Electrochemical potentials drop, i.e., the cell interior becomes less negative. Finally, cells undergo false plasmolysis and die.

Losses of electrolytes, similar to those induced by victorin, occur when sensitive tissues are treated with selective toxins produced by *Periconia circinata, Alternaria mali,* or *A. kikuchiana* (Scheffer, 1976; Hanchey and Wheeler, 1978). Similar effects, and in some cases increases in water permeability, occur in response to a number of nonselective toxins produced by a variety of pathogens (Patil, 1974; Rudolph, 1976). Somewhat more complex effects on permeability have been reported with other selective pathotoxins. The HC-toxin, produced by *Helminthosporium carbonum,* causes a sharp increase in the uptake of nitrate and certain other materials before toxin-treated tissues become leaky (Scheffer, 1976). Most, but not all, investigators have found that T-toxin, produced by *H. maydis* race T, induces losses of electrolytes from susceptible tissues. This response, however, is rather insensitive since it requires much higher concentrations of toxin than those which cause other physiological changes. Additional data on effects of T-toxin on permeability, which appear confused and in some cases contradictory, can be found in recent reviews (Strobel, 1974; Scheffer, 1976; Hanchey and Wheeler, 1978). Both T-toxin and PM-toxin, produced by *Phyllosticta maydis,* cause rapid swelling and loss of respiratory control in isolated mitochondria from corn leaves susceptible to these two pathogens.

D. Effects of Growth Regulators and Pollutants

Elevated levels of growth-regulating substances are often found in diseased plants even in the absence of obvious growth abnormalities. Auxins, cytokinins, and ethylene cause changes in cell permeability and there is some evidence for hormonal interactions with membranes (Daly, 1975). Although growth regulators produced by pathogens or by plants in response to infection probably play a role, direct effects of single plant hormones on permeability differ, both qualitatively and quantitatively, from those found in diseased plants (Wheeler, 1975). As endogenous regulators which interact among themselves and with other agents to control metabolism growth and development, plant hormones must function, at least indirectly, in the maintenance of cell permeability. It has been suggested that they may function with other agents to trigger pathogenesis (Pegg, 1976) and, by interaction with membranes, to play a role in the action of phytotoxins (Daly, 1975).

Certain oxidant air pollutants, e.g., ozone, cause permeability changes typical of those observed in infected plants. Early symptoms of ozone damage are water-soaked areas on leaves and leakiness of tissues. Evidence that ozone disrupts membranes, perhaps by reacting with the double bonds of unsaturated lipids, has been reviewed by Rich (1964).

E. Initial Events in Pathogenesis

Selective pathotoxins as disease-inducing agents which can be substituted for living pathogens, provided an opportunity to test, in a systematic way, the hypothesis that permeability changes are initial events in pathogenesis. Early results with victorin-treated tissues revealed that permeability changes, estimated from rates of electrolyte loss, could be detected within 5 min, whereas increases in respiration and other physiological effects occurred only after a lag of about 30 min (Wheeler and Black, 1963). These results were thought to reflect an initial effect on the plasmalemma followed after about 30 min by disruption of the tonoplasts. The latter, by allowing vacuolar contents (salts, phenolics, etc.) to mix with the cytoplasm, would result in changes in respiration and subsequent pathological events. These results with victorin, together with scattered data from other diseased tissues, led to the working hypothesis that changes in permeability were initial events that served to trigger pathological changes in diseased plants (Wheeler and Luke, 1963).

When first proposed, the hypothesis that permeability changes are initial events in pathogenesis was received with considerable skepticism. At that time, attention was focused on respiration and oxidative phosphorylation as energy-generating systems thought to play crucial, perhaps primary, roles in pathogenesis (Millerd and Scott, 1962). Furthermore, the prevailing concept was that changes in permeability were the result, rather than the cause, of changes in respiratory metabolism. The hypothesis gained respectability when losses of electrolytes, similar to those induced with victorin, were obtained with other toxins and with pathogen-produced pectic enzymes (Wheeler, 1975; Bateman and Basham, 1976; Scheffer, 1976). Finally, after rapid changes in cellular electrochemical potentials were found in toxin-treated and infected tissues, the hypothesis attained the rank of dogma (Novacky and Karr, 1977). More specifically, these very rapid changes in permeability were interpreted as evidence of an initial effect on the plasmalemma since this membrane would be the first important permeation barrier contacted by a pathogenic agent.

Skeptics who question popular beliefs do so at the risk of becoming themselves very unpopular. Questions about a concept by one who played a role in its advancement may be regarded as particularly dastardly. If so, the author and some of his colleagues have already stigmatized themselves by pointing out that there has been no critical test of the hypothesis that pathogens or pathogen-produced metabolites cause disease by effects on cell permeability and that, despite extensive searches, no direct evidence of an initial effect on the plasmalemma has been produced (Wheeler, 1975; Hanchey and Wheeler, 1978). Support for the hypothesis rests mainly on three lines of indirect evidence: (1) rapid, virtually instantaneous, changes in electrolyte loss or electrochemical potentials, or both, which occur in tissues exposed to pathotoxins or pectic enzymes; (2) lysis of protoplasts and cell death caused by the same agents; and (3) lethal effects of victorin on dormant oat seeds.

Evaluation of the first of these three lines of evidence led to the conclusion that the pattern of short-term changes induced by victorin was not that expected from an initial direct effect on the plasmalemma (Wheeler, 1975, 1976). During the first 5 min of exposure to victorin, the rate of electrolyte loss increases sharply, after which it levels off for a period of 20–30 min before a second and prolonged increase in the rate occurs (Saftner and Evans, 1974). The pattern of changes in electrochemical potentials is very similar except that after an initial sharp depolarization the potential actually recovers to a near-normal value before a second prolonged depolarization sets in after about 30 min (Novacky and Hanchey, 1974). It is unlikely that these changes result from sequential effects on the plasmalemma and the tonoplast. The initial period of electrolyte loss, which occurs at a rate typical of free-space equilibration, seems too brief for complete washout of the cytoplasm. Furthermore, to account for the recovery in potential, one would have to assume that, after an initial disruption, the plasmalemma was somehow temporarily repaired. Although changes in electrochemical potentials in response to other toxins or to pathogens are less rapid and of a different character than those induced by victorin, these too have been interpreted to reflect interference with the metabolic systems required for the maintenance of cell permeability rather than direct interactions with the plasmalemma (Novacky and Karr, 1977).

If the initial effect of victorin and other pathogenic agents is on the plasmalemma, changes at least as rapid and drastic as those found in toxin-treated intact tissues should occur in isolated protoplasts exposed to toxin. Results with oat protoplasts exposed to victorin have failed to provide evidence of such rapid changes. Collapse of sensitive protoplasts began only after a lag of 20–30 min and was characterized by almost simultaneous rupture of the plasmalemma and tonoplast (Samaddar and Scheffer, 1968; Easton and Hanchey, 1972). No differences between sensitive and resistant protoplasts in uptake of labeled compounds were detected until after 30 min of exposure to victorin (Rancillac et al., 1976). These results support the suggestion that initial changes induced by victorin reflect effects on cell walls and that permeability of the protoplast is disrupted only after a lag of about 30 min (Wheeler, 1975, 1976).

Lethal effects which occurred when dormant seeds of susceptible oat varieties were treated for 1 hr with victorin and then washed for 1 hr led to the conclusion that this toxin acts on resting tissues as well as on those actively metabolizing (Samaddar and Scheffer, 1970). Since uptake into the cytoplasm by dormant tissues would be most unusual, these findings implied a direct toxic effect at the protoplast surface. Reinvestigation, however, has shown that lethal effects of victorin on dormant seeds can be prevented by extended periods of washing or by brief posttreatments with alkali, which is known to deactivate this toxin (McCammon and Wheeler, 1978). These results, plus the fact that alkali posttreatments are effective on victorin-treated seeds that have been dried and stored for 2–3 weeks, indicate that dormant seeds are highly resistant to victorin and that lethal effects are caused by residual bound toxin which acts during germination. In any event, the possibility that active metabolism and uptake are required for victorin activity cannot be ruled out.

If initial rapid changes in permeability in response to pathogenic agents are not the result of effects on the plasmalemma, what do these changes signify? Any attempt to answer this question should take into consideration the fact that these very rapid changes have been obtained only with highly concentrated toxin or enzyme preparations and with massive doses of inoculum. Therefore they are artifactual, at least in the sense that they occur only under much more drastic conditions than those likely to be found in natural infections. In my laboratory, sensitive oat tissues have been treated with victorin for periods longer than those required to cause initial changes in permeability without lethal effects when the tissues were promptly posttreated to deactivate and remove adsorbed toxin. These results indicate that initial changes are not irreversible lethal events, but they do not rule out the possibility that they may be required first steps in pathogenesis.

V. MEMBRANES AS SITES OF ACTION OF PATHOGENIC AGENTS

A. Evidence of Changes in Membrane Function

Three general hypotheses have been advanced to account for permeability changes in diseased plants in terms of membrane dysfunction. One assumes that receptor sites on the outer surface of the plasmalemma interact in a highly specific fashion with pathogens or their metabolites. The second postulates effects on membrane-bound enzymes, especially ATPases, thought to be located in the plasmalemma and to function in the operation of ion pumps or other active transport systems. The third proposes that dysfunction of the plasmalemma is brought about indirectly by disruption of energy-generating systems in chloroplasts or mitochondria which provide energy for maintenance and repair of membrane function.

The receptor site hypothesis is attractive because it also accounts for the highly selective nature of certain pathogens and pathogen-produced toxins (Pringle and Scheffer, 1964). Despite extensive searches, direct support for this hypothesis is limited to results obtained with a toxin produced by *Helminthosporium sacchari,* a foliar pathogen of sugarcane. A fraction of this toxin, termed helminthosporoside, has been reported to bind specifically to a single protein present in susceptible but not in resistant clones of sugarcane. The protein is assumed to occupy a specific site on the plasmalemma and, when bound by toxin, to activate a membrane-bound ATPase. This is thought to disrupt the ionic balance of the cells and to lead to water soaking and other symptoms of disease (Strobel, 1974). The findings reported with helminthosporoside are of great interest but, in view of questions which have been raised about experimental protocol and interpretation of data (Bhullar *et al.,* 1975; Daly, 1976; Wheeler, 1975), firm conclusions must await independent confirmation. Attempts to identify specific receptors or binding sites for other pathotoxins by the methods used with helminthosporoside have, to date, yielded only negative results (Scheffer, 1976).

Support for the hypothesis that specific effects on membrane-bound enzymes are involved in permeability changes is also very limited. Other than helminthosporoside (Strobel, 1974), only T-toxin, produced by *Helminthosporium maydis* race T, has been reported to have selective effects on membrane-bound ATPase (Tipton *et al.*, 1973). Results from another laboratory, however, failed to provide confirmation of such an effect by T-toxin (Scheffer, 1976).

Although unsupported by direct evidence, most available data are consistent with the hypothesis that membrane dysfunction results from disruption of mechanisms responsible for membrane maintenance and repair. This idea was advanced to account for the increased sensitivity of aged tissues to victorin (Wheeler, 1969) and it would also account for the characteristic delay in metabolic responses to this toxin discussed in the preceding section. The light-dependent recovery of potential in cells exposed to cyanide is lost in tissues inoculated with bacteria and in those treated with certain pathotoxins (Novacky and Karr, 1977). These workers suggest that chloroplasts may alter cellular potentials either by pumping ions in or out of the cytoplasm or by providing energy to ion pumps in the plasmalemma. In either case, their results suggest that pathogenic agents interfere with energy supplied by photophosphorylation for the maintenance of membrane function. Specific binding of tentoxin, a product of *Alternaria tenuis*, by chloroplast coupling factor I, has been reported to inhibit ATPase and photophosphorylation in species sensitive to this toxin (Steele *et al.*, 1976). The selective effects of toxins produced by *Helminthosporium maydis* and *Phyllosticta maydis* on mitochondrial oxidative phosphorylation (Section IV,D) suggest that this source of energy is also required for maintenance of membrane function.

B. Evidence of Changes in Membrane Structure

Electron microscopy has yielded little evidence of early changes in the structure of the plasmalemma in tissues exposed to pathotoxins or in those infected by pathogens (Wheeler, 1975). Although changes too subtle to detect may occur in the plasmalemma, the tonoplast was the first membrane found to be disrupted in oat leaves treated with victorin (Luke *et al.*, 1966) and in corn leaves infected with *Helminthosporium maydis* (White *et al.*, 1973). A very rapid change (within 15 min) in mitochondria when susceptible corn tissues were exposed to *H. maydis* T-toxin has been reported, but this change involved a loss of density of the matrix rather than structural disruption of membranes (Aldrich *et al.*, 1977).

In contrast to the scant evidence for early structural changes in membranes, structural changes at the cell wall-plasmalemma interface appear to be characteristic early events in pathogenesis. First observed in response to victorin (Luke *et al.*, 1966), they are now known to occur in response not only to a variety of pathogenic agents but also to mechanical stress (Aist, 1976). Structurally modified cell walls in diseased tissues often appear to be greatly weakened and, in some cases, partially solubilized (Wheeler, 1975). It is possible, therefore, that lack of mechanical support by structurally modified cell walls permits uncontrolled swelling of vacuoles which brings about rupture of tonoplasts, and that this mechanism, which has been proposed to account for effects of pectic enzymes (Section IV,B), is of more general significance.

C. Evidence of Changes in Membrane Chemistry

Since details of the chemical composition of plant membranes, except chloroplast membranes, are largely unknown, any attempt to interpret changes in permeability in terms of changes in membrane chemistry would be pure speculation. The possibility that pathogen-produced enzymes degrade lipid or protein components of membranes has been considered by Bateman and Basham (1976), who conclude that there is no good evidence that these are involved in early stages of pathogenesis. Analyses of lipids during disease development, the most extensive of which has been carried out by Hoppe and Heitefuss, have revealed changes but these appear to occur too late to account for permeability changes (Hanchey and Wheeler, 1978).

VI. PHYSIOLOGICAL EFFECTS OF MEMBRANE DISRUPTION

Extensive disruption of membranes would be expected to result in complete loss of metabolic regulation and control. Energy generation and synthetic activity by highly organized, membrane-bound enzyme systems would be interrupted, and decompartmentalization would result in uncontrolled reactions between degradative enzymes and their substrates as well as inactivation of cytoplasmic enzymes by phenols or other toxic materials released from vacuoles. Such drastic events could account for the sudden granulation and collapse of protoplasts observed in tissues treated with pectic enzymes (Brown, 1965), in protoplasts exposed to pathotoxins (Easton and Hanchey, 1972), and in cells penetrated by certain pathogens (Akai, 1959). However, these changes, which probably signal cell death, do not occur until after the cells have lost most of their electrolytes (Wheeler, 1975).

In general, the physiological changes observed during pathogenesis are not those expected from extensive disruption of membranes. Uncoupling of oxidative phosphorylation, which would be expected if mitochondrial membranes were disrupted, has been postulated to account for the increases in respiration which are characteristic of diseased plant tissues (Allen, 1953), but the available evidence does not appear to be compatible with this hypothesis. Instead, increases in biosynthetic activity, geared to the increase in respiration, have been found in infected tissues (Daly, 1976) and in those exposed to pathotoxins (Rawn, 1977). Furthermore, Kosuge and Gilchrist (1976) have pointed out that uncoupling would result in a sharp drop in energy charge, which is based on the relative amounts of energy-rich phosphate bonds in cellular adenylate pools. In fully charged cells with adenylates in the form of ATP the energy charge is 1, whereas in the fully discharged condition (all adenylates in the AMP form) the energy charge is 0. In the few cases studied, energy charge in diseased tissues was only slightly lower than that in healthy tissues (Kosuge and Gilchrist, 1976).

Changes in protein and nucleic acid metabolism during early stages of pathogenesis also indicate that metabolic activities are merely redirected rather than completely disrupted. Evidence that early changes in infected tissues are similar to those induced by wounds has been summarized by Uritani (1976). Stimulation of synthesis of RNA in early stages of pathogenesis may also be a response to stress similar to that induced by wounds (Heitefuss and Wolf, 1976). If initial physiological responses to pathogenic agents reflect an activation of repair mechanisms involved in wound healing, later changes, which are similar to those found during senescence (Farkas et al., 1964), would indicate failure of repair mechanisms to operate effectively.

VII. SUMMARY

Changes in permeability are early, and, in many cases, the first detectable response of plants to pathogenic agents. Virtually instantaneous changes in permeability in tissues exposed to pathogen-produced toxins and enzymes appear to support the hypothesis that permeability changes are initial events in pathogenesis and to implicate the plasmalemma as the site at which these agents act. However, attempts to provide direct evidence of interactions with specific receptor sites or enzymes bound to the plasmalemma have, for the most part, yielded negative results.

Furthermore, the pattern of permeability changes observed in short-term experiments with pathotoxins—an initial rapid change followed by a period of recovery before a second prolonged change occurs—suggests that the initial change reflects an effect on cell walls. If this interpretation is correct, changes in protoplast permeability, assumed to be controlled by the plasmalemma, occur only after a lag period similar to that required for changes in respiration and other processes directly linked to metabolism. Thus the key question of whether permeability changes are primary causal events in pathogenesis or whether they are merely another physiological symptom of injury remains to be answered.

References

Aist, J. R. (1976). Cytology of penetration and infection-fungi. *Physiol. Plant Pathol.* **4**, 197–221.

Akai, S. (1959). Histology of defense in plants. *In* "Plant Pathology: An Advanced Treatise" (J. G. Horsfall and A. E. Dimond, eds.), Vol. 1, pp. 391–434. Academic Press, New York.

Aldrich, H. C., Gracen, V. E., York, D., Earle, E. D., and Yoder, O. C. (1977). Ultrastructural effects of *Helminthosporium maydis* race T on mitochondria of corn roots and protoplasts. *Tissue Cell* **9**, 167–177.

Allen, P. J. (1953). Toxins and tissue respiration. *Phytopathology* **43**, 221–229.

Anderson, W. P., ed. (1973). "Ion Transport in Plants." Academic Press, New York.

Bateman, D. F., and Basham, H. G. (1976). Degradation of plant cell walls and membranes by microbial enzymes. *Physiol. Plant Pathol.* **4**, 316–355.

Bhullar, B. S., Daly, J. M., and Rehfeld, D. W. (1975). Inhibition of dark CO_2 fixation and photosynthesis in leaf discs of corn susceptible to the host-specific toxin produced by *Helminthosporium maydis,* race T. *Plant Physiol.* **56**, 1–7.

Brown, W. (1965). Toxins and cell-wall dissolving enzymes in relation to plant disease. *Annu. Rev. Phytopathol.* **3**, 1–18.

Collander, R. (1959). Cell membranes: Their resistance to penetration and their capacity for transport. *In* "Plant Physiology" (F. S. Steward, ed.), Vol. 2, pp. 3–102. Academic Press, New York.

Cook, G. M. W. (1971). Membrane structure and function. *Annu. Rev. Plant Physiol.* **22**, 97–120.

Crotty, W. J., and Ledbetter, M. C. (1973). Membrane continuities involving chloroplasts and other organelles in plant cells. *Science* **182**, 839–841.

Daly, J. M. (1975). Specific interactions involving hormonal and other changes. *In* "Specificity in Plant Diseases" (R. K. S. Wood and A. Graniti, eds.), pp. 151–165. Plenum, New York.

Daly, J. M. (1976). Summary of points from discussions. *In* "Specificity in Plant Diseases" (R. K. S. Wood and A. Graniti, eds.), pp. 231–234. Plenum, New York.

Easton, C. Z., and Hanchey, P. (1972). Localization of crystals in diseased oats treated with uranyl acetate. *Plant Physiol.* **50**, 706–712.

Farkas, G. L., Dezsi, L., Horwarth, M., Kisban, K., and Udvardy, J. (1964). Common pattern of enzymatic changes in detached leaves and tissues attacked by parasites. *Phytopathol. Z.* **49**, 343–354.

Gorter, E., and Grendel, F. (1925). On bimolecular layers of lipids on the chromocytes of the blood. *J. Exp. Med.* **41,** 439–443.

Hall, J. A., and Wood, R. K. S. (1973). The killing of plant cells by pectolytic enzymes. *In* "Fungal Pathogenicity and the Plant's Response" (R. J. W. Byrde and C. V. Cutting, eds.), pp. 19–38. Academic Press, New York.

Hanchey, P., and Wheeler, H. (1969). Pathological changes in ultrastructure: False plasmolysis. *Can. J. Bot.* **47,** 675–678.

Hanchey, P., and Wheeler, H. (1978). The role of host cell membranes. *In* "Recognition Factors in Relation to Disease Specificity" (J. M. Daly and I. Uritani, eds.) (in press).

Heitefuss, R., and Wolf, G. (1976). Nucleic acids in host–parasite interactions. *Physiol. Plant Pathol.* **4,** 480–508.

Higinbotham, N. (1973). Electropotentials of plant cells. *Annu. Rev. Plant Physiol.* **24,** 25–46.

Hislop, E. C., Hoad, G. V., and Archer, S. A. (1973). The involvement of ethylene in plant diseases. *In* "Fungal Pathogenicity and the Plant's Response" (R. J. W. Byrde and C. V. Cutting, eds.), pp. 87–117. Academic Press, New York.

Hoagland, D. R., and Davis, A. R. (1923). The composition of the cell sap of the plant in relation to the absorption of ions. *J. Gen. Physiol.* **5,** 629–647.

Hughes, C. (1976). Cell surface membranes of animal cells as the sites of recognition of infectious agents and other substances. *In* "Specificity in Plant Diseases" (R. K. S. Wood and A. Graniti, eds.), pp. 77–99. Plenum, New York.

Kaars Sijpesteijn, A., and van Dijkman, A. (1973). The host–parasite interactions in resistance of tomatoes to *Cladosporium fulvum. In* "Fungal Pathogenicity and the Plant's Response" (R. J. W. Byrde and C. V. Cutting, eds.), pp. 437–448. Academic Press, New York.

Kosuge, T., and Gilchrist, D. G. (1976). Metabolic regulation in host–parasite interactions. *Physiol. Plant Pathol.* **4,** 679–702.

Luke, H. H., and Gracen, V. E., Jr. (1972). *Helminthosporium* toxins. *Microb. Toxins* **8,** 139–168.

Luke, H. H., Warmke, H. E., and Hanchey, P. (1966). Effects of the pathotoxin victorin on ultrastructure of root and leaf tissue of *Avena* species. *Phytopathology* **56,** 1178–1183.

McCammon, S., and Wheeler, H. (1978). Prevention of effects of victorin on oat seed germination. *Physiol. Plant Pathol.* **12,** 191–198.

MacRobbie, E. A. C. (1971). Fluxes and compartmentation in plant cells. *Annu. Rev. Plant Physiol.* **22,** 75–96.

MacRobbie, E. A. C. (1973). Vacuolar ion transport in *Nitella. In* "Ion Transport in Plants" (W. P. Anderson, ed.), pp. 431–446. Academic Press, New York.

Millerd, A., and Scott, K. J. (1962). Respiration of the diseased plant. *Annu. Rev. Plant Physiol.* **13,** 559–574.

Mitchell, P. (1966). "Chemiosmotic Coupling in Oxidative and Photosynthetic Phosphorylation." Bodmin, Cornwall, England.

Morré, D. J. (1975). Membrane biogenesis. *Annu. Rev. Plant Physiol.* **26,** 441–481.

Morré, D. J., and Mollenhauer, H. H. (1973). The endomembrane concept: A functional integration of endoplasmic reticulum and Golgi apparatus. *In* "Dynamics of Plant Ultrastructure" (A. W. Robards, ed.), pp. 84–131. McGraw-Hill, New York.

Mount, M. S., Bateman, D. F., and Basham, H. G. (1970). Induction of electrolyte loss, tissue maceration, and cellular death of potato tissue by an endopolygalacturonate *trans*-eliminase. *Phytopathology* **60,** 924–931.

Nathansohn, A. (1904). Ueber die Regulation der Aufnahme anorganischer Salze durch die Knollen von *Dahlia. Jahrb. Wiss. Bot.* **39**, 607–664.

Novacky, A., and Hanchey, P. (1974). Depolarization of membrane potentials in oat roots treated with victorin. *Physiol. Plant Pathol.* **4**, 161–165.

Novacky, A., and Karr, A. L. (1977). Pathological alterations in cell membrane bioelectrical properties. *In* "Regulation of Cell Membrane Activities in Plants" (E. Marré and O. Ciferri, ed.), pp. 137–144. Elsevier, Amsterdam.

Osterhout, W. J. V. (1922). Some aspects of selective absorption. *J. Gen. Physiol.* **5**, 225–231.

Page, O. T. (1972). Effect of phytotoxins on the permeability of cell membranes. *In* "Phytotoxins in Plant Diseases" (R. K. S. Wood, A. Ballio, and A. Graniti, eds.), pp. 211–255. Academic Press, New York.

Patil, S. S. (1974). Toxins produced by phytopathogenic bacteria. *Annu. Rev. Phytopathol.* **12**, 259–279.

Pegg, G. F. (1976). The involvement of ethylene in plant pathogenesis. *Physiol. Plant Pathol.* **4**, 582–591.

Poole, R. J. (1973). The H^+ pump in red beet. *In* "Ion Transport in Plants" (W. P. Anderson, ed.), pp. 129–134. Academic Press, New York.

Pringle, R. B., and Scheffer, R. P. (1964). Host-specific plant toxins. *Annu. Rev. Phytopathol.* **2**, 133–156.

Rancillac, M., Kaur-Sawhney, R., Staskawicz, B., and Galston, A. W. (1976). Effects of cycloheximide and kinetin pretreatment on responses of susceptible and resistant *Avena* leaf protoplasts to the phytotoxin victorin. *Plant Cell Physiol.* **17**, 987–995.

Rawn, C. D. (1977). Simultaneous changes in the rate and pathways of glucose oxidation in victorin-treated oat leaves. *Phytopathology* **67**, 338–343.

Rich, S. (1964). Ozone damage to plants. *Annu. Rev. Phytopathol.* **2**, 252–266.

Rudolph, K. (1976). Non-specific toxins. *Physiol. Plant Pathol.* **4**, 270–315.

Saftner, R. A., and Evans, M. L. (1974). Selective effects of victorin on growth and the auxin response in *Avena. Plant Physiol.* **53**, 382–387.

Samaddar, K. R., and Scheffer, R. P. (1968). Effect of the specific toxin in *Helminthosporium victoriae* on host cell membranes. *Plant Physiol.* **43**, 21–28.

Samaddar, K. R., and Scheffer, R. P. (1970). Effects of *Helminthosporium victoriae* toxin on germination and aleurone secretion by resistant and susceptible seeds. *Plant Physiol.* **45**, 586–590.

Scheffer, R. P. (1976). Host-specific toxins in relation to pathogenesis and disease resistance. *Physiol. Plant Pathol.* **4**, 245–269.

Scheffer, R. P., and Yoder, O. C. (1972). Host-specific toxins and selective toxicity. *In* "Phytotoxins in Plant Disaeses" (R. K. S. Wood, A. Ballio, and A. Graniti, eds.), pp. 251–272. Academic Press, New York.

Schnepf, E. (1969). Sekretion und Exkretion bei Pflanzen. *Protoplasmatologia* **8**, 1–181.

Singer, S. J., and Nicolson, G. L. (1972). The fluid mosaic model of the structure of cell membranes. *Science* **175**, 720–731.

Spanswick, R. M., and Williams, E. J. (1964). Electrical potentials of Na, K, and Cl concentration in the vacuole and cytoplasm of *Nitella translucens. J. Exp. Bot.* **15**, 193–200.

Staehelin, L. A., and Probine, M. C. (1970). Structural aspects of cell membranes. *Adv. Bot. Res.* **3**, 1–52.

Steele, J. A., Uchytil, T. F., Durbin, R. D., Bhatnager, D., and Rich, D. H. (1976).

Chloroplast coupling factor I. A species-specific receptor for tentoxin. *Proc. Natl. Acad. Sci. U.S.A.* **73,** 2245–2248.

Strobel, G. A. (1974). Phytotoxins produced by plant parasites. *Annu. Rev. Plant Physiol.* **25,** 541–566.

Thatcher, F. S. (1939). Osmotic and permeability relations in the nutrition of fungus parasites. *Am. J. Bot.* **26,** 449–458.

Thatcher, F. S. (1942). Further studies of osmotic and permeability relations in parasitism. *Can. J. Res., Sect. C* **20,** 283–311.

Tipton, C. L., Mondal, M. H., and Uhlig, J. (1973). Inhibition of the K+ stimulated ATPase of maize root microsomes by *Helminthosporium maydis* race T pathotoxin. *Biochem. Biophys. Res. Commun.* **51,** 525–528.

Uritani, I. (1976). Protein metabolism. *Physiol. Plant Pathol.* **4,** 509–525.

Van Steveninck, R. F. M. (1975). The "washing" or "aging" phenomenon in plant tissues. *Annu. Rev. Plant Physiol.* **26,** 237–258.

Wheeler, H. (1969). Genetics of pathogenesis. *Proc. Symp. Potentials Crop Prot. 1969* pp. 9–13.

Wheeler, H. (1974). Cell wall and plasmalemma modifications in diseased and injured plant tissues. *Can. J. Bot.* **52,** 1005–1009.

Wheeler, H. (1975). "Plant Pathogenesis." Springer-Verlag, Berlin and New York.

Wheeler, H. (1976). Permeability alterations in diseased plants. *Physiol. Plant Pathol.* **4,** 411–429.

Wheeler, H., and Black, H. S. (1963). Effects of *Helminthosporium victoriae* and victorin upon permeability. *Am. J. Bot.* **50,** 686–693.

Wheeler, H., and Hanchey, P. (1968). Permeability phenomena in plant disease. *Annu. Rev. Phytopathol.* **6,** 331–350.

Wheeler, H., and Luke, H. H. (1963). Microbial toxins in plant disease. *Annu. Rev. Microbiol.* **17,** 223–242.

White, J. A., Calvert, D. H., and Brown, M. F. (1973). Ultrastructural changes in corn leaves after inoculation with *Helminthosporium maydis* race T. *Phytopathology* **63,** 296–300.

Wilson, C. L. (1973). A lysosomal concept for plant pathology. *Annu. Rev. Phytopathol.* **11,** 247–272.

Wood, R. K. S. (1972). The killing of plant cells by soft rot parasites. *In* "Phytotoxins in Plant Diseases" (R. K. S. Wood, A. Ballio, and A. Graniti, eds.), pp. 273–288. Academic Press, New York.

Chapter 16

Changes in Intermediary Metabolism Caused by Disease

JOSEPH KUĆ

I. INTRODUCTION

A. Randomness versus Integration

Specificity, order, and integration of structure and metabolic activity are characteristic of living organisms. The amino acids of proteins are almost exclusively L isomers, most sugars in nature are D isomers, and most unsaturated fatty acids are *cis* isomers. Is there some reason vital to life that necessitates these configurations? What chemical and biological properties would be possessed by proteins composed of D-amino acids and polysaccharides of L-sugars? Would proteins composed of D-amino acids function as enzymes? If you cannot answer these questions, your education should not be blamed. Clearly, selectivity is evident in the organic building blocks used to construct cells and in the compounds used to provide energy and direct the metabolic activity of the cell. Death represents the ultimate in randomness, and the onset of death, therefore, represents an increase in entropy.

Metabolic controls exist at various levels: the synthesis of enzymes, activity of enzymes, and the availability of substrates, coenzymes, cofactors, and energy. Environment can profoundly affect the metabolism of living cells as it does all chemical reactions; however, environmental changes potentially have a much greater impact on living cells since they can lead to a cascade effect. A slight change in a single reaction of an integrated and closely regulated system has the potential of leading to profound and often apparently unrelated changes in many processes. Much of the energy and metabolic activity of living cells is consumed, therefore, in resisting and adjusting to changes in the environment and to maintaining the essential aspects of a metabolic and structural "status quo," i.e., maintaining a high degree of specificity, order, and integration in structure and function. Organisms resist and adjust to changes imposed upon them by the environment.

As discussed more fully by Samborski *et al.* in Chapter 17, this volume, the highly integrated and interlocked structure and metabolism of living cells is probably controlled by nucleotide sequences in the nucleic acids of the nucleus, organelles, and cytoplasm. How did this highly specific chemical code arise? Lightning bolts passing through superheated nitrogen, carbon dioxide, carbon monoxide, water vapor, methane, and ammonia during the earth's inception hardly seem adequate to the task of producing the ultimate in organization—life. If carbon, hydrogen, oxygen, and nitrogen were combined at random to yield "primitive genetic information or protein," high randomness must have been converted to order and specificity. Stability of chemical structures might have provided the capacity to resist and adjust to changes imposed by the environment. In this way, an apparently random process might have led to an ordered system with high specificity. The lightning bolts may have provided the energy for organization.

In a blade of grass, a man, an elephant, or a flea, more than 1,000 chemical reactions are simultaneously taking place—all organized, integrated, and controlled. The reactions take place at mild temperatures, approximately neutral pH, and at atmospheric pressure. A single enzyme may be remarkable in its ability to catalyze a single chemical reaction under such relatively mild conditions. The totality of enzymes, structural components, and their organization and integration can accomplish much more than the sum of all individual contributions (holistic concept—the total effect is greater than the sum of each individual contribution).

The versatility of life is dependent upon organization and integration and this requires the expenditure of considerable energy.

It is not surprising, therefore, that disruptions of this highly integrated system, either by infection or injury, lead to many metabolic alterations. Some alterations are due directly to the disruption of integration (increased entropy of the system); others are due to the defense response of the living organism and the resultant repair mechanism which tends to bring structure and metabolism back to the "status quo." In this chapter I will consider the intermediary metabolism of plants as affected by infectious agents. Infection by a virus, for example, diverts the metabolism of the host from its normal functions to the synthesis of viral nucleic acid. The viral nucleic acid assumes a major role in directing the metabolic machinery of the cell and the plant in turn resists the change and makes the necessary adjustments. The adjustments may result in various metabolic changes, some far removed from the synthesis of viral nucleic acid. These changes may result in the survival of host cells and replication of the virus or localized death of host tissue and deactivation of virus. With bacteria and fungi, the adjustment or lack of adjustment of two distinct metabolic systems—the host and the infectious agent—occurs. Each attempts to maintain its individual "status quo" and thereby maintain itself and reproduce. Both respond to each other. One or the other may survive or some sort of metabolic compromise may be reached wherein both persist to one, to the other's, or to their mutual advantage. To study these metabolic adjustments and compromises, it is well to review briefly the general mechanisms for the control of intermediary metabolism and its integration. Several recent reviews present a thorough treatment of the subject (see Preiss and Kosuge, 1976; Daly, 1976a,b; Kosuge and Gilchrist, 1976).

B. Basics of Metabolic Control

Structure and metabolism in plants and animals are regulated by enzymes which in turn are regulated by other enzymes. This apparent paradox is thought to terminate with coded information in nucleotide sequences which determine the structure and concentration of enzymes. The evolutionary origin of the nucleotide sequences is unknown, but the entire picture is further complicated by the observation that the synthesis of the nucleotides and their sequencing is dependent upon enzymes and their regulation.

What are some of the ways by which metabolism is controlled? It is important to recognize that evidence for metabolic control *in vivo* is largely circumstantial and based on "common sense" extrapolation of simple experiments largely conducted *in vitro*. Though major pathways for biosynthesis, degradations, and energy release and conversion have been elucidated, the knowledge of their regulation is only in the earliest exploratory stage.

One means of metabolic regulation is the control of two competing processes—the synthesis and degradation of enzymes. Simple metabolites can either induce or repress enzyme synthesis. The classical examples of both are found in considering the growth of *Escherichia coli*, but the principles involved undoubtedly apply to bacterial and fungal phytopathogens. Wild-type *E. coli* at best produces trace quantities of β-galactosidase, the enzyme that hydrolyzes lactose to glucose and galactose. When *E. coli* cells are placed in a culture medium containing lactose as the sole source of carbon, they respond by rapidly synthesizing large amounts of the enzyme. The increase in synthesis is evident after several minutes and can reach concentrations 1,000 times that found in cells growing on a nutrient medium containing glucose as a sole source of carbon. Aspects of nitrogen metabolism in the same bacterium provide an example of the repression of enzyme synthesis. *Escherichia coli* utilizes ammonium salts as its sole source of nitrogen and, therefore, contains all the requisite enzymes for amino acid synthesis. The addition of histidine to the medium, however, represses the synthesis of the entire sequence of nine enzymes leading to the final synthesis of histidine (co-ordinate end-product repression). In another mechanism for the repression of enzyme synthesis, glucose represses the formation of β-galactosidase by *E. coli* even when lactose is present in the culture medium (catabolite repression). There is evidence (Pastan and Perlman, 1969) to support the participation of cyclic AMP in this process. As long as glucose is available, the cyclic AMP concentration remains low, minimizing the synthesis of enzymes capable of utilizing lactose or galactose. When the glucose level drops and an alternative carbon source is available, cyclic AMP increases and binds to the appropriate catabolite gene-activator protein. The cyclic AMP activator complex then binds to a promotor site and stimulates transcription of its operon.

The regulation of gene expression is more complex in eukaryotes than in prokaryotes. The response of eukaryotic cells to inducers is generally slower and not as great as that in prokaryotes. In cells of *Neurospora,* the induction of β-galactosidase takes longer and the increase in activity is approximately tenfold rather than 1,000-fold as with *E. coli.* A sequence of enzymes catalyzing a given metabolic pathway in eukaryotes is often not coded by adjacent genes in a single chromosome but appears scattered among several chromosomes. Nevertheless, induction of enzyme synthesis and its repression are evident in plants and animals. Some key enzymes which appear inducible in plants include: nitrate reductase (Hageman and Flesher, 1960; Beevers et al., 1965), ribulose-1,5-diphosphate carboxylase and other chloroplast enzymes (Bradbeer, 1973; Huffaker et al., 1966), and phenylalanine ammonia-lyase (Zucker, 1970, 1971, 1972). The induction of polysaccharide-degrading enzymes in fungal and bacterial phytopathogens has frequently been reported (Bateman, 1976). Recent work by Cooper and Wood (1975) demonstrated that both the induction of some carbohydrases and the repression of induction are controlled by the levels of various sugars in the culture medium. In some instances, reports of the induction of enzyme synthesis in phytopathogens, however, may be the result of marked catabolite repression of constitutive enzymes (Hsu and Vaughn, 1969; Moran and Starr, 1969). Hormones in higher plants and animals have profound effects on gene expression by influencing both enzyme synthesis and activity, and it is obvious that differentiation of eukaryotes is determined by regulated gene expression at varying stages of their development. In addition to enzyme synthesis, enzyme degradation or inactivation may be inducible and involve protein synthesis (Kahl, 1974; Zucker, 1970, 1971, 1972).

A hypothesis explaining induced resistance in plants was based on the derepression of enzyme synthesis (Hadwiger and Schwochau, 1969; Schwochau and Hadwiger, 1970). Schwochau and Hadwiger proposed that low concentrations of microbial metabolites, as well as of numerous other chemicals, derepress segments of DNA that control the synthesis of regulatory enzymes necessary for the accumulation of phytoalexins. It is also possible to speculate on the existence of other mechanisms by which the synthesis of enzymes determines the specificity of the plant–parasite interaction. Microbial metabolites, or metabolites released during the interaction of host and infectious agent, could repress synthesis of specific enzymes in the host. This in turn could suppress the defenses of the host and permit development of an infectious agent. Repression or induction of enzyme synthesis in the plant or the infectious agent could explain toxin production, toxin inactivation, sensitivity to toxin, inactivation of phytoalexins, and substrate utilization.

A finer and generally more responsive control of metabolism in prokaryotes and eukaryotes is achieved by regulating the activity rather than the amount of enzyme synthesized. The regulation of enzyme activity includes feedback inhibition (allosteric control), covalent modification, and the effects of temperature, coenzymes, cofactors, activators, and inhibitors. In many biosynthetic pathways the final product of a sequence may act as a specific inhibitor of the activity of the first or an early enzyme in the pathway. The rate of the entire sequence of reactions is determined, therefore, by the steady-state concentration of the final product. A classical example of this is the inhibition by L-isoleucine of threonine deaminase, the first enzyme in the biosynthetic pathway for L-isoleucine (Umbarger, 1956). The enzyme is allosteric and has both a catalytic binding site and a site for noncovalently binding the activity-regulating molecule (modulator). Some modulators are inhibitory, e.g., L-isoleucine for threonine deaminase, whereas other modulators are stimulatory, e.g., acetyl-CoA for pyruvate carboxylase. Not all modulators are final products of a reaction sequence, e.g., acetyl-CoA modulation of pyruvate carboxylase, and in some reactions a substrate for an enzyme can function as its positive (positive cooperativity) or negative (negative cooperativity) modulator. Many enzymes concerned with energy control in cells have ATP, ADP, or AMP as allosteric modulators and, therefore, would respond to either the ATP/AMP or ATP/ADP ratios in cells. The production of energy as ATP is regulated so that it is equal to the rate of utilization. Since the pool of ATP plus ADP plus AMP is constant, the utilization of ATP results in the formation of either ADP or AMP via phosphate transfer or hydrolysis. The term "energy charge" is a term commonly used to define the energy state of the adenylate system (Atkinson and Walton, 1967).

$$\text{Energy charge} = \frac{(\text{ATP} + \frac{1}{2}\,\text{ADP})}{(\text{ATP} + \text{ADP} + \text{AMP})}$$

In general, it has been suggested that enzymes involved with generating ATP are more active at low than at high energy charge values, whereas enzymes utilizing ATP for biosynthesis are more active at high and less active at low energy charge values. Many of the changes in intermediary metabolism that are associated with infectious disease or injury may be the result of adjustments to changes in energy charge.

Enzyme activity can also be influenced by covalent modification of regulatory enzymes. A classical example of this is glycogen phosphorylase found in mammals and some fungi (Fischer et al., 1971). The enzyme catalyzes the breakdown of the reserve polysaccharide, glycogen, to glucose 1-phosphate (G-1-P) and it exists in two forms, phosphorylase a, the more active form, and phosphorylase b, the less active form. Phosphorylase a is phosphorylated, and hydrolysis of the phosphate groups by a phosphatase gives rise to two molecules of phosphorylase b. The activity of phosphorylase b can be restored by phosphorylation catalyzed by phosphorylase kinase. Another type of covalent modification is the conversion of inactive forms of enzymes (zymogens) to active forms by proteolysis. The digestive enzymes pepsin, trypsin, and chymotrypsin arise from selective proteolysis of their respective zymogens. Compared to allosteric regulation, covalent modification of enzyme activity by an enzyme introduces a marked amplification to control. One molecule of phosphorylase kinase can convert thousands of molecules of inactive phosphorylase *b* to phosphorylase *a,* which in turn can catalyze the production of thousands of molecules of G-1-P from glycogen. With zymogen activation, one molecule of pepsin catalyzes the conversion of thousands of molecules of pepsinogen to pepsin, which can in turn react with more pepsinogen.

Temperature can also have profound cascade effects on the metabolism of a host–parasite complex. Reduced or enhanced activity of even a single enzyme can disrupt metabolic balance, and it is common with microorganisms and plants to observe profound morphological changes with relatively small changes in temperature. The susceptibility and resistance of plants to infectious agents can also be influenced by small changes in temperature. Some cultivars of wheat may be resistant to races of stem rust at one temperature range and susceptible to the same races at other temperatures. Phytoalexin accumulation can also be controlled within narrow ranges of temperature. Currier and Kuć (1975) found that rishitin accumulates in potato tuber slices inoculated with incompatible races of *Phytophthora infestans* or treated with sonicates of compatible or incompatible races of the fungus at 19°C. Accumulation of the terpenoid is markedly reduced at 25° and 30°C and not evident at 14° or 37°C. Steroid glycoalkaloid accumulation, however, increases between 19° to 30°C.

The concentrations of substrates, coenzymes, cofactors, activators, and inhibitors also influence enzyme activity, and the concentrations of these are controlled by enzymes (again, we return to the paradox). In studies of plant–parasite interactions, data often have been presented to relate enzyme activity to disease reaction. Accurate interpretation of such data is extremely difficult.

Do the changes in enzyme activity determine disease reaction or are they the result of the disease reaction? It is also difficult to relate enzyme activity *in vitro* to activity *in vivo.* Are the enzymes located in cells where they come in contact with substrates and/or other enzymes in a pathway? Are allosteric modulators and other inhibitors or activators present where the enzyme is compartmentalized in a cell? To further complicate metabolic control, it is astounding to see how little is known about enzyme degradation, both mechanisms for degradation and factors influencing its rate (Huffaker and Peterson, 1974). Since the quantity of enzymes is determined by both synthesis and degradation, the latter phenomenon deserves increased attention from the scientific community.

II. INJURY AND ITS REPAIR

A. Comparison of Disease and Injury

Diseases caused by fungi, bacteria, or virus damage plants. They induce diseases with many different symptoms and many types of injury. Some fungi and bacteria rot plant tissues, and symptoms are apparent as enlarging areas of cellular disorganization which are often discolored. In other cases, diseases may show as restricted lesions which vary considerably in size depending upon the disease. Some plants wilt as a result of disease, while others show abnormal growth (stunting, excessive elongation, loss of apical dominance, uncontrolled cell proliferation). Chlorosis, enhanced pigmentation, mottling, and leaf abscission also are common symptoms. In some instances the first sign of disease is the appearance of spores on infected tissues. Many different forms of injury can give rise to similar or very different symptoms, and different infectious agents can cause similar or very different symptoms. Viruses, bacteria, and fungi can cause lesions and all three can cause abnormal growth.

Are changes in the metabolic activity of infected plants evidence that the plant is responding to the infectious agent per se (its structural presence), the extracellular products it produces, or the damage it or its products produce? Are differences in the metabolism of plants infected by different organisms due to differences in the type of injury and the attempt by the plant to repair the injury? Can studies of changes in the intermediary metabolism of infected plants yield a clue as to the metabolic alterations that are unique to the disease? Certainly many environmental stresses cause shifts in plant metabolism which are similar to those described for infected plant tissues (Wender, 1970; Dugger and Ting, 1970).

Measurements of the levels of glucose and keto substances in blood and urine are commonly used as diagnostic tests for diabetes mellitus. The elevated concentrations of these substances indicate the key metabolic processes affected by the disease but they do not elucidate the mechanism whereby the complex of hormones, receptor sites, activators, inhibitors, and enzymes interact to produce the disease. The medical literature is full of reports describing the effects of cancer on the activity of enzymes and concentrations of intermediary metabolites. Though the reports may provide information useful for the diagnosis of cancer, do they reveal a great deal about the cause of the disease? The plant physiologist cannot explain the action of indoleacetic acid despite the glut of literature reporting changes in intermediary metabolism.

Nevertheless, some changes in intermediary metabolism are associated with certain diseases, and can be used for diagnosis, as discussed by McIntyre and Sands in Volume I of this treatise. These changes may also be important in determining the susceptibility or resistance of plants to disease.

I will not cite a long list of changes and shifts in the metabolism of infected plant tissues. Many may be little more than measurements of the response of the plant to various forms of injury and the repair mechanisms of the plant, and few experiments differentiate the metabolic contribution of host and pathogen. I will discuss in detail injury responses and possible repair mechanisms in the Irish potato tuber and relate these to other plants. A comprehensive review of the effects of wounding on the metabolism of plant storage tissues is available (Kahl, 1974).

The metabolism of storage organs, such as potato tuber and sweet potato root, and probably plant tissues in general, is activated by wounding. It is likely that gene depression results in increased synthesis of structural proteins and enzymes in intact cells, and this leads to greatly enhanced metabolic activity (Kahl, 1974). Injury also results, to varying degrees, in the decompartmentalization, release, and mixing of enzymes and metabolites, creating a transitory and relatively unregulated chemical environment which is quite different from that of intact cells. It is this transitory environment which probably initiates the metabolic alterations in the intact cells adjacent to the injury.

B. Source of Carbon Skeletons

Starch is the major storage compound for carbon skeletons as well as energy for repair when potato tubers and sweet potato roots are wounded. Therefore, a very fundamental control point for metabolism exists at the level of starch hydrolysis. In other plants the hydrolysis or modification of sucrose, sorbitol, raffinose, galactomannans, glucomannans, or fructosans may function as a key control point. Starch is degraded in the tuber when it is wounded by slicing, by emerging sprouts, or by infectious agents. For example, an incompatible race of *Phytophthora infestans,* which causes a rapid but limited death of tuber or leaf tissue, also rapidly initiates starch hydrolysis and many of the subsequent changes in intermediary metabolism associated with slice wounds. On the other hand, a compatible race ramifies through tissue for several days without markedly disrupting host function or structure. Does it also mobilize starch reserves, but unlike an incompatible race, utilize the carbohydrate as a source of energy and carbon skeletons without competition from wound-activated processes in the host? This simple but basic facet of intermediary metabolism deserves more attention.

Starch degradation is evident in potato tubers within 12 hr after a slice wound, and it appears to be mainly a result of enhanced phosphorylase activity (Kahl, 1974). The sugar entering the metabolic pool is, therefore, glucose 1-phosphate (G-1-P). The G-1-P is converted to glucose by the joint actions of phosphoglucomutase and glucose-6-phosphatase (Adams and Rowan, 1970; Kahl et al., 1969a; Lange et al., 1970). The activity of phosphoglucomutase decreases after wounding but is probably still sufficient for the rapid conversion of G-1-P to G-6-P. More than 50% of the extractable phosphorus is found in the hexose fraction of potatoes 24 hr after they are wounded, and ^{32}P incorporation into G-6-P is 100% greater at this time than in freshly cut slices (Laties, 1963; Loughman, 1960). Sucrose synthesis consumes some of the G-1-P (MacDonald and De Kock, 1958; Ap Rees and Beevers, 1960a), and sucrose synthetase could catalyze the production of UDP-glucose from sucrose for the synthesis of other sugars and polysaccharides. Enhanced invertase activity in numerous wounded storage tissues leads to hydrolysis of sucrose to glucose and fructose and therefore also minimizes the accumulation of sucrose (Kahl, 1974). Carbon skeletons, derived from G-1-P, are consumed for the synthesis in the periderm of new cell wall polymers (e.g., cellulose, polygalacturonates, galactans, mannans, arabans, xylans, laminarins, lipids, and protein) as part of the repair process.

Unlike potato tuber and sweet potato root which store glucose principally as starch, photosynthesis provides a direct source of carbon in the foliage of green plants. In contrast to respiration, rates of apparent photosynthesis, with few exceptions, generally decrease in infected leaves. This is not surprising for diseases in which damage to the tissue is obvious. It has been assumed, however, that photosynthesis is crucial for the development of obligate parasites (biotrophs), but most of the reports indicate that the apparent photosynthetic rate does not increase after infection with biotrophs. This is true even with low density infections which produce green islands (Daly, 1976b).

C. Glycolysis, Pentose Phosphate Pathway, and
Tricarboxylic Acid Cycle

The rapid production of G-1-P provides carbon skeletons for both glycolysis and the pentose phosphate pathway. Glycolysis is stimulated in sliced tubers to produce increased levels of fructose 6-phosphate, fructose 1,6-diphosphate, 3-phosphoglyceraldehyde, dihydroxyacetone phosphate, 3-phosphoglycerate, phosphoenolpyruvate, and pyruvate

(Kahl *et al.*, 1969a,b; Lange *et al.*, 1970). Enhanced glycolysis also occurs in beet and carrot roots wounded by slicing (Reed and Kolattukudy, 1966; Adams, 1970; Adams and Rowan, 1970). The activities of the enzymes of glycolysis, however, are not all enhanced. Phosphoglucoisomerase, phosphofructokinase, triosephosphate isomerase, and enolase activities decrease (Kahl *et al.*, 1969a,b; Lange *et al.*, 1970; Loughman, 1960; Ricardo and Ap Rees, 1972). Phosphoglyceromutase and pyruvate kinase activities increase, and aldolase activity remains constant in potato tuber tissue after wounding. Assuming that the measurements of enzyme activity have some relation to the activity *in vivo*, it would appear that the activities of the enzymes of glycolysis in intact tubers are not in general limiting factors for an enhanced glycolysis. This would also appear valid for enhanced glycolysis arising from wounding or infection. A possible key control of glycolysis is the rapid production of G-1-P and G-6-P; the metabolic machinery is in the tissue but the raw materials are limiting. The power to drive the machinery may be derived from the drain of ATP which is consumed during the myriad of synthetic reactions associated with tissue repair. Thus the energy charge during tissue repair would favor reactions which make energy available to the cell. The question of what starts the repair process still remains. It would appear that the sudden burst of synthetic activity also starts a process for its limitation.

Activity of the pentose phosphate shunt also increases shortly after wounding and the activity reaches a maximum after 24 hr (Kahl, 1974, Ap Rees and Beevers, 1960b; Kolattukudy and Reed, 1966; Laties, 1963; Norton, 1963). This is in good agreement with the marked enhancement of the activities of glucose-6-phosphate dehydrogenase and 6-phosphogluconate dehydrogenase (Kahl *et al.*, 1969a; Lange *et al.*, 1970; Sacher *et al.*, 1972; Tomiyama *et al.*, 1967). The activity of glucose-6-phosphate dehydrogenase increases more than 100% in potato slices 48 hr after slicing.

The enhanced activity of both enzymes is also reported for wounded beet, carrot, and sweet potato tissues (Kahl, 1974), but their activities decrease in rice infected with the mycoplasma that causes yellow dwarf disease (Tschen, 1976). Intermediates of the pentose phosphate shunt provide intermediates for the reported accumulation of ascorbate (Kahl, 1974; Johnson and Schaal, 1957), and the synthesis of nucleic acids, phenylalanine, tyrosine, histidine, tryptophan, C_6–C_3 phenolics, flavonoids, isoflavonoids, and hemicellulose.

Immediately after a potato tuber is wounded, there is weak activity of the tricarboxylic acid (TCA) cycle; however, as the wound ages TCA respiration increases markedly (Laties, 1963; Romberger and Norton,

1961; Reed and Kolattukudy, 1966). The TCA cycle makes a major contribution of carbon skeletons for the synthesis of the amino acids and proteins necessary for healing (Laties, 1964). The drain of organic acids to accommodate enhanced synthesis of amino acids and proteins is compensated for by the action of two carboxylases, phosphoenolpyruvate carboxylase and malic enzyme (Clegg and Wittingham, 1970; Romberger and Norton, 1961). The first carboxylase converts phosphoenolpyruvate to oxaloacetate, whereas the second is NADP-dependent and converts pyruvate to malate.

D. Lipid Metabolism

Presumably the effect on lipid metabolism is similar in both diseased tubers and wounded tubers. Slicing potato tubers releases active acyl hydrolases and lipoxygenases and both α- and β-oxidation are active in the wounded tissue (Kahl, 1974; Galliard, 1970, 1973, 1975; Laties et al., 1972; Wardale and Galliard, 1975). Shortly after tubers are sliced, however, fatty acid metabolism becomes directed toward the synthesis of suberin and mitochondrial and cytoplasmic membranes (Kahl, 1974). This in turn requires the synthesis of saturated and unsaturated fatty acids, phospholipids, glycolipids, and sterols. Though injury releases acyl hydrolases and lipoxygenases which may be destructive to membranes, the rapid synthesis of fatty acids, phospholipids, and glycolipids and their incorporation into membranes may be a key factor in the healing mechanism that returns tissue to "normalcy."

Another aspect of lipid metabolism in a wounded potato is the rapid accumulation of steroid glycoalkaloids, principally α-solanine and α-chaconine, beneath the cut surface. A high concentration of the steroid glycoalkaloids α-solanine and α-chaconine is localized in the outer millimeter (peel) of potato tubers. Analyses of four cultivars indicated an average of 0.55 and 0.03 mg/g fresh weight in the peel and peeled tubers, respectively (Shih, 1972). Soon after removal of the peel, secondary periderm formation occurs and the steroid glycoalkaloids accumulate to a depth of 1 or 2 mm (Ishizaka and Tomiyama, 1972). The concentration which accumulates in the top millimeter of peeled tissue may reach or surpass that in the peel (Shih and Kuć, 1973; Shih et al., 1973). Slices inoculated with incompatible races of P. infestans or some nonpathogens of potato accumulate low levels of the steroid glycoalkaloids and high levels of numerous sesquiterpenoids and a norsesquiterpenoid (Kuć et al., 1976). A considerable source of carbon and a large amount of energy are required to accomplish these syntheses. The rate of starch hydrolysis, glycolysis, and pentose pathway activities would all influence the production of acetyl-CoA, and the consumption of acetyl-CoA in the tricarboxylic acid cycle and in fatty acid synthesis would limit its availability for terpenoid synthesis.

Obviously key metabolic controls of terpenoid accumulation may be those which regulate the hydrolysis of starch and the energy charge in the cell. An additional control would be the regulation of enzyme activity in pathways leading to terpenoid synthesis. These enzymes would remove products from the economy of the cell much as war removes products from a world economy. Stimulation of the economy in war or metabolism during tissue repair (defense against disease) are both false indicators of prosperity, but the stimulation is necessary for defense. What is the signal (and perhaps there is more than one) that initiates the profound metabolic changes associated with the healing and that has as one consequence the increased accumulation of terpenoids? Perhaps it is CO_2. Slicing a potato tuber has been reported to release CO_2 immediately and levels above 5% have been reported in potato tuber and sweet potato root (Boswell and Whiting, 1940; Gerhardt, 1942; Whiteman and Schomes, 1945; Burton, 1951; Kahl, 1974). The release of CO_2 is temporary and ends within 1–2 hr. At this time, the respiratory activity of wounded tuber tissue is five to ten times that of tissue within the intact tuber. The signal for terpenoid synthesis and accumulation may be the sudden release of CO_2 which removes a brake to metabolic activity and activates phosphorylase. A high concentration of CO_2, in addition to serving as a possible brake for respiratory activity, also acts as a competitive inhibitor for ethylene (Varner and Ho, 1976; Pegg, 1976). Since damage to plant tissues also enhances ethylene production, many of the metabolic alterations associated with wounding or infection may be induced by ethylene. These alterations include the accumulation of phenylpropane phenolics, coumarins, isocoumarins, flavonoids, isoflavonoids, pterocarpans, and increased activities of phenylalanine ammonia lyase, polyphenol oxidase, and peroxidase. The question remains, however, of what initiates ethylene production in damaged or infected tissues. Again, we do not know.

Fatty acid synthesis markedly declines within 10–12 hr of wounding, and if the activity of glucose metabolism via the pentose pathway and glycolysis has not declined, acetyl-CoA may accumulate and be shunted off to the synthesis of terpenoids. The pathway for synthesis of α-solanine and α-chaconine probably follows the general pathway: hydroxymethylglutaryl-CoA $\rightarrow$ mevalonate $\rightarrow$ isopentenyl PP $\rightarrow$ squalene $\rightarrow$ cycloartenol $\rightarrow$ α-solanine and α-chaconine (Goodwin, 1973; Jadhav and Salankhe, 1975). The steroid glycoalkaloids may represent a trap for acetyl-CoA until periderm around the wound is sufficient to brake metabolism, though perhaps indirectly by its effect on CO_2 and ethylene. The steroid glycoalkaloids may, however, be more than a "metabolic safety valve." Their antibiotic activity and localization at the wounded surface may make them an important part of the wound repair process (Allen and Kuć, 1968).

E. Phenylpropanoid Metabolism

Synthesis of aromatic compounds in higher plants starts with the condensation of phosphoenolpyruvate and erythrose 4-phosphate to 3-deoxy-D-heptulosonate 7-phosphate, which is rapidly converted to 5-dehydroquinate, 5-dehydroshikimate, and shikimate. The sequence leads to the synthesis of aromatic amino acids, numerous phenylpropane phenolics, lignin, coumarins, flavonoids, and isoflavonoids (Brown, 1966; Yoshida, 1969; Grisebach and Barz, 1969). Accumulation of phenolics in wounded and infected plant tissues is well documented (Uritani et al., 1967; Kuć, 1966, 1972; Stoessl, 1970; Ingham, 1972). Chlorogenic acid and isochlorogenic acid are major phenolics which commonly accumulate in wounded or diseased plants of many different families. Smaller quantities of shikimate, phenylalanine, p-coumarate, quinate, caffeate, ferulate, 3-O-p-coumaroylquinate, p-courmaroylshikimate, caffeoylquinate, caffeoylshikimate, p-coumarylglucose, umbelliferone, scopoletin, and aescutetin have also been frequently reported (Kahl, 1974). With the exception of the initiating enzyme of the shikimate pathway, 3-deoxy-D-arabinoheptulosonic acid-7-phosphate synthetase, the activities of the other enzymes of the pathway are enhanced in a wounded sweet potato root (Kahl, 1974; Uritani et al., 1967).

Phenylalanine ammonia lyase (PAL) and tyrosine ammonia lyase are important enzymes linking the shikimate pathway to the synthesis of phenylpropane phenolics. The activity of PAL is enhanced after injury or infection in foliage and other plant tissues. Its activity commonly reaches a maximum 14–30 hr after injury and then drops to levels close to those found in uninjured tissue. The fall in PAL activity after the maximum appears dependent on protein synthesis (Zucker, 1970, 1971, 1972). The metabolic role and regulation of PAL has been well reviewed (Camm and Towers, 1973; Creasy and Zucker, 1974).

We may summarize wounding and healing as follows. To accommodate the myriad of biosyntheses needed to form new cells, the synthesis of proteins and nucleic acid is enhanced (Kahl, 1974). Enhanced starch degradation, consumption of ATP, and the enhanced activity of the pentose phosphate pathway, glycolysis, and the TCA cycle result in a marked increase in the respiration of wounded storage tissues. The respiratory increase is also probably characteristic of injury in many non-storage plant tissues (Ap Rees, 1966). The wound response may also explain the enhanced respiratory and biosynthetic activity of different plants and plant organs in various stages of disease (Farkas and Király, 1962; Wender, 1970; Dugger and Ting, 1970; Kuć, 1966, 1972; Ward and Stoessl, 1976). The wound response in infected tissue may be greater than that in sliced tissue, since the latter is a transitory trauma, whereas the trauma elicited by infection and some chemical toxicants is of longer duration. In addition, some metabolic alterations of infected tissue may be unique for the interaction since two different organisms, the plant and infectious agents, are influencing each other's metabolism. The wound response need not be limited to higher plants. It is probably also characteristic of bacteria and fungi. How do infectious agents respond to the stress of an unfavorable chemical or nutritive environment in plant tissues?

III. INTERMEDIARY METABOLISM OF INTERACTION

A. Phytoalexin Accumulation and Wounding

Though enhanced metabolism is a common phenomenon observed in plants receiving a very transitory trauma such as slicing, it may vary in characteristics in different plants or plant families. Thus, although enhanced glycolysis, pentose phosphate pathway metabolism, TCA cycle metabolism, and respiration in general increase in potato tuber and sweet potato root after slicing, the accumulation of steroid glycoalkaloids is observed only in potato slices.

The presence of an infectious agent elicits a response which in many instances is inseparable from a wound response to injury of long-term duration. Many simple chemical toxicants, ultraviolet radiation, as well as low and high molecular weight products of infectious agents elicit the accumulation of phytoalexins (Anderson-Prouty and Albersheim, 1975; Ayers et al., 1976; Kuć, 1976; Kuć et al., 1976; Stoessl et al., 1976; Van Etten and Pueppke, 1976). Viruses, which cause plant injury but do not synthesize polysaccharide elicitors or extracellular peptides per se, also elicit the accumulation of high levels of many phytoalexins (Kuć, 1972, 1975 ; Van Etten and Pueppke, 1976). All of these elicitors of phytoalexin accumulation cause trauma of longer duration than simple mechanical injury and its associated rapid repair. Infectious agents and some chemical toxicants would persist in tissues serving as a prolonged source of trauma, perhaps by eliciting continued ethylene production.

In addition, fungi and bacteria would provide a persistent drain on a plant's energy and reserve of carbon skeletons.

Sesquiterpenoids and norsesquiterpenoids accumulate in plants of the Solanaceae infected with a broad spectrum of infectious agents. Similarly, isoflavonoids accumulate in the Leguminosae, furanoterpenoids in Convolvulaceae, and isocoumarins in Umbelliferea. The structure of the compounds which accumulate appears more dependent upon the family and species within a family than on the infectious agent (Kuć, 1976; Kuć et al., 1976; Stoessl et al., 1976). Rishitin accumulates in potato tuber infected by numerous fungi and bacteria but not in pepper infected by the same infectious agents. Similarly, capsidiol accumulates in infected pepper fruit and not in potato, and ipomeamarone, a furanoterpenoid, accumulates in infected sweet potato root but not in infected potato tuber or pepper fruit. On the other hand, lubimin has been detected in infected potato, eggplant, and jimson weed (Datura stramonium).

What shifts in metabolism lead to the accumulation of phytoalexins? How does the transitory trauma-induced wound response influence and interact with the response to long-term trauma, including infection, which leads to the accumulation of phytoalexins? Is it possible that transitory trauma initiates a general activation of metabolism, perhaps due to a transitory burst of ethylene production, whereas prolonged trauma initiates, in addition, the synthesis or activation of specific enzymes leading to the accumulation of the stress metabolites called phytoalexins? The enhanced metabolism common to restricted injury provides carbon skeletons and energy for the pathways leading to the accumulation of the stress metabolites. Unlike the response to injury and infection of storage tissues, some diseases, however, characteristically cause the accumulation of starch in foliage. MacDonald and Strobel (1970) reported the accumulation of starch in the chloroplasts of leaves infected with Puccinia striiformis. The authors concluded that fluctuations in levels of metabolic modulators of ADP-glucose pyrophosphorylase regulate starch accumulation. Wheeler (1975) also described the accumulation of starch in the chloroplasts of mesophyll cells of susceptible oat leaves treated with victorin, a host-specific toxin produced by Helminthosporium victoriae. Starch was observed to disappear from guard cells of leaves treated with the toxin.

Does disease resistance depend upon the activation of a mechanism in the host which is unique for its role in defense against disease and which in turn is elicited by structural components or metabolites unique to the infectious agent? Despite all the research, no one has demonstrated in plants a metabolic process which is uniquely a defense reaction against disease.

B. Terpenoid Phytoalexins

The acetate–mevalonate pathway is activated following injury or infection of potato tubers (Shih and Kuć, 1973; Stoessl *et al.*, 1976; Uritani *et al.*, 1976). In the sweet potato, the activities of many of the enzymes of the pathway are enhanced: acetoacetyl-CoA synthetase, HMG-CoA synthetase, HMG-CoA reductase, mevalonate kinase, phosphomevalonate kinase, and pyrophosphomevalonate decarboxylase. The activation of the acetate–mevalonate pathway in some stressed plants is similar to the activation of HMG-CoA reductase in stressed animals (Dietschy and Wilson, 1970; Dietschy and Brown, 1974; Weis and Dietschy, 1975). In animals this may be a key to the synthesis of steroid hormones, and in plants it may result in the increased synthesis of terpenoid phytoalexins and terpenoid hormones such as gibberellic and abscisic acids. The contribution of gibberellins and abscisic acid to disease reaction, especially in tissues accumulating terpenoid phytoalexins, deserves increased attention.

Though the pathways for the biosynthesis of the norsesquiterpenoids and sesquiterpenoids of potato and for the furanoterpenoids of sweet potato are not completely elucidated, several have been suggested. Dehydroipomeamarone may be an immediate precursor for ipomeamarone on the pathway from farnesylpyrophosphate (Oguni and Uritani, 1974). Two schemes suggested by Stoessl *et al.* (1976) accommodate spirovetiva-1(10),11-dien-2-one (solavetivone), spirovetiva-1(10),3,11-trien-2-one (anhydro-β-rotunol), lubimin, rishitin, capsidiol, and phytuberin. In a preliminary report, Kalan and Osman (1976) observed that wounded potato tubers treated with solavetivone yielded lubimin and rishitin within 24 hr. A vetispirane, isolubimin, which had not been previously detected in potatoes infected with fungi, was also isolated, and it may be an intermediate in the biosynthesis of rishitin and lubimin. The concentration of solavetivone decreased concomitantly with increased concentration of isolubimin, and isolubimin was detected prior to lubimin and rishitin and decreased with increasing rishitin and lubimin concentration. The suppression by compatible races of *Phytophthora infestans* of both the hypersensitive reaction and accumulation of rishitin, phytuberin, and lubimin (Doke, 1975; Varns and Kuć, 1971, 1972) suggests a block which may occur in the conversion of solavetivone to lubimin. An aspect of metabolism unique to interaction is the influence of compatible and incompatible races of *P. infestans* on terpenoid accumulation in tuber slices. Potato cultivars containing major R genes for resistance rapidly accumulate numerous sesquiterpenoids, including rishitin, lubimin, and phytuberin, when infected with incompatible but not compatible races of *P. infestans.* All cultivars, including those lacking major R genes, rapidly accumulate terpenoids when treated with cell-free sonicates from any race of the fungus (Tomiyama *et al.*, 1968; Varns *et al.*, 1971a,b). Rishitin, lubimin, and phytuberin accumulation are markedly suppressed, however, when inoculation with a compatible race of *P. infestans* precedes inoculation with an incompatible race or treatment with cell-free sonicates of the fungus (Varns and Kuć, 1971, 1972). A polysaccharide isolated from the zoospores or mycelium of compatible races may be the specific suppressor of terpenoid accumulation and may, therefore, determine susceptibility (Doke *et al.*, 1977; Garas *et al.*, 1977).

C. Isoflavonoid Phytoalexins

The accumulation of isoflavonoid phytoalexins after infection is probably an enhancement of the wound response. Many substances elicit the accumulation of isoflavonoid phytoalexins. An excellent review of the isoflavonoid phytoalexins in diseased plants was written by Van Etten and Pueppke (1976).

The biosynthetic pathway of isoflavonoids can be divided into three stages: (1) early reactions which are shared with other secondary metabolites, (2) reactions mutual to flavonoid and isoflavonoid biosynthesis, and (3) reactions unique to specific isoflavonoids. Only the first stage of the pathway is well understood (Van Etten and Pueppke, 1976).

The basic C_{15} molecular skeleton for isoflavonoids arises from two distinct metabolic pathways, i.e., the acetate–malonate and shikimate pathways. Radioactive labels from acetate and shikimate precursors are incorporated into the A ring and the B and heterocyclic rings, respectively (Grisebach, 1965). Both pathways are ubiquitous in higher plants and simple mechanical injury activates both pathways.

The enzymes that hydroxylate and activate *t*-cinnamic acid prior to its condensation with acetate units have been intensively studied in parsley (Grisebach and Hahlbrock, 1974; Hahlbrock *et al.*, 1971). Their activity appears modulated by light, but infection and injury also enhance activity. Condensation of the activated phenylpropanoid unit with three 2-carbon acetate-derived units is the first unique step of flavonoid biosynthesis. Although it is possible that the initial C_{15} intermediate is a polyketide, the initial C_{15} structure is a chalcone or its isomeric flavone (Grisebach, 1965; Grisebach and Barz, 1969).

It is unlikely that enzymes required early in the biosynthesis of isoflavonoids regulate the accumulation of the isoflavonoid phytoalexins in infected tissues. Partridge and Keen (1977) reported that soybean cultivars monogenically resistant or susceptible to *Phytophthora megasperma* var. *sojae* both showed activation of PAL, chalcone-flavonone isomerase, and peroxidase after wounding or inoculation with the fungus. The similarity of the enzymatic responses in wounded control and inoculated plants suggests that the enhanced enzyme activities are caused primarily by wounding. The enzymes apparently do not have a regulatory role in determining the rapid accumulation of glyceollin in the infected resistant, but not susceptible, plants. Enzyme regulation related to the magnitude of response in plant–parasite interactions, though evident with isoflavonoids such as glyceollin, daidzein, and coumesterol (Keen *et al.*, 1972), apparently occurs after the conversion of chalcone to flavonone.

D. Degradation of Phytoalexins

It has become increasingly apparent that phytoalexins, rishitin, lubimin, capsidiol, ipomeamarone, phaseollin, and probably pisatin are not stable end products of metabolism (Kuć, 1975; Haard and Weis, 1976; Horikawa *et al.*, 1976; Stoessl *et al.*, 1977). These phytoalexins generally reach a peak of accumulation 96–120 hr after infection or treatment with elicitors and then their concentrations decrease markedly until they are barely detectable 7–10 days after elicitation. The accumulation of phytoalexins and other stress metabolites, therefore, may be as much a result of decreased degradation as a result of increased synthesis. Elicitors for the accumulation of these compounds may be inhibitors of degradation or elicitors of increased synthesis. Both synthesis and degradation may be sensitive to inhibitors of protein synthesis, and both processes are likely to occur at the same time.

E. Blocking a Common Metabolic Pathway

The accumulation of ornithine in bean tissue infected with *Pseudomonas phaseolicola* is an example of interference in intermediary metabolism that is unique for a plant–pathogen interaction for which the accumulation of phytoalexins has not been reported. Patel and Walker (1963) reported that ornithine accumulated and arginine decreased in bean leaf tissues infected with compatible races of *P. phaseolicola*. In subsequent reports Patil and his co-workers (Patil, 1974, 1976) demonstrated that the extracellular toxin, phaseotoxin, produced by *P. phaseolica*, is a specific and potent inhibitor of ornithine carbamoyltransferase. This enzyme catalyzes the conversion of carbamoyl phosphate and L-ornithine to citrulline, which is an intermediate in the biosynthesis of L-arginine. The toxin appears to be a competitive inhibitor of the carbamyl site of ornithine carbamoyl transferase. The application of L-citrulline or arginine to the center of toxin-induced chlorotic haloes caused a reversal of chlorosis. These studies are unique in that a mechanism for pathogenesis has been established, and the key observation for establishment of the mechanism was the observation of a change in a common pathway in intermediary metabolism—the biosynthesis of arginine via the ornithine cycle.

IV. SUMMARY

The literature contains numerous reports of metabolic alterations in infected plants. In most instances it is not possible to determine which or whether alterations are the result of the disease, the cause of the disease, or the basis of a resistance mechanism. The metabolic responses to injury, including repair mechanisms, have many similarities to the metabolic response reported in infected plants. An important question remains unanswered: does disease resistance in plants depend upon the activation of a mechanism in the host which is unique for its role in defense against disease and in turn is elicited by structural components or metabolites unique to the infectious agent?

Acknowledgments

Journal paper no. 77–11–210 of the Kentucky Agricultural Experiment Station, Lexington, Kentucky, 40506. The author's work in this chapter was supported in part by a grant from the Herman Frasch Foundation and Grant no. 316–15–51 of the Cooperative State Research Service, United States Department of Agriculture.

References

Adams, P. B. (1970). Effect of adenine nucleotides on levels of glycolytic intermediates during the development of induced respiration in carrot slices. *Plant Physiol.* **45**, 500–504.

Adams, P. B., and Rowan, K. S. (1970). Glycolytic control of respiration during aging of carrot root tissue. *Plant Physiol.* **45**, 490–494.

Allen, E., and Kuć, J. (1968). α-Solanine and α-chaconine as fungitoxic compounds in extracts of Irish potato tubers. *Phytopathology* **58**, 776–781.

Anderson-Prouty, A., and Albersheim, P. (1975). Isolation of a pathogen-synthesized fraction rich in glucan that elicits a defense response in the pathogen's host. *Plant Physiol.* **56**, 286–291.

Ap Rees, T. (1966). Evidence for the widespread occurrence of induced respiration in slices of plant tissues. *Aust. J. Biol. Sci.* **19**, 981–990.

Ap Rees, T., and Beevers, H. (1960a). Pathways of glucose dissimilation in carrot slices. *Plant Physiol.* **35**, 830–838.

Ap Rees, T., and Beevers, H. (1960b). Pentose phosphate pathway as a major component of induced respiration of carrot and potato slices. *Plant Physiol.* **35**, 839–847.

Atkinson, D. E., and Walton, G. M. (1967). Adenosine triphosphate conservation in metabolic regulation. *J. Biol. Chem.* **242**, 3239–3241.

Ayers, A., Ebel, J., Finelli, F., Berger, N., and Albersheim, P. (1976). Quantitative assays of extracellular activity and characterization of the elicitor present in the extracellular medium of cultures of *Phytophthora megasperma* var. *sojae*. *Plant Physiol.* **57**, 751–759.

Bateman, D. F. (1976). Plant cell wall hydrolysis by pathogens. *In* "Biochemical Aspects of Plant–Parasite Relationships" (J. Friend and D. R. Threlfall, eds.), pp. 79–103. Academic Press, New York.

Beevers, L., Schrader, L. E., Flesher, D., and Hageman, R. H. (1965). The role of light and nitrate in the induction of nitrate reductase in radish cotyledons and maize seedlings. *Plant Physiol.* **40**, 691–698.

Boswell, J., and Whiting, G. (1940). Observations on the anaerobic respiration of potato tubers. *Ann. Bot.* (*London*) [N.S.] **4**, 257–268.

Bradbeer, J. W. (1973). The synthesis of chloroplast enzymes. *In* "Biosynthesis and its Control in Plants" (B. V. Milborrow, ed.), pp. 279–302. Academic Press, New York.

Brown, S. A. (1966). Lignins. *Annu. Rev. Plant Physiol.* **17**, 223–244.

Burton, W. (1951). Studies on the dormancy and sprouting of potatoes. *New Phytol.* **50**, 287–296.

Camm, E. L., and Towers, G. H. N. (1973). Phenylalanine ammonia lyase. *Phytochemistry* **12**, 961–973.

Clegg, C. J., and Wittingham, C. P. (1970). Dark CO_2 fixation by potato tuber tissue. *Phytochemistry* **9**, 279–287.

Cooper, R. M., and Wood, R. K. S. (1975). Regulation of synthesis of cell wall degrading enzymes by *Verticillium albo-atrum* and *Fusarium oxysporum* f. sp. *lycopersici*. *Physiol. Plant Pathol.* **5**, 135–156.

Creasy, L. L., and Zucker, M. (1974). Phenylalanine ammonia-lyase and phenolic metabolism. *Recent Adv. Phytochem.* **8**, 1–19.

Currier, W. W., and Kuć, J. (1975). Effect of temperature on rishitin and steroid glycoalkaloid accumulation in potato tuber. *Phytopathology* **65**, 1194–1197.

Daly, J. M. (1976a). Some aspects of host–pathogen interactions. In "Physiological Plant pathology" (R. Heitefuss and P. H. Williams, eds.) vol. **4**, pp. 27–50. Springer-Verlag, N. Y.-Heidelberg.

Daly, J. M. (1976b). The carbon balance of diseased plants: Changes in respiration, photosynthesis, and translocation. *In* "Physiological Plant Pathology" (R. Heitefuss and P. H. Williams, ed.) vol. **4**, pp. 450–479. Springer-Verlag, N. Y.-Heidelberg.

Dietschy, J. M., and Brown, M. (1974). Effect of alterations of the specific activity of the intracellular acetyl CoA pool on apparent rates of hepatic cholesterogenesis. *J. Lipid Res.* **15**, 508–516.

Dietschy, J. M., and Wilson, J. D. (1970). Regulation of cholesterol metabolism. *N. Engl. J. Med.* **282**, 1128–1138, 1179–1183, and 1241–1249.

Doke, N. (1975). Prevention of the hypersensitive reaction of potato cells to infection with an incompatible race of *Phytophthora infestans* by constituents of the zoospores. *Physiol. Plant Pathol.* **7**, 1–7.

Doke, N., Garas, N., and Kuć, J. (1977). Partial characterization and mode of action of the hypersensitivity-inhibiting factor isolated from *Phytophthora infestans*. *Program 69th Annu. Meet. Am. Phytopathol. Soc.* Abstract No. 383, p. 219.

Dugger, W. M., and Ting, I. P. (1970). Physiological and biochemical effects of air pollution oxidants in plants. *Recent Adv. Phytochem.* **3**, 31–58.

Farkas, G. L., and Király, Z. (1962). Role of phenolic compounds in the physiology of plant disease and disease resistance. *Phytopathol. Z.* **44**, 105–150.

Fischer, E. H., Heilmeyer, L. M., and Haschke, R. H. (1971). Phosphorylase and the control of glycogen degradation. *Curr. Top. Cell. Regul.* **4**, 211–251.

Galliard, T. (1970). The enzymic breakdown of lipids in potato tuber by phospholipid and galactolipid-acyl hydrolase activities and lipoxygenase. *Phytochemistry* **9**, 1725–1734.

Galliard, T. (1973). Lipids of potato tubers. *J. Sci. Food Agric.* **24**, 617–632.

Galliard, T. (1975). Degradation of plant lipids by hydrolytic and oxidative enzymes. In "Rec. Adv. Chem. and Biochem. of Lipids" (T. Galliard and E. Mercer, ed.), pp. 319–357. Academic Press, London.

Garas, N., Doke, N., and Kuć, J. (1977). Extraction of mycelial components from *Phytophthora infestans* that inhibit the accumulation of terpenoids in potato tuber tissues. *Program 69th Annu. Meet. Am. Phytopathol. Soc.* Abstract No. 399, p. 225.

Gerhardt, F. (1942). Simultaneous measurement of the carbon dioxide and organic volatiles of the internal atmosphere of fruits and vegetables. *Agric. Res.* **64**, 207–219.

Goodwin, T. W. (1973). Recent developments in the biosynthesis of plant triterpenes. *Recent Adv. Phytochem.* **6**, 97–115.

Grisebach, H. (1965). Biosynthesis of flavonoids. In "Chemistry and Biochemistry of Plant Pigments" (T. W. Goodwin, ed.), 1st ed., pp. 279–308. Academic Press, New York.

Grisebach, H., and Barz, W. (1969). Biochemic der flavonoide. *Naturwissenschaften* **56**, 538–544.

Grisebach, H., and Hahlbrock, K. (1974). Enzymology and regulation of flavonoid and lignin biosynthesis in plants and plant cell suspension cultures. *Recent Adv. Phytochem.* **8**, 21–52.

Haard, N., and Weiss, P. (1976). Influence of exogenous ethylene on ipomeamarone accumulation in black rot infected sweet potato roots. *Phytochemistry* **15**, 261–262.

Hadwiger, L. A., and Schwochau, M. E. (1969). Host resistance—an induction hypothesis. *Phytopathology* **59**, 223–227.

Hageman, R. H., and Flesher, D. (1960). Nitrate reductase activity in corn seedlings as affected by light and nitrate content of nutrient media. *Plant Physiol.* **35**, 700–708.

Hahlbrock, K., Ebel, J., Ortmann, R., Sutter, A., Wellman, E., and Grisebach, H. (1971). Regulation of enzyme activities related to the biosynthesis of flavone glycosides in cell suspension cultures of parsley (*Pentroselinum hortense*). *Biochim. Biophys. Acta.* **244**, 7–15.

Horikawa, T., Tomiyama, K., and Doke, N. (1976). Accumulation and transformation of rishitin and lubimin in potato tuber tissue infected by an incompatible race of *Phytophthora infestans*. *Phytopathology* **66**, 1186–1191.

Hsu, E. I., and Vaughn, R. H. (1969). Production and catabolite repression of the constitutive polygalacturonic acid *trans*-eliminase of *Aeromonas liquefaciens*. *J. Bacteriol.* **98**, 172–181.

Huffaker, R. C., and Peterson, L. W. (1974). Protein turnover in plants and possible means of its regulation. *Annu. Rev. Plant Physiol.* **25**, 363–392.

Huffaker, R. C., Obendorf, R. L., Keller, C. J., and Kleinkopf, G. E. (1966). Effects of light intensity on photosynthetic carboxylative phase enzymes and chlorophyll synthesis in greening leaves of *Hordeum vulgare* L. *Plant Physiol.* **41**, 913–918.

Ingham, J. (1972). Phytoalexins and other natural products as factors in plant disease resistance. *Bot. Rev.* **38**, 343–424.

Ishizaka, N., and Tomiyama, K. (1972). The effect of wounding on infection by *Phytophthora infestans* on the contents of terpenoids in potato tubers. *Plant Cell Physiol.* **13**, 1053–1063.

Jadhav, S. J., and Salankhe, P. K. (1975). Formation and control of chlorophyll and glycoalkaloid in tubers of *Solanum tuberosum* L. and evaluation of glycoalkaloid toxicity. *Adv. Food Res.* **21**, 307–354.

Johnson, S., and Schaal, L. A. (1957). Accumulation of phenolic substances and ascorbic acid in potato tuber tissue upon injury and their possible role in disease resistance. *Am. Potato J.* **34**, 200–209.

Kahl, G. (1974). Metabolism in plant storage tissue slices. *Bot. Rev.* **40**, 263–314.

Kahl, G., Lange, H., and Rosenstock, G. (1969a). Substratspiegel, enzymaktivitaten und genetische regulation nach derepression in pflanzlichen speichergeweben. *Z. Naturforsch., Teil B.* **24**, 911–918.

Kahl, G., Lange, H., and Rosenstock, G. (1969b). Regulation glykotytischen umsatzes durch synthese und abbau von enzymen. *Z. Naturforsch., Teil B.* **24**, 1544–1549.

Kalan, E. B., and Osman, S. F. (1976). Isolubimin: A possible precursor of lubimin in infected potato slices. *Phytochemistry* **15**, 775–776.

Keen, N. T., Zake, A. I., and Sims, J. J. (1972). Biosynthesis of hydroxyphaseollin and related isoflavonoids in diesase resistant soybean hypocotyls. *Phytochemistry* **11**, 1031–1039.

Kolattukudy, P. E., and Reed, D. J. (1966). Metabolism of red beet slices. II. Effects of aging. *Plant Physiol.* **41**, 661–669.

Kosuge, T., and Gilchrist, D. G. (1976). Metabolic regulation in host–parasite interactions. *In* "Physiological Plant Pathology" (R. Heitefuss and P. H. Williams, ed.), vol. 4, pp. 679–702. Springer-Verlag, N.Y.-Heidelberg.

Kuć, J. (1966). Resistance of plants to infectious agents. *Annu. Rev. Microbiol.* **20**, 337–364.

Kuć, J. (1972). Phytoalexins. *Annu. Rev. Phytopathol.* **10**, 207–232.

Kuć, J. (1975). Teratogenic constituents of potatoes. *Recent Adv. Phytochem.* **9**, 139–150.

Kuć, J. (1976). Phytoalexins and the specificity of plant–parasite interaction. *In* "Specificity in Plant Disease" (R. K. S. Wood and A. Graniti, eds.), pp. 252–271. Plenum, New York.

Kuć, J., Currier, W., and Shih, M. (1976). Terpenoid phytoalexins. *In* "Biochemical Aspects of Plant–Parasite Relationships" (J. Friend and D. Threlfall, eds.), pp. 225–257. Academic Press, New York.

Lange, H., Kahl, G., and Rosenstock, G. (1970). Enzymaktivitaten und intermediatspiegel des glucosekatabolismus bei proliferierenden und suberin synthetisieren speicherparenchymzellen von *Solanum tuberosum* L. *Planta* **91**, 18–31.

Laties, G. G. (1963). Control of respiratory quality and magnitude during development. *In* "Control Mechanisms in Respiration and Fermentation" (B. Wright, ed.), pp. 129–155. Ronald Press, New York.

Laties, G. G. (1964). The onset of tricarboxylic acid cycle activity with aging in potato slices. *Plant Physiol.* **39**, 654–663.

Laties, G. G., Hoelle, C., and Jacobson, B. (1972). α-Oxidation of endogenous fatty acids in fresh potato slices. *Phytochemistry* **11**, 3411–3413.

Loughman, B. C. (1960). Uptake and utilization of phosphate associated with respiratory changes in potato tuber slices. *Plant Physiol.* **35**, 418–424.

MacDonald, I. R., and De Kock, P. C. (1958). Temperature control and metabolic drifts in aging disks of storage tissue. *Ann. Bot.* (*London*) [N.S.] **22**, 429–448.

MacDonald, P. W., and Strobel, G. A. (1970). Adenine diphosphate glucose pyrophosphorylase control of starch accumulation in rust-infected wheat leaves. *Plant Physiol.* **46**, 126–135.

Moran, F., and Starr, M. P. (1969). Metabolic regulation of polygalacturonic acid *trans*-eliminase in *Erwinia. Eur. J. Biochem.* **11**, 291–295.

Norton, G. (1963). The respiratory pathways in potato tubers. *In* "The Growth of the Potato" (J. D. Ivins and F. L. Milthorpe, eds.), pp. 148–159. Butterworth, London.

Oguni, I., and Uritani, I. (1974). Dehydroipomeamarone as an intermediate in the biosynthesis of ipomeamarone, a phytoalexin from sweet potato root infected with *Ceratocystis fimbriata. Plant Physiol.* **53**, 649–652.

Partridge, J. E., and Keen, N. T. (1977). Soybean phytoalexins: Rates of synthesis not regulated by activation of initial enzymes in flavonoid biosynthesis. *Phytopathology* **67**, 50–55.

Pastan, I., and Perlman, R. (1969). Cyclic AMP in bacteria. *Nature* (*London*) **169**, 339–344.

Patel, P. N., and Walker, J. C. (1963). Changes in free amino acids and amide content of resistant and susceptible beans after infection with the halo blight organism. *Phytopathology* **53**, 522–528.

Patil, S. S. (1974). Toxins produced by phytopathogenic bacteria. *Annu. Rev. Phytopathol.* **12**, 259–279.

Patil, S. S. (1976). Mode of action of phaseotoxin. *In* "Biochemistry and Cytology of Plant–Parasite Interaction" (K. Tomiyama *et al.*, eds.), pp. 102–111. Am. Elsevier, New York.

Pegg, G. F. (1976). The envolvement of ethylene in plant pathogenesis. *In* "Physiological Plant Pathology" (R. Heitefuss and P. H. Williams, ed.) vol. **4**, pp. 582–591. Springer-Verlag, N.Y.-Heidelberg.

Preiss, J., and Kosuge, T. (1976). Regulation of enzyme activity in metabolic pathways. *In* "Plant Biochemistry" (J. Bonner and J. E. Varner, eds.), 3rd ed., pp. 277–336. Academic Press, New York.

Reed, D. J., and Kolattukudy, P. E. (1966). Metabolism of red beet slices I. Effects of washing. *Plant Physiol.* **41**, 653–660.

Ricardo, C. P., and Ap Rees, T. (1972). Activities of key enzymes of carbohydrate oxidation in disks of carrot storage tissue. *Phytochemistry* **11**, 623–626.

Romberger, J. A., and Norton, G. (1961). Changing respiratory pathways in potato tuber slices. *Plant Physiol.* **36**, 20–29.

Sacher, J. A., Towers, G. H., and Davies, D. D. (1972). Effect of light and aging on enzymes, particularly phenylalanine ammonia lyase in discs of storage tissue. *Phytochemistry* **11**, 2383–2391.

Schwochau, M. E., and Hadwiger, L. A. (1970). Induced host resistance—a hypothesis derived from studies of phytoalexin production. *Recent Adv. Phytochem.* **3**, 181–189.

Shih, M. (1972). The accumulation of isoprenoids and phenols and its control as related to the interaction of potato (*Solanum tuberosum* L.) with *Phytophthora infestans*. Ph.D. Thesis, Purdue University, Lafayette, Indiana.

Shih, M., and Kuć, J. (1973). Incorporation of ^{14}C from acetate and mevalonate into rishitin and steroid glycoalkaloids by potato slices inoculated with *Phytophthora infestans*. *Phytopathology* 63, 826–829.

Shih, M., Kuć, J., and Williams, E. B. (1973). Suppression of steroid glycoalkaloid accumulation as related to rishitin accumulation in potato tubers. *Phytopathology* 63, 821–826.

Stoessl, A. (1970). Antifungal compounds produced by higher plants. *Recent Adv. Phytochem.* 3, 143–180.

Stoessl, A., Stothers, J., and Ward, E. (1976). Sesquiterpenoid stress compounds of the *Solanaceae*. *Phytochemistry* 15, 855–872.

Stoessl, A., Robinson, J. R., Rock, G. L., and Ward, E. W. B. (1977). Metabolism of capsidiol by sweet pepper tissue: Some possible implications for phytoalexin studies. *Phytopathology* 67, 64–66.

Tomiyama, K., Sakai, R., Otani, Y., and Takemori, T. (1967). Phenol metabolism in relation to disease resistance of potato tuber. I. Activities of phenol oxidase and some enzymes related to glycolysis as affected by chlorogenic acid treatment. *Plant Cell Physiol.* 8, 1–13.

Tomiyama, K., Sakuma, T., Ishizaka, N., Sato, N., Katsui, N., Takasugi, M., and Masamune, T. (1968). A new antifungal substance isolated from potato tuber tissue infected by pathogens. *Phytopathology* 58, 115–116.

Tschen, J. (1976). Changes in respiration and carbohydrate metabolism of rice plants infected by mycoplasma. *In* "Biochemistry and Cytology of Plant–Parasite Interaction" (K. Tomiyama *et al.*, eds.), pp. 236–238. Am. Elsevier, New York.

Umbarger, H. E. (1956). Evidence for negative-feedback in the biosynthesis of isoleucine. *Science* 123, 848.

Uritani, I., Asahi, T., Minamikawa, T., Hyodo, H., Oshima, K., and Kouima, M. (1967). The relation of metabolic changes in infected plants to changes in enzymatic activity. *In* "The Dynamic Role of Molecular Constituents in Plant–Parasite Interaction" (C. J. Mirocha and I. Uritani, eds.), pp. 342–356. Am. Phytopathol. Soc., St. Paul, Minnesota.

Uritani, I., Oba, K., Kouima, M., Kim, W. K., Oguni, I., and Suzuki, H. (1976). Primary and secondary defense actions of sweet potato in response to infection by *Ceratocystis fimbriata* strains. *In* "Biochemistry and Cytology of Plant–Parasite Interaction" (K. Tomiyama *et al.*, eds.), pp. 239–252. Am. Elsevier, New York.

Van Etten, H. D., and Pueppke, S. G. (1976). Isoflavonoid phytoalexins. *In* "Biochemical Aspects of Plant–Parasite Relationships" (J. Friend and D. R. Threlfall, eds.), pp. 239–289. Academic Press, New York.

Varner, J. E., and Ho, D. T. (1976). Hormones. *In* "Plant Biochemistry" (J. Bonner and J. E. Varner, eds.), 3rd ed., pp. 713–723. Academic Press, New York.

Varns, J., and Kuć, J. (1971). Suppression of rishitin and phytuberin accumulation and hypersensitive response in potato by compatible races of *Phytophthora infestans*. *Phytopathology* 61, 178–181.

Varns, J., and Kuć, J. (1972). Suppression of the resistance responses as an active mechanism for susceptibility in the potato–*Phytophthora infestans* interaction.

In "Phytotoxins in Plant Diseases" (R. K. S. Wood, A. Ballio, and A. Graniti, eds.), pp. 465–468. Academic Press, New York.

Varns, J., Kuć, J., and Williams, E. B. (1971a). Terpenoid accumulation as a biochemical response of the potato tuber to *Phytophthora infestans*. *Phytopathology* **61**, 174–177.

Varns, J., Currier, W., and Kuć, J. (1971b). Specificiity of rishitin and phytuberin accumulation by potato. *Phytopathology* **61**, 968–971.

Ward, E., and Stoessl, A. (1976). On the question of "elicitors" or "inducers" in incompatible interactions between plants and fungal pathogens. *Phytopathology* **66**, 940–941.

Wardale, D., and Galliard, T. (1975). Subcellular localization of lipoxygenase and lipolytic acid and hydrolase enzymes in plants. *Phytochemistry* **14**, 2323–2329.

Weis, H. J., and Dietschy, J. M. (1975). The interaction of various control mechanisms in determining the rate of hepatic cholesterogenesis in the rat. *Biochim. Biophys. Acta* **398**, 315–324.

Wender, S. H. (1970). Effects of some environmental stress factors on certain phenolic compounds in tobacco. *Recent Adv. Phytochem.* **3**, 1–29.

Wheeler, H. (1975). *In* "Plant Pathogenesis" (G. W. Thomas, B. R. Sabey, Y. Vaadea, and L. D. Van Vleck, eds.), p. 54. Springer-Verlag, Berlin and New York.

Whiteman, T., and Schomes, H. (1945). Respiration and internal gas content of injured sweet potato roots. *Plant Physiol.* **20**, 171–182.

Yoshida, S. (1969). Biosynthesis and conversion of aromatic amino acids in plants. *Annu. Rev. Plant Physiol.* **20**, 41–62.

Zucker, M. (1970). Rate of phenylalanine ammonia-lyase synthesis in darkness. *Biochim. Biophys. Acta* **208**, 331–333.

Zucker, M. (1971). Induction of phenylalanine ammonia-lyase in *Xanthium* leaf discs. *Plant Physiol.* **47**, 442–444.

Zucker, M. (1972). Light and enzymes. *Annu. Rev. Plant Physiol.* **23**, 133–156.

Chapter 17

Transcription and Translation in Diseased Plants

D. J. SAMBORSKI, R. ROHRINGER, AND W. K. KIM

I. INTRODUCTION

All living cells hold genetic information in DNA molecules which can be transcribed into RNA and translated into polypeptides according to the code of the particular gene. This chapter presents a general outline of the biochemical processes involved in transcription and translation and considers how they are affected in plants after infection with pathogens. It is probably through studies of such processes that specific gene interactions leading to compatibility and incompatibility will be understood and more sophisticated control measures of plant diseases developed.

Mechanisms involved in transcription and translation are now fairly well understood, at least in prokaryotes. Transcription is the process by which genetic information encoded in DNA is copied during the process of mRNA synthesis. Translation is the process by which the information is used to specify certain amino acid sequences during peptide synthesis. These processes are illustrated in Fig. 1, redrawn from Karlson (1975).

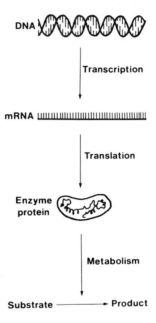

DNA

Transcription

mRNA

Translation

Enzyme protein

Metabolism

Substrate ⟶ Product

Fig. 1.   Mechanisms involved in transcription and translation.

For more detailed accounts of transcription and translation that are not too highly specialized, see Lehninger (1975) and Karlson (1975), or for a popular account, Gore *et al.* (1976). Much of the information on these processes has come through the study of *Escherichia coli* and *E. coli* infected with bacteriophages. We have therefore included a brief description of transcription and translation in this system, which, because of its simplicity, can serve as a model for other, more complex systems. In studies on transcription and translation in plant diseases, those dealing with plant diseases caused by viruses have yielded most of the information available in this field. Although many studies have considered the effect of fungal plant pathogens on nucleic acid and protein metabolism, few have dealt with the specific mechanisms involved. In bacterial diseases, studies have been largely carried out on crown gall.

Many host–parasite systems involve gene-for-gene interactions in which a specific gene for resistance in the host corresponds to a specific gene for avirulence in the parasite. It has been suggested that gene-for-gene interactions are superimposed on a basic system of compatibility in a host–parasite system and that the specificity in such systems resides in the incompatible interaction that is conditioned by the gene for resistance and the gene for avirulence (Ellingboe, 1976). Speculations based on the information derived from studies on transcription and translation must correspond to these known genetic facts.

II. GENERAL FEATURES OF RNA AND PROTEIN METABOLISM

A. Control of Gene Activity

Control of transcription is based on the specific interaction of molecules directly with chromatin, the DNA–protein complex. In prokaryotes gene activity is controlled by a product (repressor) of a regulator gene which binds specifically to a region of DNA (operator) that controls the action of an associated structural gene. Attachment of the repressor blocks the access of RNA polymerase to promoter regions, the initial sites of attachment for this enzyme. Absence of the repressor makes the promoter accessible to RNA polymerase and transcription of the structural gene begins. In addition, gene activity may be regulated by the presence of inducers or corepressors (enzyme induction, end-product inhibition) or by changes in the binding specificity of RNA polymerases.

Regulation of gene activity in eukaryotic cells is more complex and the mechanism of induction and repression, which is very common in prokaryotes, is seldom used by eukaryotes. Where it is used, it is slow. This fundamental difference reflects the fact that eukaryote cells have a more constant environment. Chromosomes of eukaryotes contain, in addition to DNA, a large variety of proteins and apparently some RNA. Two classes of chromatin-associated proteins are known: histones and non-histone protein. Histones appear to be involved in nonspecific repression of transcription, while non-histone protein and possibly chromosomal RNA appear to effect histone displacement and gene derepression. Both polypeptides and nucleotides may alter the specificity and activity of RNA polymerase in eukaryotes. Although cyclic AMP is known to have an important regulatory function in animal tissues and many prokaryotes (Goldberg et al., 1976), there is little information on its presence and role in higher plants (Amrhein, 1977).

B. Transcription and Processing of Transcribed RNA

Synthesis of RNA containing the genetic information encoded in the base sequence of DNA involves a series of biochemical events mediated by the enzyme RNA polymerase. The DNA-directed synthesis of RNA by this enzyme includes a number of steps: (1) template-binding RNA polymerase attaches to the DNA template and locates a specific site at which chain initiation can occur; (2) the RNA chain is initiated; (3) the RNA chain elongates; (4) the RNA chain is terminated and releases RNA polymerase from the DNA template.

RNA polymerase synthesizes a single-stranded RNA molecule which is complementary to one of the strands of DNA. In prokaryotes, a variety of RNA molecules are synthesized by one species of RNA polymerase. The holoenzyme contains four major subunits β, β', σ and α, which are present in the molar ratio 1:1:1:2. While the core enzyme is sufficient for polymerization of RNA, an additional factor (called the σ factor) is essential for selection of the site for attachment to the promoter. In prokaryotes, transcription and translation are coupled closely; translation usually is initiated on the newly formed mRNA before its synthesis is completed. This contrasts with the situation found in eukaryotes, where sites of transcription and translation are spatially isolated and where mRNA is translocated to the sites of peptide synthesis.

Eukaryotes contain several RNA polymerases which differ in their chromatographic behavior, intracellular localization, α-amanitine sensitivity, subunit composition, cation requirements, and in the type of RNA they synthesize. Polymerase I is required for synthesis of rRNA. Similarly, polymerase II is required for the production of mRNA, and polymerase III for that of tRNA. All of the newly synthesized polynucleotides need further processing before they become functional. This processing includes the following: (1) nucleolytic reactions such as cleavage and trimming of large precursor molecules by RNAase; (2) terminal addition of nucleotides such as 3'-polyadenylation of mRNA and addition of –CCA to the 3'-terminus of tRNA; (3) nucleoside modifications such as base or ribose methylation and the conversion of uridine to pseudouridine.

C. Translation and Posttranslational Modification of Protein

Translation of mRNA takes place on ribosomes. These subcellular particles consist of two subunits of different size. Each subunit is composed of various proteins and an RNA species of high molecular weight. In addition, the larger of the two subunits contains a low molecular weight RNA (5 S RNA). In eukaryotes, 80 S ribosomes are either free or membrane bound, while in prokaryotes the smaller 70 S ribosomes are mostly free because of the absence of a clearly defined endoplasmic reticulum. Ribosomes and ribosomal subunits of prokaryotes and eukaryotes differ in size, but the process of translation is similar in both and can conveniently be divided into four stages:

1. Amino acids are activated and aminoacyl-tRNA is synthesized.
2. Peptide synthesis is initiated when mRNA and methionyl-tRNA are bound to the 40 S ribosomal subunit in eukaryotes (formyl-methionyl-tRNA to the 30 S ribosomal subunit in prokaryotes) through mediation of several initiation factors. The 60 S ribosomal subunit of eukaryotes (or the 50 S ribosomal subunit of prokaryotes) attaches to this complex to form a functional ribosome initiation complex (80 S in eukaryotes, 70 S in prokaryotes).
3. Elongation of the polypeptide chain occurs through enzymatic transfer of new aminoacyl residues from aminoacyl-tRNA esters, each of which is bound to the ribosome in response to a specific codon or base triplet in the mRNA. Each tRNA is specific for a particular amino acid and can exist in multiple forms termed isoaccepting tRNA. Between 60 and 80 different molecular species of tRNA are presently known to be involved in the transfer of the 20 amino acids that occur in protein.
4. Termination of the polypeptide chain takes place when appropriate termination signals in the mRNA are reached during the process of translation and the product is released from the ribosome.

A polypeptide formed in this manner may undergo chemical modification before it attains its biologically active form. This can occur through methylation, acetylation, thiolation, adenylation, and phosphorylation. Proteolysis is also known to be involved in the processing of newly synthesized protein.

III. EFFECT OF INFECTION ON SYNTHESIS OF RNA AND PROTEIN

A. Bacteriophage Infections

Transcription of the genetic information in the phage genome begins after the viral nucleic acid has entered the bacterial host. Infection is accomplished in several ways but in essence it involves host-specific attachment of the bacteriophage to the bacterial cell wall and entry of the nucleic acid portion of the phage into the bacterium. The specific binding depends on a close complementary fit between the molecules of the phage tail fibers or capsid and those of the receptor on the bacterial cell wall. In the case of the T5 phage, the receptor is known to be a lipoprotein molecule of high molecular weight (Braun *et al.*, 1973).

After infection, phage nucleic acid serves as a template for the replication of more phage nucleic acid and for the production of mRNA (transcription) to make phage enzymes and coat protein (translation). Host components are needed to carry out these synthetic processes. The key step in transcription is mediated by RNA polymerase. After infection, there are three possible ways in which the specificity of this important enzyme can be altered. First, a new enzyme synthesized with information encoded in the phage genome can be produced. This mechanism is represented in T7 phage infections, where the new RNA polymerase is a single peptide with a molecular weight of 107,000 daltons (Chamberlin *et al.*, 1970). Second, the host RNA polymerase can be altered by posttranslational modifications of the enzyme, as is the case after infection with T4 phage. Here, no new enzyme is produced, and specificity is altered by modifications such as addition of one adenine nucleotide to the host enzyme (Goff, 1974). Third, the RNA polymerase is composed of subunits derived from the host genome and others derived from the viral genome. This is the case after infection of *E. coli* with QB, a phage with an RNA genome. Here, the enzyme contains three subunits of host origin and one subunit derived from the viral genome (Kamen, 1970).

Another important step in the interaction between phage and bacterium is the inhibition of transcription of the host genome. This serves to facilitate phage production and is accomplished, in bacteria infected with T7 phage, by inhibition of the holoenzyme of host RNA polymerase by a new protein (molecular weight 66,000 daltons) derived from the phage genome. This protein does not inhibit the T7 RNA polymerase and thus specifically inhibits transcription of the host genome (Hesselbach *et al.*, 1974). After infection with T4 phage, RNA and protein synthesis are inhibited, apparently by an unidentified product of the phage genome (Travers, 1976).

Transcription of the viral genome yields mRNA that is translated to specific viral proteins. Some of these, like a new T7-specific RNA polymerase, have been mentioned. Others are coat proteins and lysozymes. The lysozymes are a later product of transcription from the phage genome and serve to lyse the host cells after coat protein and nucleic acid of the virus have been assembled.

Of the phages considered above, only QB phage contains an RNA genome and it serves directly as mRNA in the host organism for the synthesis of phage protein and as template for its own replication. In phages possessing a DNA genome, mRNA must be synthesized from the phage DNA after infection.

B. Viral Plant Diseases

In a group of plant viruses, such as brome mosaic virus (BMV) and tobacco mosaic virus (TMV), the RNA genome can function directly as mRNA after infection, much like that of the QB phage. In other plant viruses, the viral nucleic acid must first be transcribed to yield mRNA, either from an RNA or a DNA template. The classification by Hamilton (1974) is based on this distinction. Infection with viruses typified by BMV results in the formation of an RNA-dependent RNA polymerase (also called replicase), with a subunit structure that has been postulated to contain peptides derived from both host and BMV genomes (Hamilton, 1974). This, too, is analogous to the situation existing in *E. coli* infected with QB phage. In other types of plant viruses containing RNA (those with single-stranded RNA, as in lettuce necrotic yellow virus, or with double-stranded RNA, as in wound tumor virus) the virus-specific, RNA-dependent RNA polymerase is present in the complete virion and transcribes the viral RNA to mRNA after infection (Franki and Randles, 1972).

The major products of the viral genome are coat protein and enzyme or enzyme subunits that participate in the transcription of the viral genome. Other proteins have been detected after infection, but it is not established whether these are of host or viral origin. In plants containing genes for the hypersensitive response, several low molecular weight proteins unrelated to TMV coat protein have been detected (Zaitlin and Hariharasubramanian, 1972). Inclusions consisting of proteins that are observed after infection with viruses of the potato virus Y group have a molecular weight twice that of the coat protein (Hiebert and McDonald, 1973). Commonly, higher activities of ribonuclease are observed in virus-infected tissues. This enzyme may play an important role in the processing of host or viral RNA after infection and could be responsible for the degradation of host RNA, furnishing nucleotides for the synthesis of viral RNA. After infection with cucumber mosaic virus, the composition of the host chromatin is altered because of an increase in arginine-rich histone (Kato and Misawa, 1971). This may affect transcription of the host genome.

Since the plant hosts of the virus are eukaryotes possessing a highly structured organization, the question may be asked where synthesis of viral RNA and viral protein occur in the host cell. It does not seem to be possible to make a general statement on this subject since little is known about sites of virus synthesis and because different viruses may well be transcribed in different cell compartments. The RNA of turnip yellow mosaic virus appears to be synthesized in the chloroplasts and, more specifically, in vesicles that arise in chloroplasts of infected plants (Hamilton, 1974). These vesicles, in turn, originate from membrane structures of the chloroplasts as a consequence of infection. Significantly, the vesicle membrane differs in structure from the membrane from which it is thought to be derived. An ultrastructural analysis of plants infected with barley yellow dwarf virus suggested that, although the virus particles of one strain were assembled around the nucleolus, replication of the viral RNA might be associated with vesicles enclosed by endoplasmic reticulum (Gill and Chong, 1976). In fact, Hamilton (1974) suggested in a recent review that vesiculate membranes associated with viral infection may play an important role in the synthesis and assembly of virus components.

C. Bacterial Plant Diseases

The literature contains few reports on the effect of nontumerogenic bacterial plant pathogens on transcription and translation in the host. There are numerous reports on the occurrence and role of degradative enzymes of bacterial origin in plant diseases, but such studies generally have not considered the possible effect on host transcription and translation and are therefore outside the scope of this chapter.

Crown gall caused by *Agrobacterium tumefaciens* has attracted much attention from molecular biologists because the tumerous tissue, once transformed by the pathogen, continues unregulated growth in the absence of the bacterium. After entry through wounds, the bacterial cells attach to specific sites on the cell surface (Lippincott and Lippincott, 1969). Biologically active DNA molecules move out of the pathogen and integrate into the host DNA. However, it is not clear whether the tumor-inducing principle is in the form of a part of the bacterial genome or whether it is bacterial plasmid DNA (Kado, 1976). The DNA of bacterial origin replicates along with the host DNA, probably with the help of host DNA polymerase.

Crown gall cells, as compared to normal cells, have been shown to contain different amounts of nucleic acid and enzymes associated with nucleic acid metabolism. Apparently, a new RNA methylase (7-methylguanine tRNA methylase), not present in healthy plants, occurs in tumor cells of *Parthenocissus tricuspidata* (Dubois and Weil, 1977). Changes also have been observed in the amounts of other proteins and enzymes in crown gall tissue. Several isozymes that are not present in healthy cells have been detected in crown gall cells, but it is not clear whether they are derived from the bacterial or host genome (Reddy and Stahmann, 1973).

D. Fungal Plant Diseases

Studies on transcription and translation in diseased plants infected by fungi are complicated because they involve two eukaryotic organisms, each possessing its own mechanisms for these biochemical processes and because these in turn occur in specific organelles or compartments in each organism. Consequently, it is often difficult to distinguish between the contributions each of these makes to changes observed in the analyses. Although each organism can be studied separately, the constituents produced by the fungus in artificial culture may not necessarily be the same as those produced by it in the host.

Mechanisms for resistance and susceptibility will not be considered here unless such a discussion contributes to the understanding of transcription and translation in host–parasite systems. In host–parasite systems where gene-for-gene interactions determine resistance or susceptibility, transfer of information must occur from one organism to the other, presumably through mediation by macromolecules. Some of the recent concepts on information exchange between partners of a parasitic relationship require either recognition at the plasma membrane or transfer of macromolecules across the plasma membrane and perhaps into specific organelles of the host or parasite where they can carry out their function.

All parasites, while growing parasitically, are dependent on their hosts for nutrients. Historically, it was considered that obligate parasites, such as the rusts and mildews, may have become so dependent on the host that they require functional macromolecules of host origin to complete their normal life cycle. This extreme dependence does not appear likely, since at least some of the rusts have been grown axenically on simple media.

One of the features common to all infective agents is their capability to reproduce in the host and at the expense of the host. To do this they alter the metabolism of their host in specific ways. One way in which bacteriophages and plant viruses alter host metabolism to their advantage is by production of a new virus-specific RNA polymerase or by changing the specificity of host RNA polymerases. It is therefore reasonable to expect that the specificity of RNA polymerases will also be altered in the more complex plant–fungus interactions, although such

altered specificity must allow continued host protein synthesis in interactions involving biotrophic parasites, where continued survival of the
host is advantageous to the pathogen. However, there has been little
research on RNA polymerases in fungal diseases of plants. In fact, we
know of only one published report on this subject. It was published in
two separate abstracts (Flynn *et al.*, 1976; Scott *et al.*, 1976) and has
been discussed in two review articles by one of the authors (Chakravorty
and Shaw, 1977a,b). RNA polymerases associated with chromatin isolated from healthy and rust-infected wheat leaves produced different
products *in vitro*. The authors concluded that infection induced changes
in the specificity of these important enzymes.

Most of the research that has a bearing on transcription and translation
in plant–fungus interactions has described quantitative changes in RNA
and protein and changes in the activity of certain enzymes.

The total RNA content has been shown to increase in various kinds of
susceptible, reacting plants after infection with rust (Heitefuss and Wolf,
1976). At least some of this increase can be attributed to fungal RNA.
Although the effect of infection on RNA content is usually observed 6 to
8 days after inoculation, a qualitative change in RNA composition has
been determined 24 to 48 hr after infection. The $A + U/G + C$ nucleotide ratio of salt-soluble RNA isolated from flax cotyledons increased
after infection with flax rust (Chakravorty and Shaw, 1971). Possibly
this reflects synthesis of new species of tRNA that are not present in
healthy wheat leaves or in rust spores, although it may represent other
functional RNA species that are present in the salt-soluble fraction. Other
workers have noted increases in soluble RNA after infection of plants
with rusts or mildew, but no further fractionations were carried out.
Although mRNA would be included in the salt-soluble fraction, no specific studies on this species of RNA appear to have been carried out in
mycoses, probably because of the technical difficulties involved. These
are readily apparent when one considers that, in contrast to bacteriophages and viruses where the mRNA is polygenic, each mRNA from
eukaryotes codes for a single protein, necessitating a multiplicity of
mRNA species.

Studies on rRNA in the host–parasite complex are also difficult to
interpret because cytoplasmic rRNA of host and fungal origin have similar sedimentation coefficients and cannot be easily separated. In general,
what has been found is an increase of cytoplasmic rRNA in susceptible-
reacting plants following infection with rust. This undoubtedly reflects
the increased metabolic activity of the host after infection and the synthesis of ribosomes in the parasite.

The ribonucleases are translation products that have an important role in processing products of transcription. In general, RNAase activity increases after infection. In a number of host–rust systems the infected tissues were shown to contain an RNAase with a substrate specificity that differed from that of host or parasite ribonuclease (Chakravorty et al., 1974a,b; Harvey et al., 1974). It was suggested that this altered specificity is due to the addition of a fungus-derived subunit to the multimeric enzyme. This would require transfer of either the appropriate mRNA or of a polypeptide across the host–parasite interface.

Many other changes in enzyme activities accompanying fungal infections have been described. These changes were either deduced after following the metabolism of particular substrates in vivo or by determining enzyme activities in tissue extracts of various levels of purification. More direct evidence was obtained by separating proteins with advanced techniques such as gel chromatography and electrophoresis, sometimes coupled with immunochemical methods, in attempts to determine whether the proteins of interest were of host or of fungal origin. In some cases, the enzymes studied were represented by a mixture of isozymes.

It was commonly noted in a number of susceptible-reacting tissues that there was an increase in the activity of enzymes involved in photosynthesis, the pentose phosphate pathway, glycolysis, the TCA cycle, oxidative phosphorylation, and in the shikimate pathway (Uritani, 1976). Most of the reactions catalyzed by these enzymes concern energy-producing processes. The increase in their activity probably reflects the heightened metabolic activity associated with fungal growth in these interactions. In resistant-reacting tissues, where much less fungal growth occurs, much smaller increases have generally been observed in activities of these enzymes. Only a relatively few enzymes have been studied in resistant interactions. The enzymes that received most attention were phenylalanine ammonia lyase, tyrosine ammonia lyase, peroxidase, and polyphenoloxidase. The resistant response was generally accompanied by an increase in the activity of these enzymes, or of certain isozymes of these enzymes. These proteins would not be translation products of genes for resistance, but represent secondary changes of protein synthesis that are also often observed in mechanically injured tissues.

Several research groups have attempted to detect translation products of genes for resistance without having to make an arbitrary decision to look for certain enzyme activities. Von Broembsen and Hadwiger (1972), using a double labeling technique to distinguish proteins synthesized in noninfected flax leaves from those synthesized in rust-infected flax leaves, found that the rate of protein synthesis increased in the incompatible interaction very early after infection, but remained constant or decreased in the compatible interaction. For each of several gene interactions distinctive patterns were observed. However, their fractionation methods did not permit the isolation of single proteins. The isotope technique used by Yamamoto *et al.* (1976) on oat leaves infected with crown rust did not reveal differences in incorporation of activity into protein fractions from noninfected and resistant-reacting leaves. Using immunoelectrophoresis, they did, however, detect what might be a new protein of host origin in extracts from the incompatible interaction. On the other hand, Frič and Heitefuss (1970), also using immunoelectrophoresis methods, could not detect any new host proteins in either compatible or incompatible interactions between wheat and stem rust of wheat.

E. Significance of Observed Changes in RNA and Protein Metabolism in Infected Hosts

DNA-dependent RNA polymerase is the key enzyme involved in transcription. In interactions involving bacteriophages, plant viruses, and their hosts, where the infective agent is genetically relatively simple, changes in the specificity of this enzyme have been documented. These changes facilitate replication of the pathogen. A preliminary report (Flynn *et al.*, 1976; Scott *et al.*, 1976) suggests that such altered specificity of this enzyme also occurs in the more complex rust–wheat system. In our opinion this would be a factor in the basic compatibility between wheat and stem rust of wheat, and is probably not involved in specific gene-for-gene interactions. The altered specificity of this enzyme in the wheat–rust system was observed in the compatible interaction. We agree with Ellingboe (1976) that specificity in gene-for-gene interactions resides in the incompatible interaction and therefore these systems must be studied to determine biochemical specificity.

Changes in protein synthesis have been commonly observed in studies on interactions between infective agents and their hosts. Most of the studies report increases in the activities of particular enzymes. Plant growth substances may well play an important role in this pathogen-induced protein synthesis since many plant diseases are accompanied by increases in the concentration of growth substances. It has been shown that both indoleacetic acid and cytokinins affect transcription by activation of chromatin-bound RNA polymerase (Guilfoyle and Hanson, 1974), or through a number of other mechanisms that are not well understood (Jacobsen, 1977). The possible relationship between growth substances and transcription in diseased plants should be investigated.

The appearance of a new protein may result from derepression of a specific host cistron or from enhanced protein synthesis leading to detectable concentrations of the enzyme after infection. A new protein isolated from a host–parasite complex may be of parasite origin and may not be detected when the parasite is grown *in vitro*. Generally, most of the studies on protein synthesis in plant–parasite systems have involved soluble proteins, but specificities such as in gene-for-gene interactions may reside in insoluble peptide or glycoprotein structures associated with cell surfaces. Specific recognition phenomena involving receptors on the cell surface have been documented in interactions with bacteriophages, viruses, and crown gall. While there is no direct biochemical evidence for specific recognition between host and fungal parasites, histological studies on the wheat–rust system (Samborski *et al.*, 1977) suggested that gene-specific recognition occurs between the fungal cell wall and the host plasma membrane.

IV. PROMISING APPROACHES TO FUTURE RESEARCH

Most workers in plant pathology now recognize the advantages of genetically well-defined material. Host lines isogenic for a particular gene for resistance are often used but isogenic lines of the parasite often are not available. Use of such genetically defined materials would facilitate the detection of products unique to each specific gene interaction.

If the gene product is a novel protein, new methods of protein fractionation (O'Farrell, 1975), with the potential of resolving up to 7,000 individual proteins, should increase the chance for its detection. This procedure involves separation in two dimensions, using isoelectric focusing in one dimension and separation by molecular weight in the second dimension.

Affinity chromatography may also be useful for the detection and isolation of specific gene products. Products of genes for resistance could be more readily identified if hosts with functional alleles were available since the products of such alleles would be more closely related to each other than the products of two unrelated genes for resistance. The expected minor differences in structure should simplify recognition with modern analytical techniques. However, there has been very little work done on the analysis of the fine structure of genes for resistance; we know of only the study by Shepherd and Mayo (1972) on the flax–rust system.

Although it may be possible to detect specific gene products with the above methods, determination of their function will probably require larger amounts than can be isolated with conventional techniques. It is possible that the recently developed methods of gene amplification by DNA cloning can be used to produce the required amounts. While this method does not appear to have been used successfully with DNA from higher plants, it is probable that appropriate techniques will be developed in the future.

Acknowledgments

The authors wish to thank Drs. A. K. Chakravorty and M. Shaw for allowing us to see two manuscripts in press. The figure was reproduced by courtesy of Dr. P. Karlson and Academic Press. This chapter is contribution No. 810, Research Station, Winnipeg.

References

Amrhein, N. (1977). The current status of cyclic AMP in higher plants. *Annu. Rev. Plant Physiol.* **28**, 123–132.

Braun, V., Schaller, K., and Wolff, H. (1973). A common receptor protein for phage T5 and colicin M in the outer membrane of *Escherichia coli* B. *Biochim. Biophys. Acta* **323**, 87–97.

Chakravorty, A. K., and Shaw, M. (1971). Changes in the transcription pattern of flax cotyledons after inoculation with flax rust. *Biochem. J.* **122**, 551–557.

Chakravorty, A. K., and Shaw, M. (1977a). The role of RNA in host–parasite specificity. *Annu. Rev. Phytopathol.* **15**, 135–151.

Chakravorty, A. K., and Shaw, M. (1977b). A plausible molecular basis for obligate host pathogen interactions. *Biol. Rev. Cambridge Philos. Soc.* **52**, 147–179.

Chakravorty, A. K., Shaw, M., and Scrubb, L. A. (1974a). Ribonuclease activity of wheat leaves and rust infection. *Nature (London)* **247**, 577–580.

Chakravorty, A. K., Shaw, M., and Scrubb, L. A. (1974b). Changes in ribonuclease activity during rust infection. II. Purification and properties of ribonuclease from healthy and infected flax cotyledons. *Physiol. Plant Pathol.* **4**, 335–358.

Chamberlin, M., McGrath, J., and Waskell, L. (1970). New RNA polymerase from *Escherichia coli* infected with bacteriophage T7. *Nature (London)* **228**, 227–231.

Dubois, E. G., and Weil, J. H. (1977). A 7-methylguanine tRNA methylase present in crown gall cells but absent in normal *Parthenocissus tricuspidata*. *Plant Sci. Lett.* **8**, 385–394.

Ellingboe, A. H. (1976). Genetics of host–parasite interactions. *In* "Physiological Plant Pathology," Encyclopedia of Plant Physiology, Vol. 4 (R. Heitefuss and P. H. Williams, eds.), pp. 760–778. Springer Verlag, New York.

Flynn, J. G., Chakravorty, A. K., and Scott, K. J. (1976). Changes in the transcription pattern of wheat leaves during the early stages of infection by *Puccinia graminis tritici. Proc. Aust. Biochem. Soc.* **9**, 44.

Franki, R. I. B., and Randles, J. W. (1972). RNA-dependent RNA polymerase associated with particles of lettuce necrotic yellow virus. *Virology* **47**, 270–275.

Frič, F., and Heitefuss, R. (1970). Immunochemische und elektrophoretische Untersuchungen der Proteine von Weizenblättern nach Infektion mit *Puccinia graminis tritici. Phytopathol. Z.* **69**, 236–246.

Gill, C. C., and Chong, J. (1976). Differences in cellular ultrastructural alterations between variants of barley yellow dwarf virus. *Virology* **75**, 33–47.

Goff, C. G. (1974). Chemical structure of a modification of the *E. coli* RNA polymerase α polypeptides induced by bacteriophage T4 infection. *J. Biol. Chem.* **249**, 6181–6190.

Goldberg, R. F., Deeley, R. G., and Mullinix, K. P. (1976). Regulation of gene expression in prokaryotic organisms. *Adv. Genet.* **18**, 1–67.

Gore, R., Dale, B., and Meltzer, D. (1976). The awesome worlds within a cell. *Nat. Geogr.* **150**, 355–395.

Guilfoyle, T. J., and Hanson, J. B. (1974). Greater length of ribonucleic acid synthesized by chromatin-bound polymerase from auxin-treated soybean hypocotyls. *Plant Physiol.* **53**, 110–113.

Hamilton, R. I. (1974). Replication of plant viruses. *Annu. Rev. Phytopathol.* **12**, 223–245.

Harvey, A. E., Chakravorty, A. K., Shaw, M., and Scrubb, L. A. (1974). Changes in ribonuclease activity in *Ribes* leaves and pine tissue culture infected with blister rust, *Cronartium ribicola. Physiol. Plant Pathol.* **4**, 359–371.

Heitefuss, R., and Wolf, G. (1976). Nucleic acids in host–parasite interactions. *In* "Physiological Plant Pathology," Encyclopedia of Plant Physiology, Vol. 4 (R. Heitefuss and P. H. Williams, eds.), pp. 480–508. Springer Verlag, New York.

Hesselbach, B. A., Yamada, Y., and Nakada, D. (1974). Isolation of an inhibitor protein of *E. coli* RNA polymerase from T7 phage infected cell. *Nature (London)* **252**, 71–74.

Hiebert, E., and McDonald, J. G. (1973). Characterization of some proteins associated with viruses in the potato virus Y group. *Virology* **56**, 349–361.

Jacobsen, J. V. (1977). Regulation of ribonucleic acid metabolism by plant hormones. *Annu. Rev. Plant Physiol.* **28**, 537–564.

Kado, C. I. (1976). The tumor-inducing substance of *Agrobacterium tumefaciens. Annu. Rev. Phytopathol.* **14**, 265–308.

Kamen, R. (1970). Characterization of the subunits of QB replicase. *Nature (London)* **228**, 527–533.

Karlson, P. (1975). "Introduction to Modern Biochemistry," 4th ed. Academic Press, New York.

Kato, S., and Misawa, T. (1971). Studies on the infection and the multiplication of plant viruses. *Ann. Phytopathol. Soc. Jpn.* **37**, 272–282.

Lehninger, A. L. (1975). "Biochemistry." Worth Publ., New York.

Lippincott, B. B., and Lippincott, J. A. (1969). Bacterial attachment to a specific wound site as an essential stage in tumor initiation by *Agrobacterium tumefaciens. J. Bacteriol.* **97**, 620–628.

O'Farrell, P. H. (1975). High resolution two-dimensional electrophoresis of proteins. *J. Biol. Chem.* **250**, 4007–4021.

Reddy, M. N., and Stahmann, M. A. (1973). A comparison of isozyme patterns of crown gall and bacteria-free gall tissue cultures with non-infected stems and non-infected tissue cultures of sunflower. *Phytopathol. Z.* **78**, 301–313.

Samborski, D. J., Kim, W. K., Rohringer, R., Howes, N. K., and Baker, R. J. (1977). Histological studies on host-cell necrosis conditioned by the *Sr6* gene for resistance in wheat to stem rust. *Can. J. Bot.* **55**, 1445–1452.

Scott, K. J., Chakravorty, A. K., and Flynn, J. G. (1976). Biochemical aspects of gene expression in wheat and barley after infection with rust and mildew fungi. *Aust. Plant Pathol. Soc. Newsl.* **5**, Suppl., 9.

Shepherd, K. W., and Mayo, G. M. E. (1972). Genes conferring specific plant disease resistance. *Science* **175**, 375–380.

Travers, A. (1976). RNA polymerase specificity and the control of growth. *Nature (London)* **263**, 641–646.

Uritani, I. (1976). Protein metabolism. *In* "Physiological Plant Pathology," Encyclopedia of Plant Physiology, Vol. 4 (R. Heitefuss and P. H. Williams, eds.), pp. 509–525. Springer Verlag, New York.

von Broembsen, S. L., and Hadwiger, L. A. (1972). Characterization of disease resistance responses in certain gene-for-gene interactions betweeen flax and *Melampsora lini. Physiol. Plant Pathol.* **2**, 207–215.

Yamamoto, H., Tani, T., and Hokin, H. (1976). Protein synthesis linked with resistance of oat leaves to crown rust fungus. *Annu. Phytopathol. Soc. Jpn.* **42**, 583–590.

Zaitlin, M., and Hariharasubramanian, V. (1972). A gel electrophoretic analysis of proteins from plants infected with tobacco mosaic and potato spindle tuber viruses. *Virology* **47**, 296–305.

Chapter 18

Senescence and Plant Disease

G. L. FARKAS

I. WHAT IS SENESCENCE?

If we are to elucidate senescence, we must elucidate aging because I think that we must look upon senescence as the outer end of a continuum that moves imperceptibly from young through middle age to old age and eventually to senescence. The word "age" applies to any point in life, say 10 years of age, but the verb form, aged or aging, means old or to become old. A person that is aged is considered to be senescent. Some of his functions and faculties have begun to diminish. When he loses a significant number of his faculties, he is senile. We do not normally use the word "senile" to describe plants, but perhaps we should.

Our medical colleagues generally consider senescence as the clinical aspects of aging. They have gone so far as to delineate a special medical practice called geriatrics, the doctoring of old age.

As to the plant kingdom, Leopold (1964) says that "The deteriorative processes which naturally terminate the functional life of an organ or an organism, are collectively called senescence." Some other authors also distinguish the "late developmental processes" (aging) from the

391

"deteriorative" ones (senescence). However, pathogens do attack plants during their whole life span and although they do "recognize" age differences they do not recognize a sharp boundary between aging and senescence. One fades into the other. Therefore, in discussing senescence and plant disease, senescence should be regarded as a developmental phase and, wherever possible, should be discussed in context with earlier phases of development.

In this chapter, I propose to examine the diseases of old age, i.e., diseases that induce senescence and diseases that senescence induces.

II. IS SENESCENCE RELATED TO STRESS?

A new fad is abroad in the biological world—stress physiology. There are those who would put disease-induced senescence among the stress-induced phenomena. I do not really have any quarrel with that except that the concept of stress is vague. It involves mechanical damage, chemical damage, and the metabolic load associated with the presence of the pathogen, etc. None of these is a well defined phenomenon at the biochemical level. However, these factors do play a role in most diseased tissues and they are important from the point of view of the plant pathologist because they lead to striking changes in metabolism; they dominate the metabolic patterns of diseased tissues. Consequently, they can easily obscure the more subtle, specific changes. As will be discussed later, it is a formidable task to identify primary reactions in the presence of such a "background." But for us this background is also very interesting because these types of metabolic changes are often related to symptom development. They are responsible for the fact that some symptoms, e.g., senescence, are so widespread in diseased plants.

III. GENERAL RELATION OF AGE TO DISEASE

Although the general problem of age-related diseases will be discussed in Volume IV of this treatise, a few statements are in order for this chapter. This seems necessary as a prelude to the subsequent discussion of diseases of old age.

A. How Old Is Old?

As always in science there is the problem of measurement. How old is old? Chronological age is generally unsatisfactory because plants grow faster or slower with temperature and other factors.

Erickson and Michelini (1957) have devised a plastochron index that can lead to more comparable and exact studies of age-related processes in many plants. It is based on the growth of the leaf and the time that elapses between the setting of new leaves. It is surprising how little advantage has been taken of this index.

The "physiological leaf age" depends both on the age of the entire plant and on the "absolute age" of a particular leaf, as related to the other. This is exactly what the plastochron indexes are supposed to express. The most revealing investigations of the use of the plastochron in diseases are those of Takahashi (1971, 1972).

The multitudinous cases in which disease intensity and/or symptom expression depend on the age of plant or plant organ are well known to plant pathologists. The emerging picture is kaleidoscopelike and it makes no sense to compile even a tentative list. Instead, we shall discuss a few examples, as case histories, which might be more instructive.

B. Case Histories

1. Virus-Induced Local Lesions

It is customary to take the number of virus-induced local lesions as a measure of susceptibility. If we express lesion number as a function of increasing leaf age (leaf plastochron index), susceptibility increases to a high point and then decreases, showing an almost symmetrical maximum curve. The wide range of conflicting data published on the age dependence of local lesion number in virus-infected plants could probably be reconciled if the plastochron approach of Takahashi (1971, 1972) were applied.

The response of lesion number to cytokinin treatment of tissues and other manipulations resulting in the rejuvenation of leaves (Király et al., 1968) also suggests that lesion number depends on the age of leaves. It is, of course, a question how far lesion number (actually disease intensity) can be regarded as a measure of "resistance." A recent work directs attention to the need for a reinvestigation of these results because the number of visible lesions is not necessarily a measure of infection sites (Balázs et al., 1976). One-cell lesions are known to arise as a result of the hypersensitive reaction induced by bacteria (Turner and Novacky, 1974) and their occurrence in kinetin-treated, virus-infected leaves has also been claimed (Balázs et al., 1976). In these types of studies the vital staining technique applied by Turner and Novacky (1974) should be used instead of simple lesion counting.

It should be recalled that a number of metabolic parameters exhibit a maximum curve, and relatively few (e.g., peroxidase activity) show more or less linear changes as a function of leaf insertion level. The maxima of several parameters tend to coincide with the full expansion of the leaf. The processes associated with maturation are interdependent. They are the reflection rather than the cause of a complex developmental process. Consequently, all theories that single out fashionable parameters (e.g., the rate of RNA or protein synthesis) as processes directly responsible for resistance or susceptibility must be treated with reservation.

The size of virus-induced local lesions is also dependent on leaf age. Lesion size, as a function of leaf insertion level, shows a maximum curve with a linear part. Lesions tend to be greatest in the younger leaves of the linear part of the curve (Weststeijn, 1976).

Peroxidase activity is well known to increase upon lesion formation. A rough inverse correlation is usually obtained between peroxidase activity and lesion size (Van Loon, 1976). The inverse correlation is, however, quite good if only the peroxidase activity of the immediate, 2-mm zone surrounding the lesions is taken into consideration. This is the area in which the most intensive response is observed.

Another physiological parameter that drastically changes with lesion development is the permeability of the affected cells. Age dependence of the increase in permeability upon virus infection, as well as after wounding, has been found in the bean by Thilo and Nienhaus (1975). Middle leaves gave the most intensive response. Also, the time required for the collapse of the cells, after transfer to 24°C of "Samsun NN" tobacco leaves, previously infected and kept at 32°C, is strictly dependent on leaf age. The middle leaves react most rapidly (Takahashi, 1975).

2. Symptom Expression in Virus-Infected Systemic Hosts

An instructive example that shows how symptom expression depends on the age at which the virus invades a particular leaf has been analyzed by Atkinson and Matthews (1970). After infection of a lower leaf of a systemic host, tobacco mosaic virus (TMV) must reach the young leaves in a well-defined developmental stage to induce the typical mosaic pattern with dark green areas. Similarly, vein clearing symptoms in TMV-infected tobacco are obtained only if the virus invades the upper leaves in an age which is well defined by a given range of leaf plastochron indexes (Takahashi, 1971).

3. Biochemical Changes in Virus-Infected Leaves

The biochemical changes induced by infection also depend on the age of the infected leaf. For example, in systemically infected tobacco, the younger the secondarily infected leaf, the longer the duration of TMV RNA synthesis, and the higher the amount of virus accumulated. In very young virus-invaded leaves the accumulation of rRNA and tRNA is strongly inhibited. In older but still expanding ones, only the accumulation of chloroplast rRNA is affected, whereas in the primarily infected leaves, entering the period of senescence, the expected decrease of rRNA and tRNA is retarded by infection (Fraser, 1972).

These examples show that both symptom expression and the biochemical alterations in the diseased tissues are age dependent. Perhaps the explanation of systemic acquired resistance in virus-infected plants will follow along these lines. Balázs *et al.* (1977) have reported that the cytokinin content of the upper leaves is increased when they have acquired resistance to viruses. If so, the associated changes in the developmental stage of these leaves can easily lead to apparent changes in resistance.

IV. DISEASE-INDUCED SENESCENCE

Before turning to senescence-induced diseases, we shall deal with disease-induced senescence. The invading pathogen sets in motion a series of changes that reduce the functions (and "faculties") of the tissue. In short, the tissue senesces. A substantial portion of the literature of physiological plant pathology has a direct or indirect bearing on this problem. However, many of the research techniques that have been used to investigate this phenomenon are somewhat obsolete, as shown below.

A. Analytical Techniques—The Challenge To Identify the Primary Reactions

1. Nucleic Acids

The analytical techniques so far used for detecting alterations in the nucleic acids as the disease progresses can and do show changes in the overall drift: change in the total amounts, but not the specific or primary reactions as the disease begins to induce senescence.

The idea of information transfer between host and parasite has often been entertained but has hardly been subjected to experimental test. However, the recent progress in the detection and separation of polyadenylic acid [poly (A)]-containing mRNA's opens up new possibilities in plant pathology (Yoshikawa *et al.*, 1977).

2. Proteins

Changes in proteins in the diseased plant have been the focus of interest of plant pathologists for almost two decades. However, the limits set by the present-day rapid separation techniques are two orders of magnitude lower than the actual number of proteins to be separated. These techniques have favored the detection of major, nonspecific changes in proteins triggered by mechanical injury, and/or associated with degradative processes.

Oddly enough, due to the inadequate level of separation techniques we know much about the nonspecific changes that accompany parasitically induced injury. As we shall see later, these nonspecific changes have a major role in the induction of senescence.

The specific changes associated with the onset and progress of disease are likely to occur in the minor fractions. Therefore, simple screening procedures would be necessary to precede the classical separation techniques. It remains to be seen whether the recent high-resolution separation techniques, developed for *Escherichia coli* by O'Farrell (1975), can be applied in plant pathology.

3. The Domain of Small Molecules

In the domain of small molecules such as toxins, hormones, and phytoalexins, the classical chemical methods have been adequate and fruitful in the development of senescence physiology. Some of these compounds are elicitors of symptoms in plant–pathogen interactions and occasionally can be actual determinants of host–parasite specificity (e.g., host–specific toxins).

4. The Importance of Kinetic Studies

The earlier a change can be detected in a diseased tissue, the higher is the probability that the change is specific. For example, detailed kinetic studies suggest that such processes as parasitically increased respiration or accumulation of phenolics are not specific. Consequently, these alterations can be regarded only as typical stress phenomena (Daly, 1972).

B. The General Features of Metabolism Leading to Senescence

A critical survey of the metabolic characteristics associated with disease-induced senescence has shown that senescence, in a large number of cases, is causally correlated with the high metabolic rate of the host–parasite complex at the infection centers. The high overall rate of metabolism is due, on the one hand, to a significant contribution of the parasite and, on the other, to the "overheated" metabolism of the host itself. As a result, metabolic sinks arise at the infection centers and lead to a depletion of metabolites and senescence of nearby tissues. The general aspects of these phenomena leading to senescence are discussed in this section followed by a detailed description of the metabolic characteristics of senescing tissues in Section IV,C.

1. Contribution of the Parasite

One often overlooked special feature of plant diseases is the large amount of nonhost "material" within the host–parasite complex, especially in the susceptible combinations. The dramatic increase in respiratory rate in rust- and mildew-infected leaves is associated with the presence of large amounts of hyphae and spores in the invaded tissues. The number of bacterial cells can also reach high levels in susceptible combinations. In TMV-infected tobacco 50% of the nitrogen content of the leaf may be of viral origin. This explains, at least in part, why the metabolic activity (respiration and nucleic acid and protein synthesis) of an infected tissue can be several times higher than that of the uninfected one.

2. "Overheated" Metabolism of the Host

In addition to the contribution of the parasite, there is good evidence for an increased metabolic rate in the vicinity of infection in the healthy cells. This phenomenon seems to be associated with injury caused by the penetration of parasite and/or lesion formation after its establishment. The most unequivocal case for high metabolic activity in the uninfected cells of the host is the increase in respiratory rate, mitochondria, ribosomes, endoplasmic reticulum, and a number of enzyme activities around virus-induced lesions in hypersensitive host–virus combinations (Ross, 1961; Solymosy and Farkas, 1963). Simultaneously, respiratory substrates and low molecular weight building blocks of macromolecules accumulate in the affected area, partly as a result of energy-dependent transport.

3. Depletion of Nearby Tissues

The infection centers are metabolic sinks that rob the surrounding tissues of low molecular weight metabolites like sugars. The infection centers are comparable to the young organs of a plant that also accumulate and utilize low molecular weight substances at the expense of older tissues, thereby accelerating their senescence. An explanation for the development of the infection center sinks could be the accumulation of cytokininlike substances in them (Király et al., 1966; Dekhuijzen and Staples, 1968). Cytokinins, if applied locally, induce a directed transport of metabolites to the site of application. However, the rearrangement of the translocation pattern in diseased plants might also be due to a preferential retention of transportable material in the infected leaves (Zaki and

Durbin, 1965). The flow of food is discussed in more detail by Zimmermann and McDonough in this volume, Chapter 5.

The significance of depletion as a component of the stress syndrome is less clear-cut with necrotrophic parasites. In this case, the infection centers do not represent sites of intense syntheses and consequently the sink effect is less pronounced.

C. Metabolism of the Senescent Plant

1. Photosynthesis

Since chloroplasts show the earliest signs of damage, and their optimal activity requires functioning that is coordinated with other cell compartments (cytoplasm, peroxisomes, mitochondria), it is not surprising that in most diseased tissues a decrease in photosynthetic activity has been found. Diseased leaves tend to be chlorotic. However, not all cases of chlorophyll loss can be equated with disease-induced senescence. Perhaps the most conspicuous case of typical, disease-induced senescence is the yellowing of leaves infected by obligate parasites, perhaps more accurately called biotrophic parasites and/or viruses inducing local lesions. Small increases in photosynthesis have often been reported for the early phase of such infections but in the light of more recent studies it seems likely that the photosynthetic apparatus remains, at best, relatively intact and functional (Montalbini and Buchanan, 1974). The preservation of structure is logical since the biotrophic parasites require a continuous supply of metabolites for their own synthetic processes and even the form in which they receive these substances may be important.

Even the green island formation around the infection centers might represent a site of maintained photosynthetic activity rather than a site of increased chlorophyll synthesis and CO_2 fixation (Harding et al., 1968). Also, high rates of incorporation into the macromolecular components in green islands do not necessarily indicate higher rates of synthesis but might reflect a change in endogenous precursor pools, altered turnover rates, and/or altered uptake processes. At any rate, a significant difference can be noted between the preservation of chloroplast structure and activity in green islands and the rapidly senescing parts of the same leaves. For example, one-half of a leaf moderately infected by rust accumulates more dry matter and grows more intensively than the slowly senescing noninfected half (El Hammady et al., 1968). Maintained photosynthetic activity; cytokinin-directed, long-distance metabolite transport; inhibition of metabolite efflux; stimulation of RNA and protein synthesis; and inhibition of protein breakdown might equally play a role. The relative contribution of these factors will have to be elucidated.

2. Respiration

One of the most widespread responses of host tissues to infection is an increase in respiratory rate. This increase is not directly related to senescence. It can be detected relatively early after infection. This is true for both virus-infected local lesion hosts and tissues attacked by biotrophic parasites. In this latter case the contributions of host and parasite are difficult to separate, the overall increase being very much confined to the infection centers.

The rise in respiratory rate reflects the general anabolic nature of metabolism in infection centers. It is generally agreed that the preponderance of synthetic processes, both in host and parasite, and the associated high ATP-consumption, providing suitable phosphate acceptor (adenosine diphosphate) level, is the main controlling factor responsible for the high respiratory rates in tissues infected with biotrophic parasites. Especially during the sporulation phase there is a great demand for substrates and easily convertible energy. This requirement is met, at least partly, at the expense of noninfected areas of the leaf or other parts of the plants, leading thereby to their premature senescence. These tissues are characterized by a low respiratory rate and loss of the respiratory control of their mitochondria.

Senescent tissues tend to be rich in peroxidases. Disease and damage add to the high peroxidase level. Although there is no proof for the *in vivo* functioning of peroxidases, at the later stages of infection and/or senescence the breakdown of cellular compartmentation might lead to peroxidative processes contributing to the overall O_2 uptake. Death is not an immediate event. Between a healthy and a totally collapsed cell a number of transition states occur with ample opportunity for the pathological operation of enzymes (peroxidase, polyphenol oxidase, ascorbic oxidase) with obscure *in vivo* function. These processes are, of course, secondary. Nevertheless, they can contribute to an autocatalytic speedup of aging and lesion formation, and can also affect symptom expression. The overall change can affect the spread of the parasite.

Respiratory increase in the infected host tissues is a common phenomenon with necrotrophic parasites as well and can be elicited by the application of pure toxins. Since toxin treatment results in a rapid loss of ions from the tissues, the idea of uncoupling of phosphorylation resulting in a lesion in energy-driven ion uptake and retention emerged. However, the uncoupling might be secondary since ion leakage precedes the onset of increase in respiration (Arntzen et al., 1973). This aspect of parasitically induced respiratory increase has not been studied in relation to senescence.

3. Nucleic Acid Metabolism

To measure accurately the rates of synthesis of macromolecules like nucleic acids in higher plants is a difficult task because of the changing precursor pools and turnover rates. With fungal and bacterial diseases the extra precursor pools of the parasites complicate the methodology. *In sensu stricto*, none of the works so far carried out on host–parasite complexes meets the present-day requirements. Nevertheless, at least with the biotrophic parasites, there is evidence that at the infection centers nucleic acid synthesis is enhanced: (1) nuclei and nucleoli are enlarged, a phenomenon indicative of enhanced RNA (and protein) synthesis; (2) cytophotometric studies suggest an increase in RNA content within the nucleus and/or nucleolus; (3) biochemical analysis shows an increased RNA content in the infected leaf area; and (4) incorporation of labeled precursors into fungal as well as host cells is increased in, and in the vicinity of, the infection center (Bhattacharya *et al.*, 1965; Williams *et al.*, 1973). Whatever the reason and mechanism of this increased synthetic capacity is, its consequences for leaf juvenility and/or senescence are obvious. If viewed under the electron microscope, the infected cell, in the early phase of disease, appears more juvenile than a mature one (Ehrlich and Ehrlich, 1971). The increased synthetic activity is, however, restricted to the infected area and its immediate vicinity. Later, even in the infected cell, the synthetic capacities start to decline but the parasite, especially in the compatible combinations, continues to accumulate RNA and proteins of its own. In this phase, the compartmentation of the metabolism of host and parasite becomes increasingly evident. Even the various compartments of the host cell behave differently. The chloroplasts show the first indications of damage. Correspondingly, the chloroplast RNAs start to incorporate less label from precursors, and their amount also decreases preferentially (Tani *et al.*, 1973a). This phenomenon is typical for aging, healthy leaves as well. Therefore, a careful analysis of the ribosomal RNAs can provide one of the best criteria to detect disease-induced, premature senescence. In work of this type a cautious approach must be used, since too heavy an infection, especially in an early stage of leaf development, is likely to result in an inhibition rather than a stimulation of RNA synthesis.

Ribosomes accumulate in tissues surrounding virus-induced local lesions. Lasting injury and the presence of neighboring healthy cells are necessary for such effects. If the damage is systemic or the toxic effect is too rapid, such changes do not take place or if they do, they are too fast to be observed. Thus, bacterial toxins can induce early senescence without previously (or locally) detectable synthetic phase (Lovrekovich *et al.,* 1964). High polysome/monosome ratios are indicators of high synthetic rates. Stress conditions (drought, osmotic stress) are known to decrease the polysome/monosome ratio. It fits into the general picture that infection of barley with powdery mildew results in a preferential decrease of chloroplast polysomes (Dyer and Scott, 1972).

Our knowledge of alterations in specific tRNAs by disease in higher plant cells is limited. This is surprising in view of the great interest in changes of the tRNA isoacceptor spectrum in relation to developmental processes, including senescence. In senescent tissues the decrease in the amount of total tRNA is generally less than that of the total rRNA/unit of DNA. This probably occurs in disease-induced senescence as well.

There has been much speculation on the role of mRNAs in host–parasite relationships. The limited evidence based on the appearance of DNA-like, rapidly labeled RNA exhibiting high turnover rate and template activity suggests that the increase in template-active mRNA synthesis might be one factor regulating the overall increase in synthetic activities (Tani *et al.,* 1973b). One recent work reports the isolation of poly(A)-containing mRNA from diseased tissues (Yoshikawa *et al.,* 1977).

The turnover and breakdown of nucleic acids have received less attention, although it is becoming increasingly evident that without determining these parameters it might be difficult to assess the synthetic rates. Thus, early work attributed a major, if not exclusive, role to increased RNA and protein synthesis when kinetin retards leaf senescence.

However, inhibition by kinetin of the breakdown of RNA and protein might be more important in maintaining steady RNA and protein levels. This important aspect of nucleic acid and protein metabolism has not been studied by plant pathologists. This is surprising because a possible connection between high synthetic activity in infection centers and local accumulation of cytokinins has repeatedly been suggested (El Hammady *et al.,* 1968).

In contrast to breakdown of RNA per se, much attention has been devoted to the study of increased nuclease levels, an ubiquitous phenomenon in disease and senescence. RNAases are known to be localized mainly in special cell compartments similar to the lysosomes of mammalian cells and the cellular RNAs are protected against them (Matile, 1975). This explains that increase in RNAase level can occur concomitantly with an increase in total RNA. Therefore, the role of nucleases in host–parasite physiology remains to be elucidated. In the late stages of infection the breakdown of cellular organization might release the nucleases from their compartments and they can lead to a rapid degradation of nucleic acids, leading finally to cell death. However, it is imperative to emphasize that aging is an orderly process which cannot be based on a random degradation of macromolecules.

4. Protein Metabolism

The accumulation of ribosomes involves the synthesis of ribosomal protein and usually reflects a general stimulation of protein synthesis. Therefore, the pattern of protein synthesis in infected plants is similar to that already described for nucleic acids. In addition to the accumulation of ribosomes, the formation of mitochondria has also been reported, especially in the healthy cells surrounding the site of damage. The overall view is indicative of a synthetic metabolism geared to a higher level, supplied at the same time with the necessary energy-yielding machinery. The eliciting factors are not known. Probably no single wound hormone exists. One component of the "wound hormone complex" is ethylene which is invariably synthesized in injured cells and has been shown to regulate RNA and protein (enzyme) synthesis. However, ethylene stimulates the synthesis only of some enzymes and cannot be made responsible for the overall changes in protein level.

In contrast to nucleolytic enzymes, the proteases and peptidases have hardly been studied by plant pathologists. Although these enzymes have often been implicated in aging, their role in protein breakdown and/or turnover is not known.

5. Abnormal Proteins

Aging in microbial and mammalian cells is often explained by the formation of abnormal proteins unable to carry out normal functions. The synthesis of such proteins could be due to somatic mutation, as well as to transcriptional or translational error. These changes are supposed to accumulate during ontogenesis and result in metabolic disturbances. The cell has built-in safety machinery to eliminate the abnormal proteins since they are more vulnerable to proteolytic attack because of their altered tertiary structure. All this is well documented for bacterial and animal cells. With plants one would be inclined to see a phylogenetic adaptation in the fact that, at least in some cases, proteolytic enzymes accumulate in leaf tissues as senescence proceeds. There appears to be the greatest need to eliminate aberrant proteins in this period. However, these bulk proteases are separated from the cellular sites where the safety machinery should work and probably play just the opposite role: they take part in the breakdown of total proteins at the onset of very advanced stages of senescence. Our knowledge about proteases and their role in protein breakdown in plants is meager. The idea of aberrant proteins and a study of built-in defense mechanisms to eliminate them has not received any attention in relation to plant diseases and senescence, although such an approach might be highly rewarding.

6. Enzyme Patterns in Senescing Tissues

Since senescence is accelerated in detached leaves, this system has been widely used for studying aging phenomena. A whole set of enzyme activities is changed in detached leaves. Even if the different systems vary to some extent, on the basis of comparative studies we have put forward the idea that there is a "common pattern of enzymatic changes in detached leaves and tissues attacked by parasites" (Farkas et al., 1964; Farkas and Lovrekovich, 1965). Ultrastructural studies have corroborated this view (Shaw and Manocha, 1965). It follows logically from this observation that a number of major enzyme changes in diseased tissues reflect the accelerated aging associated with disease.

The detached leaf system has some drawbacks. Bacterial contamination might induce enzyme changes attributed to detachment but no experimental evidence has been presented to support this view. A more serious problem is that leaf detachment or the use of leaf disks is invariably associated with mechanical damage, and injury-induced changes have to be distinguished from disease-induced changes. This drawback, however, can be turned into an advantage. Since senescence is not immediate, the early changes are more likely to be due to cellular injury, whereas in the later phases senescence might predominate. The ubiquitous increase in RNAase activity in injured, senescent, and diseased tissues might serve as a model example. In a number of disease situations a biphasic increase in RNAase activity has been observed. With rusts the two peaks seem to correspond to different enzymes (Chakravorty et al., 1974). Logically, the enzyme responsible for the early rise could be an RNAase induced by the reaction of the host to shock or injury, and the other to the presence of large amounts of fungal cells or induced senescence. Indeed, we have shown that the overall increase in RNAase activity in excised leaves can be resolved into two components at least. The early rise is due to the accumulation of a relatively purine-specific RNAase, widespread in plant tissues (Wyen et al., 1969), whereas in the later stages another, aging-specific nuclease accumulates, which exhibits relative adenine specificity (Wyen et al., 1971a; Udvardy and Farkas, 1972). This process might well mimic the pathological situations, as suggested by the observation that in virus-infected local lesion hosts only the "injury-specific" nuclease increases under conditions when no senescence of the tissues is apparent (Wyen et al., 1972). The term "injury" should be used in a broad sense, not restricted to mechanical damage, because the level of the same nuclease has been shown to increase dramatically in tobacco protoplasts exposed to osmotic stress (Lázár *et al.*, 1973; Premecz *et al.*, 1977).

Similarly instructive is the investigation of peroxidases. Peroxidase activity also tends to increase in aging or injured tissue as well as in diseased tissue. The total peroxidase activity can be resolved into several components. With peroxidases the chance is especially high of isolating artefacts (a point often overlooked by those who never tried to purify peroxidases and "separated" peroxidase isozymes by gel assay only). Therefore, reports on a very high number of peroxidase isozymes in crude extracts from infected tissues must be treated with reservation. Nevertheless, it seems possible to show, also in purified preparations, that senescence is associated with the accumulation of well-defined peroxidases. One such peroxidase appears prematurely in virus-infected young bean leaves (Farkas and Stahmann, 1966), an observation which gave experimental support to the common pattern hypothesis. However, in addition to aging, necrobiosis (lesion formation) per se has also a profound effect on peroxidases. In a number of host–virus combinations definite, predictable changes in peroxidase isozyme patterns were observed whenever lesions appeared. The changes depended on the host and not on the virus which induced the lesions (Solymosy *et al.*, 1967). This suggests that necrobiosis affects the formation of specific enzymes. Elegant studies on nearly isogenic, rust-infected wheats have also shown that the necrogenic reaction is associated with the increase in one particular peroxidase isozyme, whereas the susceptible reaction favors the accumulation of a different one (Seevers *et al.*, 1971).

The level of peroxidases is quite stable once it has reached a peak. This is shown by the continued increase of peroxidase activity in the advanced stages of senescence, when the level of other "stress enzymes" starts to decline along with the excessive breakdown of total proteins (Lázár and Farkas, 1970). A further proof is the maintenance of the level of one peroxidase isozyme, specifically induced during resistant reaction, when the reaction type of the leaf, in which the resistant reaction had already developed, was shifted by experimental manipulation to a susceptible one, a reaction normally not associated with the formation of this particular isozyme (Seevers *et al.*, 1971).

Further typical stress enzymes which are generally present in injured and diseased tissues include the first two dehydrogenases of the hexose monophosphate shunt, enzymes involved in aromatic biosynthesis, and polyphenol oxidase. The reports are too numerous to list here.

As with nucleases and peroxidases, efforts have been made to see whether the elevated enzyme levels are associated with a preferential increase of specific molecular forms. Except for polyphenol oxidases in injured sweet potato (Hyodo and Uritani, 1967), the results are not very convincing.

The analysis of changed enzyme patterns is complicated by the greatly different turnover rates of different proteins. Thus, injury can induce the formation of enzyme–inactivating systems and these might lead to a decrease in enzyme level after an initial rise (Tanaka *et al.*, 1974). The stability of some other enzymes appears to be high. Ribulose-1,5-P-carboxylase, e.g., under normal conditions, has hardly any turnover. However, it is broken down in disease and senescence.

7. Lysosomes

Since nucleic acids and proteins are quite stable, enzymes have to play a role in their breakdown. Specific processing by hydrolytic enzymes of nucleic acids and proteins is well documented for bacterial and animal cells. With plants the progress in this field is slow. One reason is the high background of nonspecific enzyme activities. The bulk of hydrolytic enzymes (nucleases, proteases, esterases) are contained in special cell compartments, lysosomes (Matile, 1975), and mostly sequestrated from plant metabolism. There is no evident reason why the plant needs such a large reservoir of hydrolytic enzymes that are seldom used during normal cell functioning. However, there must be some reason. We only know that the amount of lysosomal enzymes is greatly increased during aging and probably during disease. Perhaps, in plant disease and senescence, lysosomes temporarily play a protective role by compensating for a regulatory system that went wrong because of a defect in disease and/or senescence, and overproduced hydrolytic enzymes.

The lysosomal aspects of alterations in hydrolytic enzymes in diseased plant tissues have hardly been studied. Still, the type of enzyme changes (Wyen *et al.*, 1971b) leaves little doubt that the injury- and senescence-associated changes in hydrolytic enzymes occur at the lysosomal level (see also Pitt, 1973). The study of lysosomes is a very fashionable and fruitful field of human pathology. The relevance of the results to the pathological state in general should receive more attention.

8. Secondary or Stress Metabolites

In the diseased tissues secondary or stress metabolites accumulate. With some oversimplification, two major pathways are responsible for the synthesis of these compounds: (1) aromatic biosynthesis and (2) terpenoid biosynthesis. Only the first has a major impact on senescence. Stimulated aromatic biosynthesis has long been known to be associated with necrobiosis. It is not clear why aging leads to a similar phenomenon. The difference, however, might be more phenomenological than basic. An essential attribute of senescence is the gradual breakdown of cellular structure. This is injury at the submicroscopic level. A plant organ does not have to be sliced or heavily infected to notice damage. The mostly forgotten observation that simple shaking of a leaf leads to respiratory increase can be interpreted as wound respiration due to systemic "mini" damage. It is even more remarkable that plants in outer space respond to the stress of weightlessness with an increased peroxidase activity.

The enzymatic machinery responsible for the accumulation of aromatic compounds in injured and/or infected storage tissues has been elucidated in detail. Such data, however, are almost completely lacking in relation to senescence.

9. Hormone Balance

The high frequency of hormone changes in infected plants is odd, and is in contrast to the majority of animal and human diseases. The reason is in the peculiarities of hormonal regulation in the plant kingdom. On the one hand, several plant hormones simultaneously affect certain physiological phenomena. On the other hand, plant hormones are multi-target compounds, and one hormone can influence simultaneously (or sequentially) a large variety of different physiological processes. This network of hormonal regulation by a relatively low number of hormones explains that almost any change at any place in the network will lead to coupled changes in the interconnected processes.

Hormone changes in infected tissues are too numerous to be listed here. We shall discuss only some cases which have a bearing on parasitically induced senescence.

Since infection often leads either to rejuvenation or to induced senescence, it is logical to explain these phenomena by alterations in the hormone levels. As already discussed, the large hormone pools (indoleacetic acid and cytokinin) accumulating in the infection centers of leaves attacked by biotrophic parasites are probably responsible for the transport of metabolites to the infection site, the formation of green islands, and the early senescence of the other parts of the leaf. Stimulation of ethylene formation is also a typical stress-induced process. The enhanced synthesis of ethylene induces fruit and leaf senescence, and a number of associated processes. Perhaps the most remarkable is the dramatic increase in peroxidase level and the induction of phenylalanine ammonia lyase (PAL), an enzyme which might control the rate of aromatic biosynthesis. It has not been investigated whether the accumulation of phenolics during leaf senescence can be correlated with a more intense ethylene formation and a resulting increase in PAL activity. The available limited evidence suggests a rise and a later decline of PAL activity during seedling growth. However, PAL is a labile enzyme especially in the presence of phenolics and their oxidases. The detection of lower activities in old tissues does not necessarily mean lower PAL levels. This aspect of natural and disease-induced senescence is worth reinvestigating.

10. Increased Permeability

Diseased plant tissues, especially those reacting hypersensitively, exhibit increased permeability suggesting damage to the cell membranes. The ubiquitous increase in permeability in infected plants seems to be due to a general wound stimulus (Thilo and Nienhaus, 1975). The analysis of membrane lipid components after infection also reveals little specificity (Hoppe and Heitefuss, 1975).

Since ethylene is known to increase permeability, it might be a factor involved in permeability changes in senescence and disease. However, its mode of action is unknown and might be indirect. We only know that the permeability of cell membranes tends to increase with aging.

V. SENESCENCE-INDUCED DISEASES

Every coin has its obverse. If we have disease-induced senescence, we must have senescence-induced diseases—and we have. A few examples will suffice to illustrate the latter point.

As pointed out earlier, loss of chlorophyll and, hence, yellowing are classic symptoms of senescence. The mild yellowing virus of sugar beet sharply increases the intensity of attack by downy mildew caused by *Peronospora farinosa* (Russell, 1966a). Russell shows that the result is probably due to lowered sugar levels in the senescent tissue. Maintaining leaves in the dark lowers the sugars, brings on senescence, and increases the damage from the disease. Similarly, the primary leaves on the sugar beet are lower in sugar and are damaged more by the pathogen than the secondary leaves.

Russell (1966b) also shows that the *Alternaria* sp. attack the yellowed, senescent, low-sugar leaves of the beet more than they do the healthy leaves.

This brings to mind the effects of *Alternaria solani* on solonaceous plants. The intensity of attack by *Alternaria* is more severe on the old senescent leaves of potato or tomato than on young ones. According to Rowell (1953) the fungus invades leaves of all ages, but is more damaging to the old ones. He also showed that the hexose content of the leaves falls as the leaves grow older and that disease intensity is correlated with the hexose level. Apparently the hexoses move out of the old leaves into the young leaves and fruit. Sands and Lukens (1974) have shown that the sugar inhibits the cell-degrading enzymes, thereby reducing the damaging effect of *Alternaria*. Apparently *Alternaria* attacks the cells for the sugar it needs only when the sugar pool runs low.

Two other examples of senescence-induced diseases will be sufficient. When barley yellow dwarf was first observed in California in 1951, the fusarial root rot was so severe on the plants, that it was first suspected as the cause of the disease. We now know that the root rot is the result and not the cause of the disease.

According to Watson and Guthrie (1964) red clover roots are severely damaged by root rot if the foliage is diseased with the clover yellow mosaic disease.

VI. EPILOGUE: DID PHYSIOLOGICAL PLANT PATHOLOGY TURN INTO A BRANCH OF STRESS PHYSIOLOGY?

Are the results described in this chapter discouraging? Did physiological plant pathology become a branch of stress physiology? These are questions which would probably be answered in a variety of ways by different research workers. The enthusiastic types (sometimes called "the naives") will continue to regard almost any change in any parameter to be specific and will "explain" resistance, susceptibility, and specificity by the particular change they happened to observe. Some others will be more cautious. Personally, I do not look disapprovingly at the enthusiasts—within limits, of course. Perhaps it is my own "age-dependent defense reaction" that I consider it important that research workers continue to disprove the redox hypothesis of virus-induced hypersensitive reaction (Farkas et al., 1960), even after almost two decades.

We learned that, for a number of reasons outlined in this chapter, one has to be increasingly careful in claiming specificity in host–parasite relationships. But this is equally true for stress physiology. To claim that a change in a diseased plant is specifically due to stress and/or senescence, if not well documented, is the same mistake as to claim other types of unproved specificities. Therefore, the "common pattern" hypothesis of enzyme changes in senescence and disease, and the other ideas entertained in the present chapter, describe, at best, trends that have many exceptions. The most important warning we can give to stress physiologists is that not everything is stress in plant pathology.

References

Arntzen, C. J., Haugh, H. F., and Bobick, S. (1973). Induction of stomatal closure by *Helminthosporium maydis* pathotoxin. *Plant Physiol.* **52**, 569–574.

Atkinson, P. H., and Matthews, R. E. F. (1970). On the origin of dark green tissue in tobacco leaves infected with tobacco mosaic virus. *Virology* **40**, 344–356.

Balázs, E., Barna, B., and Király, Z. (1976). Effect of kinetin on lesion development and infection sites in Xanthi nc. tobacco infected by TMV: Single cell local lesions. *Acta Phytopathol. Acad. Sci. Hung.* **11**, 1–9.

Balázs, E., Sziráki, I., and Király, Z. (1977). The role of cytokinins in the systemic acquired resistance of tobacco hypersensitive to tobacco mosaic virus. *Physiol. Plant Pathol.* **11**, 29–37.

Bhattacharya, P. K., Naylor, J. M., and Shaw, M. (1965). Nucleic acid and protein changes in wheat leaf nuclei during rust infection. *Science* **150**, 1605–1607.

Chakravorty, A. K., Shaw, M., and Scrubb, L. A. (1974). Changes in ribonuclease activity during rust infection. II. Purification and properties of ribonuclease from healthy and infected flax cotyledons. *Physiol. Plant Pathol.* **4**, 335–358.

Daly, J. M. (1972). The use of near-isogenic lines in biochemical studies of the resistance of wheat to stem rust. *Phytopathology* **62**, 392–400.

Dekhuijzen, H. M., and Staples, R. C. (1968). Mobilization factors in uredospores and bean leaves infected with bean rust fungus. *Contrib. Boyce Thompson Inst.* **24**, 39–52.

Dyer, T. A., and Scott, K. J. (1972). Decrease in chloroplast polysome content of barley leaves infected with powdery mildew. *Nature (London)* **236**, 237–238.

Ehrlich, M. A., and Ehrlich, H. G. (1971). Fine structure of *Puccinia graminis* and the transfer of C-14 from uredospores to *Triticum vulgare*. In "Morphological and Biochemical Events in Host-Parasite Interactions" (S. Akai and S. Ouchi, eds.), pp. 279–307. Phytopathol. Soc. Jpn., Tokyo.

El Hammady, M., Pozsár, B. I., and Király, Z. (1968). Increased leaf growth regulated by rust infections, cytokinins and removal of the terminal bud. *Acta Phytopathol. Acad. Sci. Hung.* **3**, 157–164.

Erickson, R. O., and Michelini, F. J. (1957). The plastochron index. *Am. J. Bot.* **44**, 297–305.

Farkas, G. L., and Lovrekovich, L. (1965). Enzyme levels in tobacco leaf tissues affected by the wildfire toxin. *Phytopathology* **55**, 519–524.

Farkas, G. L., and Stahmann, M. A. (1966). On the nature of changes in peroxidase isoenzymes in bean leaves infected by southern bean mosaic virus. *Phytopathology* **56**, 669–677.

Farkas, G. L., Király, Z., and Solymosy, F. (1960). Role of oxidative metabolism in the localization of plant viruses. *Virology* **12**, 408–421.

Farkas, G. L., Dézsi, L., Horváth, M., Kisbán, K., and Udvardy, J. (1964). Common pattern of enzymatic changes in detached leaves and tissues attacked by parasites. *Phytopathol. Z.* **49**, 343–354.

Fraser, R. S. S. (1972). Effect of two strains of tobacco mosaic virus on growth and RNA content of tobacco leaves. *Virology* **47**, 261–269.

Harding, H., Williams, P. H., and McNabola, S. S. (1968). Chlorophyll changes, photosynthesis and ultratsructure of chloroplasts in *Albugo candida*-induced "green islands" on detached *Brassica juncea* cotyledons. *Can. J. Bot.* **46**, 1229–1234.

Hoppe, H. H., and Heitefuss, R. (1975). Permeability and membrane lipid metabolism of *Phaseolus vulgaris* infected with *Uromyces phaseoli*. IV. Phospholipids and phospholipid fatty acids in healthy and rust-infected bean leaves resistant and susceptible to *Uromyces phaseoli*. *Physiol. Plant Pathol.* **5**, 263–271.

Hyōdo, H., and Uritani, I. (1967). Properties of polyphenol oxidases produced in sweet potato tissue after wounding. *Arch. Biochem. Biophys.* **122**, 299–309.

Király, Z., Pozsár, B. I., and El Hammady, M. (1966). Cytokinin activity in rust-infected plants: Juvenility and senescence in diseased leaf tissues. *Acta Phytopathol. Acad. Sci. Hung.* **1**, 29–37.

Király, Z., El Hammady, M., and Pozsár, B. I. (1968). Susceptibility to tobacco mosaic virus in relation to RNA and protein synthesis in tobacco and bean plants. *Phytopathol. Z.* **63**, 47–63.

Lázár, G., and Farkas, G. L. (1970). Patterns of enzyme changes during leaf senescence. *Acta Biol. Acad. Sci. Hung.* **21**, 389–396.

Lázár, G., Borbély, G., Udvardy, J., Premecz, G., and Farkas, G. L. (1973). Osmotic shock triggers an increase in ribonuclease level in photoplasts isolated from tobacco leaves. *Plant Sci. Lett.* **1**, 53–57.

Leopold, A. C. (1964). "Plant Growth and Development." McGraw-Hill, New York.

Lovrekovich, L., Klement, Z., and Farkas, G. L. (1964). Toxic effect of *Pseudomonas tabaci* on RNA metabolism in tobacco and its counteraction by kinetin. *Science* **145**, 165.

Matile, P. (1975). "The Lytic Compartment of Plant Cells." Springer-Verlag, Berlin and New York.

Montalbini, P., and Buchanan, B. B. (1974). Effect of rust-infection on photophosphorylation by isolated chloroplasts. *Physiol. Plant Pathol.* **4**, 191–196.

O'Farrell, P. H. (1975). High resolution two-dimensional electrophoresis of proteins. *J. Biol. Chem.* **250**, 4007–4021.

Pitt, D. (1973). Solubilization of molecular forms of lysosomal acid phosphatase of *Solanum tuberosum* L. leaves during infection by *Phytophthora infestans* (Mont.), de Bary. *J. Gen. Microbiol.* **77**, 117–126.

Premecz, G., Oláh, T., Gulyás, A., Nyitrai, Á., Pálfi, G., and Farkas, G. L. (1977). Is the increase in ribonuclease level in isolated tobacco protoplasts due to osmotic stress? *Plant Sci. Lett.* **9**, 195–200.

Ross, A. F. (1961). Localized acquired resistance of plant virus infection in hypersensitive hosts. *Virology* **14**, 329–339.

Rowell, J. B. (1953). Leaf blight of tomato and potato plants. *R.I., Agric. Exp. Stn., Bull.* **320**, 1–29.

Russell, G. E. (1966a). Some effect of inoculating with yellowing viruses on the susceptibility of sugar beets to fungal pathogens. I. Susceptibility to *Peronospora farinosa*. *Trans. Br. Mycol. Soc.* **49**, 611–619.

Russell, G. E. (1966b). The control of *Alternaria* spp. on leaves of sugar beet infected with yellowing viruses. II. Experiments with two yellowing viruses and virus-tolerant sugar beet. *Ann. Appl. Biol.* **57**, 425–434.

Sands, D. C., and Lukens, R. L. (1974). Effect of glucose and adenosine phosphates on production of extracellular carbohydrates of *Alternaria solani*. *Plant Physiol.* **54**, 666–669.

Seevers, P. M., Daly, J. M., and Catedral, F. F. (1971). The role of peroxidase isozymes in resistance to wheat stem rust disease. *Plant Physiol.* **48**, 353–360.

Shaw, M., and Manocha, M. S. (1965). The physiology of host–parasite relations. Fine structure in rust-infected wheat leaves. *Can. J. Bot.* **43**, 1285–1295.

Solymosy, F., and Farkas, G. L. (1963). Metabolic characteristics at the enzymatic level of tobacco tissues exhibiting localized acquired resistance to viral infection. *Virology* **21**, 210–221.

Solymosy, F., Szirmai, J., Beczner, L., and Farkas, G. L. (1967). Changes in peroxidase-isozyme patterns induced by virus-infection. *Virology* **32**, 117–121.

Takahashi, T. (1971). Studies on viral pathogenesis in plant hosts. I. Relation between host leaf age and the formation of systemic symptoms induced by tobacco mosaic virus. *Phytopathol. Z.* **71**, 275–284.

Takahashi, T. (1972). Studies on viral pathogenesis in plant hosts. III. Leaf age-dependent susceptibility to tobacco mosaic virus infection in "Samsun NN" and "Samsun" tobacco plants. *Phytopathol. Z.* **75**, 140–155.

Takahashi, T. (1975). Studies on viral pathogenesis in plant hosts. VIII. Systemic virus-invasion and localization of infection on "Samsun NN" tobacco plants resulting from tobacco mosaic virus infection. *Phytopathol. Z.* **84**, 75–87.

Tanaka, Y., Kojima, M., and Uritani, I. (1974). Properties, development and cellular localization of cinnamic hydroxylase in cut-injured sweet potato. *Plant Cell Physiol.* **15**, 843–854.

Tani, T., Yoshikawa, M., and Naito, N. (1973a). Effect of rust infection of oat leaves on cytoplasmic and chloroplast ribosomal nucleic acids. *Phytopathology* **63**, 491–494.

Tani, T., Yoshikawa, M., and Naito, N. (1973b). Template activity of ribonucleic acid extracted from oat leaves infected by *Puccinia coronata*. *Ann. Phytopathol. Soc. Jpn.* **39**, 7–13.

Thilo, H. J., and Nienhaus, F. (1975). Stoffwechselphysiologische Veränderungen in der Pflanze nach Virus-Infektion unter Einfluss von Wundreiz. 5. Einfluss des Pflanzenalters auf Permeabilitätsänderungen. *Phytopathol. Z.* **82**, 291–296.

Turner, J. G., and Novacky, A. (1974). The quantitative relation between plant and bacterial cells in the hypersensitive reaction. *Phytopathology* **64**, 885–890.

Udvardy, J., and Farkas, G. L. (1972). Abscisic acid stimulation of the formation of an ageing-specific nuclease in *Avena* leaves. *J. Exp. Bot.* **23**, 914–920.

Van Loon, L. C. (1976). Systemic acquired resistance, peroxidase activity and lesion size in tobacco reacting hypersensitively to tobacco mosaic virus. *Physiol. Plant Pathol.* **8**, 231–242.

Watson, R. D., and Guthrie, J. W. (1964). Virus-fungus interrelationships in a root rot complex in red clover. *Plant Dis. Rep.* **48**, 723–727.

Weststeijn, E. A. (1976). Peroxidase activity in leaves of *Nicotiana tabacum* var. Xanthi nc. before and after infection with tobacco mosaic virus. *Physiol. Plant Pathol.* **8**, 63–71.

Williams, P. H., Aist, S. J., and Bhattacharya, P. K. (1973). Host–parasite relations in cabbage clubroot. *In* "Fungal Pathogenicity and the Plant's Response" (R. J. W. Byrde and C. V. Cutting, eds.), pp. 141–158. Academic Press, New York.

Wyen, N. V., Udvardy, J., Solymosy, F., Marré, E., and Farkas, G. L. (1969). Purification and properties of a ribonuclease from *Avena* leaf tissues. *Biochim. Biophys. Acta.* **191**, 588–597.

Wyen, N. V., Erdei, S., and Farkas, G. L. (1971a). Isolation from *Avena* leaf tissues of a nuclease with the same type of specificity towards RNA and DNA. Accumulation of the enzyme during leaf senescence. *Biochim. Biophys. Acta* **232**, 427–283.

Wyen, N. V., Udvardy, J., and Farkas, G. L. (1971b). Changes in the level of acid phosphatases in *Avena* leaves in response to cellular injury. *Phytochemistry* **10**, 765–770.

Wyen, N. V., Udvardy, J., Erdei, S., and Farkas, G. L. (1972). The level of a relatively purine specific ribonuclease increases in virus-infected hypersensitive or mechanically injured tobacco leaves. *Virology* **48**, 337–341.

Yoshikawa, M., Masago, H., and Keen, N. T. (1977). Activated synthesis of poly(A)-containing messenger RNA in soybean hypocotyls inoculated with *Phytophthora megasperma* var. sojae. *Physiol. Plant Pathol.* **10**, 125–138.

Zaki, A. J., and Durbin, R. D. (1965). Effect of bean rust on the translocation of photosynthetic products from diseased leaves. *Phytopathology* **55**, 528–529.

Chapter 19

Relation between Biological Rhythms and Disease

T. W. TIBBITTS

I. INTRODUCTION

Among the subtle, yet apparently vital physiological processes of plants are the endogenous rhythms. These physiological activities of plants, although often unrecognized, are ubiquitous in plants and are expressed in many different ways: leaf movements, flowering, photosynthesis, respiration, and cell division. The significance of endogenous rhythms for plants is a long-standing question that has not been adequately explained, but the universality of rhythms in all species of plants indicates very strongly that rhythms increase the chances that the plant will survive and reproduce. This chapter will provide an assessment of rhythms in plants in relation to possible impacts that disease infection may have upon rhythms.

The term "biological rhythms" is, by definition, limited to those fluctuating cycles of activity in plants that occur and are free-running for several repeating cycles when "all" environmental conditions as light, temperature, humidity, etc., are maintained at a constant level (Sweeney, 1969). These rhythms are commonly termed "endogenous rhythms" because the "motor" for these rhythms and the regulation of these rhythms is apparently within the plant. Many reports will be found using the term "rhythm" to describe the fluctuations in different physiological processes over the course of solar days, lunar cycles, or seasonal periods.

However, these cannot be termed rhythms and should be termed the fluctuations or cycles in each process.

Some chronobiologists have questioned the concept of endogenous rhythms, for they suggest that it is impossible to provide a constant environment in any controlled environment facility. They have demonstrated that certain geophysical parameters in the natural environment, such as electromagnetic fields, gravitational forces, and air ions, are not being held constant (Brown, 1968). These scientists suggest that these uncontrolled forces act to phase rhythms in plants in the absence of light, temperature, or moisture fluctuations. Experiments to demonstrate the regulation of rhythms by these subtle geophysical factors have not been able to be consistently reproduced and, therefore, it is generally accepted that the regulation of rhythms is endogenous or within the plants.

II. TYPES OF RHYTHMS

The living plant has been shown to exhibit rhythmic fluctuations in most living functions and processes (Bunning, 1967; Hillman, 1976; Queiroz, 1974; Sweeney, 1969). The most obvious and most common type of rhythm is the leaf movement rhythm of plants (Sweeney, 1969). Other growth and development rhythms such as rhythms in germination, tissue enlargement, and flowering have also been recorded, along with rhythms in all aspects of metabolism within the plant, i.e., photosynthesis, respiration, synthesis of amino acids, and associated activity of enzymes (Chia-Looi and Cummings, 1972; Heide, 1977; Mergenhagen and Schwerger, 1975a; Pallas et al., 1974; Satter and Galston, 1973; Steer, 1973, 1974; Wagner, 1976).

Rhythms have also been documented for changes in cell permeability that cause fluctuations in water and nutrient uptake (Ehrler et al., 1965; Leo et al., 1968) and stomatal opening (Hopmans, 1971; Martin and Meidner, 1975). Of considerable interest has been the evidence for rhythms in the transcription of RNA by DNA and the translation of proteins by RNA (Ehret, 1974; Mergenhagen and Schwerger, 1975b), for these processes are basic to the growth and functioning of the whole plant.

III. NATURE OF RHYTHMS

One might logically ask if the rhythms in different processes are in synchrony, suggesting a common rhythmic control, or if they are each phased separately, suggesting the possibility of separate control mechanisms in each process. The available data suggest that rhythms for different processes within the plant appear to be phased together or phased successively with similar periodicities, suggesting a direct linkage between the separate processes.

The particular phase that a rhythm assumes, the time of the maximum and minimum of each cycle, is controlled principally by the time of the onset of light or dark. Apparently all rhythms are phased and the rhythm can be reset to a different phase by the onset of light or dark (Bunning, 1967; Sweeney, 1969). Thus, anytime a diurnal light–dark cycle is present, an endogenous rhythm will be phased with this diurnal cycle. If plants in controlled environments are moved to a growing room having a different time for onset of light and dark, the plants will alter the phase of endogenous rhythms to phase with the new cycle. However, biologists should be aware of the fact that it takes 2 or more days for complete rephasing and during this interval plant responses may be erratic and follow a "disturbed" pattern.

IV. PERIODICITY OF RHYTHMS

The most commonly documented rhythm in plants is the diurnal rhythm. This periodicity in response has been monitored for many of the different physiological reactions discussed in previous chapters of this volume (Sweeney, 1969). The rhythmic response is documented by maintaining the plant under continuous light or continuous dark. Under continuous light, the rhythm in certain activities has been sustained for more than a week and under continuous dark for several days (Sweeney, 1969; Alford and Tibbitts, 1970). Rhythms tend to damp out more rapidly under continuous dark, apparently because of the lack of photosynthetic energy supply. The diurnal rhythm is commonly termed a "circadian rhythm" because, under continuous light or continuous dark, the cycle length of the rhythmic response is rarely a period of 24 hr, but rather differs ±3 hr. This variation in cycle length has never been adequately explained, and is altered to only a small extent by the particular environment under which plants are maintained.

Rhythms of duration exceeding 24 hr have been reported. Lunar rhythms of 29.5 days and semilunar rhythms of 14.8 days have been reported for the movement of algae (Sweeney, 1969). Evidence indicates that these rhythms are only present in algae that have evolved in seas or waterways subject to lunar tides. Rhythms in fruiting and branching of certain fungi have been found with periodicities of 4 to 24 days (Feldman and Hoyle, 1974; Manachire and Robert, 1972). Annual rhythms in plant activity have been often reported (Sweeney, 1969). These include rhythms in seed germination, bud growth, and coleoptile growth. Some of the most convincing evidence for annual rhythms is in a report of induction of fluctuations in radish flowering over a 2-year period, as reported by Yoo and Uemoto (1976). Although evidence for annual rhythms is quite convincing, there has not been sufficient monitoring in most studies to document whether environmental conditions have been adequately controlled to ensure that there are not environmental fluctuations triggering these cycles. For instance, the CO_2 level of a laboratory is usually higher in the winter, when there are larger amounts of recirculated air, compared to summer. Also, the dew point of laboratory air varies greatly between seasons, particularly in areas with extremes in temperature between winter and summer. Annual "rhythms" in plants developing under field or natural conditions cannot be established with any confidence because of the ever-present annual fluctuations in exogenous factors such as temperature, light intensity, day length and moisture stress.

Short period rhythms of 30 to 240 min are commonly seen in the growth and elongation of stems, leaves, and roots (Alford and Tibbitts, 1971; Baillaud, 1957; Spurny, 1966). These result from cycling of cell division and cell enlargement across the growing tissue. Rhythmic fluctuations in stomatal opening and closing, with periodicities between 20 and 60 min, are well documented (Hopmans, 1971). These short-period rhythms in stomatal aperture have been shown to be initiated by a sudden environmental change that increases or decreases the moisture or CO_2 gradient at the stomatal pore. These induced fluctuations decrease in amplitude with each successive cycle and will damp out if the environmental conditions are maintained constant.

V. SIGNIFICANCE OF RHYTHMS

Rhythms are an ever-present aspect of the biological functioning of plants. Their usefulness is apparently associated with survival for living plants. At the start of growth, they probably aid in ensuring germination and establishment at the most optimum time of day and season of the year. During plant growth, rhythms ensure a more constant and vigorous growth under changing environmental conditions during each day and over successive days. In flowering of photoperiodically sensitive plants, rhythms apparently provide a desirable timing mechanism for photoperiodic control in order to ensure seed production at a favorable time of the year.

Superimposed upon all rhythmic plant responses are fluctuations in these responses that result from both sudden and daily changes in the environment. Plants respond rapidly to alterations in radiation intensity as the sun goes under a cloud, to the sudden increase in wind that increases transpiration and decreases leaf temperature, or to changes in CO_2 level. From one day to the next, plants respond to the level of sunlight, extent of moisture deficit in the air, and the temperature level. Thus, the rate observed for any particular plant process is a composite result of both the ambient environmental conditions and endogenous rhythmic factors. When changing environmental conditions induce a change in rate of a plant response, that is in the opposite "direction" as the rhythmic changes, the net effect will appear as an intensified rhythmic response. However, when changing environmental conditions induce a change in rate of a plant response, that is in the opposite "direction" as the rhythmic changes, the net effect will be a decreased rhythmic response or the rhythmic response may be damped out completely for particular periods. Thus, under field environments, rhythmic responses of plants are often difficult to identify. Only under controlled environments within a greenhouse, and more particularly within growth chambers, can rhythmic responses be identified and quantified.

VI. DISEASE EFFECTS UPON RHYTHMS

Little research has been directed toward the effect of disease upon rhythms. A single report by Novak (1972) demonstrated that tobacco mosaic virus did not inhibit the circadian cycle in photosynthesis of leaves of tobacco plants. It is to be expected that plant rhythms, particularly circadian rhythms, would not be affected very greatly by plant diseases. This assumption is based on the fact that application of toxic chemicals to plants has done little to either cause the cessation of rhythms or to reset rhythms to a different phase (Hillman, 1976). Many different specific enzyme inhibitors have been applied to plants with little effect upon circadian rhythms. Protein synthesis inhibitors have been found to inhibit oxygen evolution rhythms when in contact with the cells, but the rhythm returned and maintained the original phase when the inhibitors were removed (Mergenhagen and Schwerger, 1975b). The author has watched a unifoliate bean leaf, injured by application of a toxic chemical, maintain a diurnal leaf movement during a 5-day period in which the leaf slowly shriveled and became necrotic. Movement abruptly stopped only when the leaf was completely necrotic. The opposite unifoliate leaf maintained a very regular circadian movement prior to and after death of the affected leaf.

Thus diseases will not likely cause the cessation of circadian rhythms in plants unless the pathogen produces a specific inhibitor of the rhythmic controlling mechanisms within the plant. Identification of diseases that would completely inhibit plant rhythms or alter rhythms by shifting the phase or changing the amplitude of the rhythm would provide important clues toward understanding the basic controlling mechanism of rhythms in plants.

Although diseases probably do not alter circadian rhythms, diseases are more likely to have a controlling effect upon rhythms of other periodicities. The periodicity of both the growth and development rhythms of many plants and the branching and fruiting rhythms of fungi are significantly affected by temperature changes (Baillaud, 1957; Manachire and Robert, 1972), and thus may be regulated by the stress of a disease. Similarly, the rhythm in stomatal apertures can be stopped by drying of the soil or humidification of the air. Thus, diseases that alter the water balance of the plant may have a significant controlling effect on certain rhythms.

If a disease reduces the total vigor of the plant through effects upon any overall plant process such as photosynthesis, nutrient uptake, or translocation, presumably the amplitude of rhythmic response of both circadian and other rhythms would be reduced. The presence of disease thus may make it more difficult to observe a rhythmic response in a plant process.

VII. CONCLUSION

Diseases obviously have some effect upon rhythms in plants. The greatest effect from diseases will be expected upon rhythms that are of a periodicity other than circadian. The effect of rhythm modification will likely be of reduced vigor and adaptation as indicated through the assumption of Queiroz (1974) that "rhythms are necessary for normal growth and longevity" of plants. Reduced growth and reduced reproductive capacity could result from alterations in one or more of the physiological or biochemical processes in the plants known to have rhythmic cycles of activity. However, the possibility of determining the significance of the negative impacts from disease-rhythm interactions under natural conditions is not likely. It would be extremely difficult to monitor changes in rhythms for they always are superimposed upon diurnal and other fluctuations induced by the natural environment.

References

Alford, D. K., and Tibbitts, T. W. (1970). Circadian rhythms of leaves of *Phaseolus angularis* plants grown in a controlled carbon dioxide and humidity environment. *Plant Physiol.* **46,** 99–102.

Alford, D. K., and Tibbitts, T. W. (1971). Endogenous short-period rhythms in the movements of unifoliate leaves of *Phaseolus angularis. Plant Physiol.* **47,** 68–70.

Baillaud, L. (1957). Recherches sur les mouvements spontanes des plantes grimpantes. *Ann. Sci. Univ. Bes.* Ser. 2. Bot. 235 pp.

Brown, F. A., Jr. (1968). A hypothesis for extrinsic timing of circadian rhythms. *Can. J. Bot.* **47,** 287–298.

Bunning, E. (1967). "The Physiological Clock." Springer-Verlag, Berlin and New York.

Chia-Looi, A., and Cummings, B. G. (1972). Circadian rhythms of dark respiration, flowering, net photosynthesis, chlorophyll content, and dry weight changes in *Chenopodium rubrum. Can. J. Bot.* **50,** 2219–2226.

Ehret, C. F. (1974). The sense of time: Evidence for its molecular basis in the eukaryotic gene-action system. *Adv. Biol. Med. Phys.* **15,** 47–77.

Ehrler, W. L., Nakayama, F. S., and von Bavel, C. H. M. (1965). Cyclic changes in water balance and transpiration of cotton leaves in steady environment. *Physiol. Plant.* **18,** 766–775.

Feldman, J. F., and Hoyle, M. N. (1974). A direct comparison between circadian and non-circadian rhythms in *Neurospora crassa. Plant Physiol.* **53,** 928–930.

Heide, O. M. (1977). Photoperiodism in higher plants. An interaction of phytochrome and circadian rhythms. *Physiol. Plant.* **39,** 25–32.

Hillman, W. S. (1976). Biological rhythms and physiological timing. *Annu. Rev. Plant Physiol.* **27,** 159–179.

Hopmans, P. A. M. (1971). Rhythms in stomatal opening of bean leaves. *Meded. Landbouwhogesch. Wageningen* 71–3; 86 pp.

Leo, M. W. M., Wallace, A., and Joven, C. B. (1968). Root pressure exudation in tobacco. *Adv. Front. Plant Sci.* **21,** 45–91.

Manachire, G., and Robert, J. C. (1972). Fruiting rhythm of a basidiomycete mushroom, *Coprinus congregatus* Bull. ex Fr: Physiological significance of the variation of rhythm periods in relation to culture temperature. *J. Interdiscip. Cycle Res.* **3,** 135–143.

Martin, E. S., and Meidner, H. (1975). The influence of night length on stomatal behavior in *Tradescantia virginiana. New Phytol.* **75,** 507–511.

Mergenhagen, D., and Schwerger, H. G. (1975a). Circadian rhythm of oxygen evolution in cell fragments of *Acetabularia mediterranea. Exp. Cell Res.* **92,** 127–130.

Mergenhagen, D., and Schwerger, H. G. (1975b). The effect of different inhibitors of transcription and translation on the expression and control of circadian rhythm in individual cells of *Acetabularia. Exp. Cell Res.* **94,** 321–326.

Novak, J. (1972). Wachstumsanderungen der Tabakpflanze *Nicotiana tabacum* cv. Samsun nach der Inokulation des Tabakmosaikvirus. *Acta Sc. Nat. Brno* 9(8), 1–41.

Pallas, J. E., Jr., Samish, Y. B., and Willmer, C. M. (1974). Endogenous rhythmic activity of photosynthesis, transpiration, dark respiration, and carbon dioxide compensation point of peanut leaves. *Plant Physiol.* **53,** 907–911.

Queiroz, O. (1974). Circadian rhythms and metabolic patterns. *Annu. Rev. Plant Physiol.* **25,** 115–134.

Satter, R. L., and Galston, A. W. (1973). Leaf movements: Rosetta stone of plant behavior. *BioScience* **23**, 407–416.

Spurny, M. (1966). Spiral feedback oscillations of growing hypocotyl with radicle in *Pisum sativum* L. *Biol. Plant.* (Praha) **8**(5), 381–392.

Steer, B. T. (1973). Diurnal variations in photosynthetic products and nitrogen metabolism in expanding leaves. *Plant Physiol.* **51**, 744–748.

Steer, B. T. (1974). Control of diurnal variations in photosynthetic products. *Plant Physiol.* **54**, 758–65.

Sweeney, B. M. (1969). "Rhythmic Phenomena in Plants." Academic Press, New York.

Wagner, E. (1976). Kinetics in metabolic control of time measurement in photoperiodism. *J. Interdiscip. Cycle Res.* **7**, 313–332.

Yoo, K. C., and Uemoto, S. (1976). Studies on the physiology and flowering in *Raphanus sativus* L. II. Annual rhythm in readiness to flower in Japanese radish, cultivar 'Wase-shijunichi.' *Plant Cell Physiol.* **17**, 863–865.

Author Index

Numbers in italics refer to the pages on which the complete references are listed.

A

Adams, P. B., 358, 359, *368*
Afanasiev, M. M., 271, *276*
Aho, P. E., 292, *293*
Ahrens, J. R., 234, *251*
Aist, J. R., 226, *228*, 284, *293*, 342, *344*
Aist, S. J., 400, *411*
Aiyar, S. P., 168, 171, *178*
Akai, S., 342, *344*
Akazawa, T., 99, *113*
Albersheim, P., 280, 281, 282, 283, 284, *293, 294, 296, 297*, 363, *368, 369*
Albion, R. G., 318, *322*
Aldrich, H. C., 341, *344*
Alexander, M., 223, 229, 243, *252, 253*
Alford, D. K., 415, 416, *419*
Allen, E., 362, *368*
Allen, E. K., 217, *228*
Allen, O. N., 217, *228*, 243, *252*
Allen, P. J., 187, *197*, 343, *344*
Alvim, P. de T., 27, *47*
Amrhein, N., 377, *388*
Anderson, D. B., 22, *49*
Anderson, G. R., 176, *179*
Anderson, I. C., 243, *254*
Anderson, W. P., 331, 332, 333, *344*
Anderson-Prouty, A., 363, *368*
Ap Rees, T., 358, 359, 362, *368, 372*
Aragaki, M., 265, *278*
Arber, A., 53, *82*

B

Archer, S. A., 329, *345*
Armstrong, P. B., 219, 220, *228*
Arntzen, C. J., 399, *409*
Arny, D. C., 42, *47*
Asahi, T., 362, *373*
Aspinall, G. O., 280, 281, *293*
Atkinson, D. E., 354, *369*
Atkinson, P. H., 394, *409*
Ayers, A., 363, *369*
Ayers, P., 142, 145, 151, *162*
Ayres, P. G., 95, 103, *113, 114*

B

Bacon, G. A., 66, *82*
Bailey, I. W., 119, *137*
Bailiss, K. W., 188, *197*
Baillaud, L., 416, 418, *419*
Baker, K. F., 130, *137*, 267, 268, 272, *275*
Baker, R. J., 387, *388*
Balázs, E., 190, *200*, 395, *409*
Ballio, A., 78, *83*
Balmer, E., 173, *179*
Balodis, V., 118, *140*
Bamburg, J. R., 188, *198*
Banfield, W. M., 133, *137*
Barash, I., 263, *275*
Barham, R. O., 247, *251*
Barker, K. R., 169, 171, 176, *178, 180*, 240, 246, 248, *251, 253, 254, 256*

421

Y

Yabuta, T., 7, *18*
Yamada, Y., 380, *389*
Yamadori, S., 171, *179*
Yamamoto, H., 386, *390*
Yang, A. F., 270, *278*
Yang, S. F., 185, 193, *198, 200*
Yarwood, C. E., 178, *181*
Yates, M., 66, *82*
Yoder, O. C., 76, *83*, 328, 341, *344, 346*
Yoo, K. C., 416, *420*
York, D., 341, *344*
Yoshida, S., 362, *374*
Yoshihara, O., 283, *295*
Yoshikawa, M., 395, 400, 401, *411, 412*
Younis, M. A., 174, *181*

Z

Zadoks, J. C., 15, *18*
Zaenen, I., 202, 205, *213*
Zaitlin, M., 6, *17*, 94, *116*, 186, *197*, 381, *390*
Zake, A. I., 367, *371*
Zaki, A. J., 397, 398, *412*
Zaumeyer, W. J., 273, *278*
Zelitch, C. P., 22, 24, *51*
Zelitch, I., 88, 91, 93, 95, 102, 112, *116*
Ziegler, H., 123, 124, *140*
Zimmerman, M. H., 22, *51*, 118, 119, 120, 121, 122, 123, 126, 129, *140*
Zuber, M. S., 309, 310, 311, 322, *325*
Zucker, M., 353, 362, *369, 374*

Subject Index

among plant parts, 184
in plant teratomas, 218
Comparative pathology, science, 54
Comparative studies, resistant and susceptible varieties, 76
Compartments, in xylem, 120
Compartmentalization, 227
of tree stems, 160
Compartmentation, altered by disease, 106
of cells, 329
effect of disease, 99, 100
of enzymes, 342, 343
of hydrolytic enzymes, 405
loss, 93
metabolic systems, 90
of metabolism, 400
of plant cells, 332
Compatibility, 2, 77
in host-parasite systems, 376
understanding of, 375
Compatibility phenomena, 280
Compatible combinations, of host and parasite, 400
Compensation strategies, of symbiotic associations, 239, 240
Competition, during establishment of symbionts, 250
with mycorrhizal fungi, 246–247
among roots, 231
between symbionts, 241, 242
Components, of yield, 90
Compression wood, formation in healthy plants, 38
Compressive forces, 305, 306
Concentration, of potassium, in cells, 328
Conceptual theory, of disease, 196
of pathology, 6, 8, 9
Conflict, components, 65
Conjugation, of Ti plasmid, 206
Constitutive components, of host resistance, 75
Construction, proper use of wood, 288, 289
Continued irritation, importance in definition of disease, 59
Control, of disease, 55, 56, 292
of gene activity, 377
strategies, 55, 56, 79
tactics, 56, 79

Cooperation, components, 65
in disease physiology and plant physiology, 195
between plant pathologists and agronomists, 322
between plant pathologists and foresters, 322
between plant pathologists and horticulturalists, 322
between plant pathologists and plant physiologists, 45, 46, 314
with scientists in other disciplines, need, 113
Cooperative research, value, 311
Copper, deficiency, 166, 167
effects on symbionts, 244
role in healthy plants, 29
Corepressors, role in regulation of genes, 377
Coriolus versicolor, see *Polyporus versicolor*, 290
Cork formation, induced by pathogen, 176
Corn, *Colletotrichum graminicola*, 275
dwarfism, 187
Gibberella zeae, 266
growth, 183
lodging, 309–311
Pythium aphanidermatum, 171
Rhizoctonia, 171
Sphacelotheca, reiliana on, 261
stalk rots, role in lodging, 309, 310
susceptibility to frost, 42
Ustilago, 191
Corn smut, 191
Cortex, damage, 150
role, in darkening of roots, 147
in resistance to pathogens, 150
Cortical cells, role in healthy plants, 33
Cortical pathogens, immobilization of nutrients, 174
Coruneum beijerinckii, on cherry, 314
Corynebacterium, 174
fascians, 192
Cost, of crop production, 249
of disease management, 136
of laboratory versus field research, 47
of research, 80
Cotton, *Fusarium oxysporum*, 173
weather-induced shedding of bolls, 314

E

Early leaves, in perennial plants, 43
Early-wood vessels, 125
Ear rot, of cereals, 266
Ear rot fungi, resistance of high lysine corn, 266
Economics, of crop production, 249
  importance, in definitions of disease, 58
Economic criteria, treatment of plant disease, 136
Ecosystem, study, 55
Ecology, role in biology, 54
Ecological niche, occupation of, 257
Ectomycorrhizae, 239
  energy cost, 234, 235
  longevity, 232
  protection against *Phytophthora cinnamomi*, 233
  role, in protection against feeder root pathogens, 247
    in uptake of nutrients, 232
  suppression of chlamydospore formation in *Thielaviopsis basicola*, 233
Efficiency, of plants, 4
Eggplant, metabolic response to pathogens, 364
Elastic limit, definition, 305
  of plants, 302
Electrical circuit model, of water system, 144, 161, 162
Electrical neutrality, in plants, 163, 164
Electrochemical potentials, effect, on permeability of plants, 336
  of victorin, 338
  role in permeability, 329, 332
Electrolytes, leakage, 336
  from plant cells, 286
  from plant tissues, 329
  loss, rate of increases, 338
  rate of loss, 337
Electrophoresis, use in separating proteins, 385
Embden-Meyerhof pathway, 104
Embolism, conducting units, 124–126
  major cause of wilt in plants, 125
  permanent, 126
  role in wilting, 153
  of vessels, 119–121
  without injury, 126

Embryo transmission, of viruses, 270–272
Empiricism, role in plant pathology, 55
Encapsulation, of bacteria, 220
Encapsulation matrix, 221
Endodermis, role, in darkening of roots, 147
  in water movement, 143
  in water uptake, 150
Endogenous rhythms, 413, 414
Endoglucanases, 283
Endol, 4-*B*-mannanase, 283
Endomycorrhizae, ericaceous, role in uptake of nutrients, 232
Endopectate lyases, role in tissue maceration, 285
Endopectin methyltranseliminase, of *Sclerotinia fructigena*, 285
Endoplasmic reticulum, role in virus-infected plants, 382
  structure, 333
Endophyte, onion, 236
  nodule-forming, 220
Endopolygalacturonases, role in tissue maceration, 284, 285, 334
Endopolygalacturonate transeliminase, 285
Endothia parasitica, 130
Energy capture, effect of disease, 112
  in healthy plants, 86–92
Energy charge, of plant cells, 354, 359, 361
Energy cost, of ectomycorrhizae, 234, 235
  of symbiosis, 231, 233–236
  management of, 251
  vesicular arbuscular mycorrhizae, 235
Energy loss, loss of reducing power, 109
  through respiration, 104–107
  storage organs, 109
Energy utilization, in diseased plants, 59
  effect of disease, 108, 109
  in healthy plants, 86–92
Energy, competition, in diseased plants, 187
  by host and pathogen, 108
  metabolic, 329, 350
  role in plant disease, 86–116
  use in diseased plants, 11
  waste, disease induced respiration, 104–108

Rosette, symptom of mineral deficiency, 165

Rosetting, caused by mineral deficiency, 31

Rotted roots, function in water uptake, 148

Rubber trees, root rot, 317

Rubbery wood disease, 300, 311, 312
  of apples, 311

Rubidium, uptake by plants, 34

Rust, bean, effect on starch accumulation, 100
  effect, on CO_2 uptake, 96
    on lifting force, 143
    on loading of phloem, 131
    on photosynthesis, 94
    on membrane permeability, 334
    on transport, of photoassimilates, 102
    on water status, 146, 158
  infection, 397
  obligate parasites, 383
  role in RNAase, 385
  of wheat, 384
    necrogenic reaction, 404

Rust galls, as meristematic sinks, 176

Rye, ergot of, 262

S

Safety, of phloem, 122

Safety, of water system, 121, 124–126, 136, 154

Safflower, *Fusarium oxysporum*, 274
  phyllody, 261
  Verticillium dahliae, 274
  Verticillium wilt, 273

Salinity, effect on symbionts, 243, 244
  influence on mycorrhizae, 250

Salivary glands, of insects, 224

Salix, senescence, 36

Sand blasting, losses, 322
  of young seedlings, 303

Saprophytic ability, of soil pathogens, 148

Saprophytism, 63
  distinct from parasitism, 64

Sapwood, decay susceptibility, 289
  function, 287
  transformation to heartwood, 287

SBMV, *see* Southern bean mosaic virus

Scab disease, of barley, 266

of wheat, 266

Science, of biology, 53
  usefulness, 15

Scientists, education, in plant pathology, 56

Scientists, perspectives, in plant pathology, 56

Sclerenchyma cells, in plants, 307

Sclerospora graminicola, on millet, 261

Sclerotinia fructigena, 285
  role in damping off, 313
  sclerotiorum, 315
    on celery, 175
    on flowers, 267

Sclerotium rolfsii, 64

Seasonal change, in plant growth, 39, 41, 42

Secondary cell wall, 280

Seed, lethal effects of victorin, 339

Seed coat transmission, of viruses, 269

Seed contamination, by bacterial pathogens, 273

Seed destruction, 265, 266

Seed dormancy, in healthy plants, 35

Seed germination, 192
  rhythms, 416

Seed quality, pathogenic effects, 274

Seed transmission, of viruses, 268–272

Seed treatment, role in plant quarantine, 267

Seed viability, influence of pathogens, 267–275

Seed-borne pathogens, 267–274
  bacterial diseases, 272, 273
  biological implications, 267, 268
  effect on germinability of seed, 274, 275
  fungal diseases, 273, 274
  mechanisms for transfer, from mother plant, 268
  vascular, 274
  virus diseases, 268–272
    importance and frequency, 268
    mechanisms of transmission, 269–272

Seedling diseases, 274, 275

Seedling growth, 31

Seedling metabolism, prevention, 4

Selection, of microbial symbionts, 250
  of pathogen strains, 267